Introduction to Neural and Cognitive Modeling

Daniel S. Levine, Ph. D.
University of Texas at Arlington

LEA LAWRENCE ERLBAUM ASSOCIATES, PUBLISHERS
Hillsdale, New Jersey   London

Illustrations for this book were done by Mr. Wesley Elsberry, using CorelDraw 1.21.

Lawrence Erlbaum Associates, Inc., Publishers
365 Broadway
Hillsdale, New Jersey 07642

Library of Congress Cataloging-in-Publication Data

Levine, Daniel S.
 Introduction to neural and cognitive modeling / Daniel S. Levine.
        p.   cm.
 Includes bibliographical references and index.
 ISBN 0-8058-0267-3. -- ISBN 0-8058-0268-1 (pbk.)
 1. Neural networks (Computer science) 2. Brain--Computer
simulation. 3. Cognition--Computer simulation. I. Title.
QP363.3.L48   1990
612.8'2'0113--dc20   90-43390
                                                                    CIP

Printed in the United States of America
10 9 8 7 6 5 4 3 2

Contents

CONTENTS

CONTENTS

PREFACE

How do you eat an elephant? One bite at a time.

Beverly Johnson, after climbing the rock face of
El Capitan, Yosemite National Park

Neural networks, which were relatively unknown to the educated public until the early 1980's, have recently catapulted into a major place in the consciousness of the scientific and technical community. There are at least two reasons for the sudden growth of interest in this field. One reason is the rapid progress both of neural network theory itself and of experimental neurobiology. Another reason is the increasing complexity of technical problems encountered in various industrial applications of knowledge-based computers, combined with some inadequacies found in traditional artificial intelligence approaches.

Yet after more than forty years of research, neural networks are still surrounded by mythology. Are these networks clever computers that can perform cognitive tasks with greater flexibility than expert systems? Or are they models of actual human or animal brains? Or are they abstract physical systems with certain interconnections, laws, and state transitions? In reality, all of these descriptions are partially valid but incomplete. The field of neural networks is now being investigated by researchers from as wide a range of disciplines as any field in the recent history of knowledge. These researchers include, in no particular order, biologists, psychologists, cognitive scientists, computer scientists, mathematicians, engineers, physicists, and even, at an early stage, a few social scientists and philosophers.

This book concentrates more on the theory of neural networks than on the applications. Some good reviews of neural network applications are found in Hecht-Nielsen (1986, 1988, 1990), Lippmann (1987), the DARPA study (1988), and Miller, Walker, and Ryan (1989). Theoretical principles, and their implementation in widely known models, are discussed here in a manner intended to be useful to a broad class of students, designers, and researchers. Neuroscientists can use these principles to move toward a stronger theoretical foundation for brain science, including clinical neurology and psychiatry, and suggest further experiments. Modelers can use these principles to help provide a rational basis for building networks. Computer scientists and engineers can use them for inspiration in building devices. Mathematicians and physicists can find interesting theoretical problems in dynamical systems derived from neural or cognitive models. Psychologists and cognitive scientists can achieve more coherence in their understanding of interrelationships among cognitive phenomena. Social scientists can gain a better understanding of human decision processes, which should ultimately yield predictions relevant to their own disciplines (for a discussion, see Leven, 1988). Finally, philosophers can find insights into age-old epistemological issues, thereby adding scientific yet non-reductionistic bases to existing qualitative theories of the mind (see, *e.g.*, Maslow, 1968, 1972; Powers, 1973; Turner, 1981).

This book is aimed toward the varied audience of the neural network field, and is intended as a textbook for a graduate or advanced undergraduate course in the area. The focus will be on the common interest that all these researchers share, namely, the relationship between neural structure and cognitive function. The cognitive functions to be discussed include learning, perception, attention, memory, pattern recognition, categorization, and motor control, for examples. The neural structures will at times incorporate organizing principles such as competition, association, and opponent processing, principles which can be suggested either by the exigencies of modeling psychological data or by the description of known neuroanatomical structures. These principles will be developed in early chapters and will appear throughout the book.

In keeping with the goal of accessibility to a varied audience, technical prerequisites in any one discipline are kept to a minimum. Recent advances in computing make the field accessible to many more people than before. Hence, for students, access to either a personal or mainframe computer is assumed. For those needing additional background in neurobiology or in mathematics, appendices in those fields are included; the appendices also list sources for more detailed coverage.

A word should be said here about equations. The last section of each chapter includes differential or difference equations for some of the networks discussed in that chapter, so that the reader can gain hands-on experience in computer simulation of the networks. On first reading, the student without mathematical background can skip these equations and follow the development of networks by means of the figures. On second reading, the same student can turn to Appendix 2 for explanations of how equations reflect the qualitative relationships in networks, and simple algorithms for simulating such equations. All but the last two sections of Appendix 2 are written so as not to require previous background in differential equations; notions needed from elementary calculus are redefined and motivated in the context of neural network applications.

All chapters except the first and last contain both thought experiments and computer simulation exercises pertaining to various neural network models. The exercises herein are a small sampling of the possible questions that can be asked about the material discussed in the book, and the instructor is encouraged to supplement them as he or she sees fit. Many of the questions asked here do not have right and wrong answers, only a variety of better and worse answers. The reader should approach the field with at least as much intellectual flexibility and curiosity as possessed by the systems we model.

The following diagram illustrates how the understanding of each chapter depends on previous chapters:

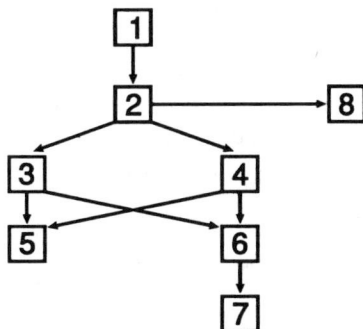

After the introduction in Chapter 1 and the historical account in Chapter 2, Chapters 3 through 7 reflect a hierarchy from simpler to more complex neural and cognitive processes. Chapters 3 and 4 discuss various neural network rules for associative learning and for competition, respectively. Small networks incorporating some of these rules are then embedded in larger networks that model complex cognitive processes such as conditioning and attention (Chapter 5) and coding and categorization (Chapter 6). Chapter 7, building on earlier chapters, concludes this hierarchy with networks designed to deal with problems that are still largely open (such as optimization, control, decision making, and knowledge representation). Finally, Chapter 8 gives a brief survey of recent advances in both neurobiology and artificial neural systems.

The study of mind is a densely interconnected subject. Hence, the organization of topics in this book is not the only possible one, and the boundaries between classifications that I use are far from rigid. But if sense is to be made of the multiplicity of cognitive processes, and models of these xii

processes, it seems natural to try to isolate important sub-processes and thereby proceed from simpler to more complex cognitive functions.

I shall close by acknowledging and thanking the many people who helped make possible the undertaking of writing this book. The Editors from Lawrence Erlbaum Associates (LEA) at different stages, Julia Hough and Judi Amsel, both gave the book project enthusiastic support, and were tireless in working to smooth out administrative details. Major assistance was also provided by other members of the staff at LEA, in particular Lawrence Erlbaum, the company President; Joseph Petrowski, from the Publicity Department; and Arthur Lizza, from Production.

Wesley Elsberry provided months of technical assistance which allowed for delivery of camera-ready copy, expediting the book's production considerably. His expertise, and good humor, with computer software covered word processing, graphics, and the running of homework exercises. The text was done in WordPerfect 5.1 and the graphics in CorelDraw 1.21. The computer and printer on which the final copy was made are located at the Automation and Robotics Research Institute (ARRI) at the University of Texas at Arlington. Charles Lindahl, William Ford, and other staff members at ARRI provided generous assistance with their facilities. Raju Bapi, Nilendu Jani, and Carey Weathers also helped with graphics and with testing of homework exercises.

Several of my colleagues and former students made suggestions on earlier drafts of the book. Gregory Stone and Richard Golden reviewed the entire manuscript carefully and helped to shape the revisions. Others contributing valuable comments on sections of the manuscript included Manuel Aparicio, Raju Bapi, Daniel Bullock, Heather Cate, David Hestenes, Peter Killeen, Samuel Leven, Wing-Kwong Mak, Ennio Mingolla, Haluk Ogmen, David Olson, Alice O'Toole, Paul Prueitt, and David Stork. Their comments considerably improved the book's style and brought to it a variety of professional perspectives.

Finally, my wife, Lorraine Levine, lived patiently with the highs and the rebound lows associated with the book's composition. She combined an appreciation of the project's value with a warm sense of humor that kept me on course but helped me avoid the perils of overly grim determination.

Daniel S. Levine
Arlington, Texas

Some Technical Conventions Used in This Book

In figures, the following symbols are used for connections between nodes:

Filled semicircle for a modifiable connection;
Filled arrow for a non-modifiable connection;
"+" for an excitatory connection;
"-" for an inhibitory connection.

In equations, I have attempted to make the terminology as uniform as possible across different authors. Hence, node activities are generally denoted by "x" and "y," with subscripts, and connection weights by "w," with subscripts, regardless of what letters were used by the equation's originator. A few judicious violations of these conventions are explained in footnotes. Positive constants defined as system parameters are usually represented by other small Roman letters.

In exercises, I have not included answers in the back of the book. This is because most of the exercises are either thought experiments with no definite right or wrong answer (*e.g.*, "can the model of Anderson, 1968, be extended to a model of selective attention?") or computer simulations where the object is to come as close as possible to reproducing results that are described either in the text or the original source. In some cases, I have included expected results in the problem statement itself. More detail about the exercises will be contained in an accompanying instructor's manual. As for the symbols at the left margin of exercises, a single star (*) means that the problem involves a relatively difficult computer simulation. A double star (**) means that the problem requires a higher degree of mathematical knowledge than others. An open circle (o) means that the problem is an open-ended thought experiment or modeling exercise.

1

Brain and Machine: The Same Principles?

My mind to me a kingdom is,
Such perfect joy therein I find
As far exceeds all earthly bliss
That God or nature hath assigned.

Edward Dyer

What is mind? No matter. What is matter? Never mind.

Thomas Hewitt Key (epigram in *Punch*)

What Are Neural Networks?

The late 1980's and early 1990's are seeing what may be the start of a major intellectual revolution. The rapid development of neurobiology and of experimental psychology has led us closer to an understanding of biological cognitive functioning than most of us had thought possible a short time ago. At the same time, the expansion of cognitive or adaptive capabilities in industrial applications of computers has proceeded even more rapidly. As the two fields of neurobiology and artificial intelligence develop further, it is natural to look for common organizing principles to both.

The belief that there are common quantitative foundations for both brain science and artificial intelligence has come and gone and come again. In the 1940's and 1950's, the notion that neurons are digital "on-off" switches (either firing or not firing), and thus that brains and the newly emerging digital computers had similar structural organizations, captured the imagination of scientists. Eventually, biologists discovered that the digital metaphor was an inadequate one for capturing what was known about neurobiology and psychology. It was found necessary to understand the graded (or analog, or gray scale) as well as the all-or-none (or digital) components of neuron responses (see, *e.g.*, Thompson, 1967, Ch. 1). Concurrently, artificial intelligence moved in the direction of writing digital computer programs to perform narrowly specified cognitive tasks without much reference to how those same tasks are performed by humans or animals (see, *e.g.*, Newell & Simon, 1972; Winston, 1977).

Recent years have seen a partial reunification of the two fields of neurobiological modeling and artificial intelligence. Designers of machines for performing cognitive functions have taken a renewed interest in learning how the brain performs those functions. Consequently, such machines have been built like simulated brain regions, with nodes corresponding to neurons or neuron populations, and connections between the nodes; at times, their designers have borrowed ideas from recent experimental results on the brain's analog responses. This development is often called *connectionism* (*e.g.*, Feldman & Ballard, 1982; Rumelhart & McClelland, 1986; articles in Volume 9 of *Cognitive Science*); the industrial applications of connectionist theory are often called *artificial neural systems* (*e.g.*, Hecht-Nielsen, 1986). Both of these terms are of recent coinage; the much older term *neural networks* is usually considered to encompass both theoretical and applied models. These models may or may not be designed as theories of actual brain organization, but always include nodes[1] and connections. The DARPA study (1988) gives a reasonable definition of the term:

---

[1] The functional units in neural networks have alternatively been called "nodes," "units," "cells," and "populations." I prefer the first two terms because they do not commit the user to an assumption that units correspond to either single or multiple neurons. This book most often uses the term "node," but sometimes uses "unit" since that usage has been popularized by Rumelhart and McClelland (1986). In Section 2.1, the word "cells" or "neurons" is used for units in the network of McCulloch and Pitts (1943), for reasons explained in that section.

a neural network is a system composed of many simple processing elements operating in parallel whose function is determined by network structure, connection strengths, and the processing performed at computing elements or nodes. ... Neural network architectures are inspired by the architecture of biological nervous systems, which use many simple processing elements operating in parallel to obtain high computation rates (p. 60).

Many factors have contributed to the recent renaissance of neural networks. First, in applied areas such as knowledge processing, robotic control, pattern classification, speech synthesis and recognition, and machine vision, computer engineers have encountered problems not easily amenable to the symbolic processing programs of mainstream artificial intelligence (for examples, see Hewitt, 1986; Minsky, 1986; Winograd & Flores, 1987). Second, neurobiological experimental methods and data analysis have advanced greatly. Techniques such as recording with electrodes from up to fifty neurons at once and taking tomographic scans of the entire brain have made neurophysiology more amenable to quantification. At the same time, advances in computing (running the gamut from personal computers to supercomputers) have made simulation of biological data easier and more practical. Third, a few publications, such as the article of Hopfield (1982) and the two-volume book edited by Rumelhart and McClelland (1986), have brought neural networks to the attention of mainstream scientists in such disciplines as physics and computer science.

The rapid surge in popularity of the neural network field conceals the field's maturity. The history of neural network models, which was summarized (up to 1983) in the review article of Levine (1983b) and is discussed further in Chapter 2, shows that most modern ideas in network design have much earlier antecedents. For example, the current distinction between input, hidden, and output units (Rumelhart & McClelland, 1986) owes much to the early work of Rosenblatt (1962) on networks with sensory, associative, and response units (though modern networks go far beyond Rosenblatt's perceptrons). Rosenblatt, in turn, combined extensions of the linear threshold law due to McCulloch and Pitts (1943) with extensions of the learning law due to Hebb (1949). Moreover, several current leaders in neural network research, such as Shun-ichi Amari, James Anderson, Walter Freeman, Stephen Grossberg, and Teuvo Kohonen, have been publishing in the field since the late 1960's or early 1970's. Many ideas that these investigators laid out in their early work remain fruitful today.

What Are Some Principles of Neural Network Theory?

As the neural network literature grows, it is essential to find criteria for making distinctions among competing models. Let us first seek guidance from the notion of connectionism. Connectionism means that models are based on activities of nodes, strengths of connections between nodes, and laws (defined typically by difference or differential equations) for the changes over time of these activities and connection strengths. In artificial intelligence and cognitive science circles, connectionist approaches are typically contrasted with previous approaches based on serial computer programs with intricate instructions but no nodes. But the connectionist versus non-connectionist distinction gives little information about the validity of a particular network as a brain model, or about the network's utility for practical computing applications. Once we have said that a particular model is connectionist, it is important also to describe the structure of the model's network connections and to identify some of its useful subsystems.

The importance of subsystem identification is suggested by the parable of the watchmakers (Simon, 1969, pp. 90-93). One watchmaker tries to fashion a whole watch simply from fitting parts together. Another watchmaker instead starts with the same parts but puts some of them together into subsystems. Not until the subsystems are working does he then join them into a watch. The second watchmaker prospers, while the first has to start all over again whenever he is interrupted.

Likewise, any major cognitive process needs to be analyzed into subprocesses. The understanding of the subprocesses will then suggest principles that can also be used in models of a wide variety of other processes. Agreement on the principles of how to organize neural networks into subnetworks is far from universal. In part, this reflects the incredible variety of ways in which any given cognitive function is organized across many biological species (indeed, phyla) and individuals within these species. In part, it reflects the variety of theoretical perspectives brought to bear on these problems. Yet it is possible to see through this diversity a few subnetwork organizing themes that are common to network models arising from many sources.

Two general examples of analyzing a cognitive process into subprocesses are now given. Both relate to main topics of later chapters in this book.

The first example, discussed more fully in Chapter 6, is the categorization of sensory patterns. Besides being one of the most common commercial applications of artificial neural networks, categorization is necessary for natural neural networks to make sense of their environments. For definiteness, say that the network is processing handprinted characters and attempting to match each one to a known letter of the alphabet. The simplest (although not the only) way to understand this process is to include in the network some units ("feature nodes") that respond to presence or absence of writing at particular locations, and other units ("category nodes") that respond to patterns of feature node activation representing particular letters (see Figure 1.1).

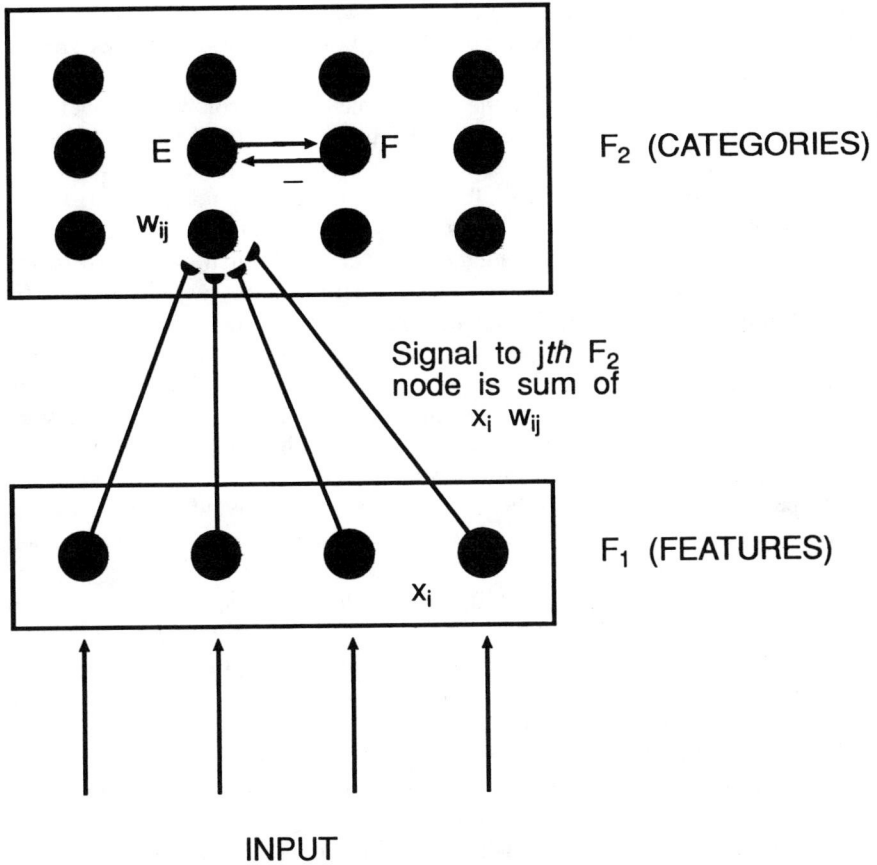

Figure 1.1. Generic categorization network combining associative learning and competition. (Modified from Levine, 1989c, with permission of Miller Freeman Publishers.)

To carry our thought experiment further, how are feature nodes and category nodes likely to be connected? If these connections were hard-wired, the network could not adjust to significant changes in its patterns of activation — for example, receiving inputs of Japanese or Russian instead of Roman characters. Hence, it is usually desired that the strength of the connection between any specific feature node and any specific category node be allowed to change over time, as a result of repeated activation of the connection. Such a change is often accomplished by a principle called *associative learning*; the various different types of associative learning laws in neural networks are the main topic of Chapter 3.

Hence, associative learning has been one of the common subnetwork organizing principles in neural networks from the early 1960's to the present. But many categorization models also depend on another organizing principle, commonly known as *competition*. To motivate the idea of neural competition, consider a sloppily written letter that is ambiguous — say, it could either be an E or an F. The network needs a method for deciding which of the two letters is the most likely one. The feature nodes are activated, to varying degrees, by the incoming letter, and in turn activate their category nodes via the internode connections. As shown in Figure 1.1, there is mutual inhibition between the category nodes. Thus, if both the "E" and "F" nodes are activated but the activation of the "E" node is greater than that of the "F" node, the system makes a definite decision that the letter is an E. Inhibition between nodes at the same level of the network (in this case, the category level) is often regarded as competition between the different cognitive entities coded by those nodes; the various competition laws in neural networks are the main topic of Chapter 4.

Associative and competitive principles are likely to combine in many other cognitive processes besides categorization. Our second example is attentional modulation of conditioning, which is discussed more fully in Chapter 5. Suppose that a neutral stimulus such as the sound of a bell has become associated with food. Then how does the animal learn to pay more attention to the bell than to other neutral stimuli in its environment?

Again, the psychological results suggest the neural network principle of competition. If different nodes develop representations of sensory stimuli, then the bell node should somehow "win" a competition with other sensory nodes for storage in short-term memory.

But why is the competition among sensory representations biased in favor of the bell? One plausible answer suggests the neural network principle of associative learning. Other things being equal, the bell node tends to be activated because of prior association between the bell and food, or, more abstractly, between the bell and satisfaction of the hunger drive (see Figure 1.2). Hence, if there is another node representing the primary reinforcer of food or the hunger drive itself, repeated pairing of the bell and food tends to activate the bell representation if the animal is hungry. Strengthening pathways based on pairing of stimuli is an example of associative learning.

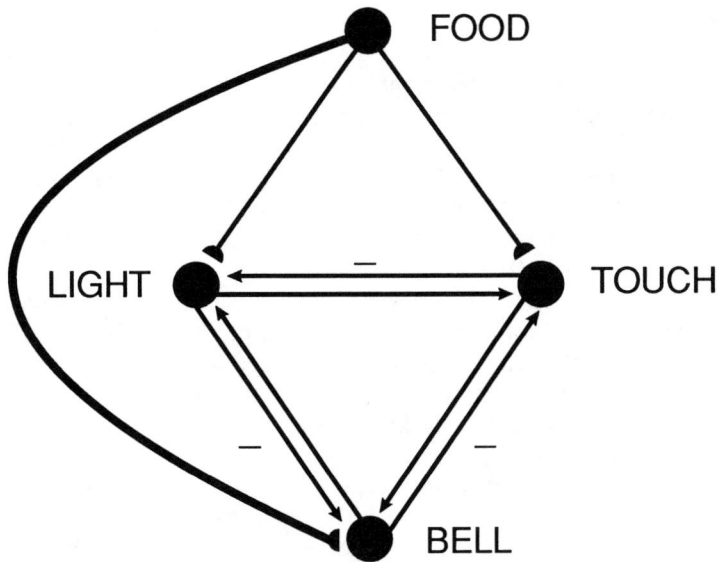

Figure 1.2. Generic selective attention network combining associative learning and competition. Darker line from food to bell represents a connection strengthened by learning, based on pairing of the bell with food. This biases competition among light, touch, and bell nodes in the bell's favor. (Modified from Levine, 1989c, with permission of Miller Freeman Publishers.)

In the course of this exposition, a few other major subnetwork organizing principles will emerge besides association and competition. One of these principles, particularly used by Grossberg and his co-workers, is *opponent processing*. This means that neural architectures are organized into pairs of pathways with opposite significance (*e.g.*, light and dark, or reward and punishment) in such a manner that a sudden decrease in activity of one pathway activates the opposing pathway. Like association and competition, opponent processing plays a role in some of the attention models discussed in Chapter 5 and some of the categorization models discussed in Chapter 6.

All these principles have been suggested by a heterogeneous database that is partly physiological and partly psychological. In some cases, the theory was first suggested by psychological results and later, at least qualitatively, supported neurophysiologically. An example is associative learning theory. Hebb (1949), inspired by Pavlovian conditioning data, proposed that if one neuron connects to another via a synapse, and the firing of the first neuron is repeatedly followed by firing of the second, then the synapse should become strengthened. As is seen in the early part of Chapter 3 below, examples of neuronal behavior consistent with this general hypothesis (though not in the precise form that Hebb had proposed) were discovered first in invertebrates (for example, Kandel & Tauc, 1965) and later in vertebrates (for example, Bliss & Lomo, 1973). More recent experimental studies, moreover, have partially supported mathematical variations of Hebb's learning law proposed by modelers in the 1960's and 1970's.

In this book, I will speak rather freely of "representation" by network nodes of features, categories, or other concepts. I largely skirt the controversy over whether this is a true representation (see, *e.g.*, Stone & VanOrden, 1989). My own position is that the concept of representation is a vast oversimplification, but a useful one for modeling purposes. Typically, one models a small part of a cognitive process at a time. A complete model of how particular sensory events are transformed into particular movement sequences depends on concatenation of several networks, each modeling a different level of the process. For example, one network might transform raw visual data (*cf.* Section 4.3 of this book), another might associate the same data with other co-occurring visual (or auditory) data, and yet another might classify these data. So while the "representations" in any particular neural network are not equivalent to the entity being represented, that should not invalidate the concept of representation because the part should not be confused with the whole.

Methodological Considerations

Subnetworks incorporating principles such as associative learning, competition, and others to emerge in later chapters can be thought of as part of the neural network modeler's "tool kit." So can larger networks that have been developed by major researchers in the field. For a given application, the adaptive resonance or back propagation or brain-state-in-a-box theory (to give the names of some popular theories we see more of later) might be suitable in some ways but not in others. As needed, one can readily add or subtract connections in a model or concatenate subnetworks from different models. A few researchers have already developed hybrids of previous network architectures for particular applications (see, *e.g.*, Hecht-Nielsen, 1987; Elsberry, 1989; Li & Wee, 1990).

Hence, when neural networks are used in brain modeling, new discoveries on the brain or on biological cognition may force modifications of a theory rather than abandonment of the entire structure. Also, a general qualitative theory may be widely applicable but the detailed instantiation of that theory may vary enormously between individuals and species.

The models discussed in this book vary in how faithful they are to known neuroanatomy and neurophysiology. The units, or nodes, in neural networks are often regarded as populations of neurons that are unified in some functional sense (*e.g.*, those cells responsive to light in a particular part of the visual field, or to the hunger drive). Analogies of these units to averaged cell types in particular brain areas are quite close at times, more remote at other times. Hence, the book title uses the broad term *neural and cognitive modeling* to encompass models with different degrees of fidelity to real brain structure. The boundary between more and less "brain-faithful" networks is a fluid one. In fact, as the above example of associative learning shows, models that are more "cognitive" than "neural" sometimes lead to "neural" predictions that are later supported by data.

As neural networks have become popular, it is often asked how much neural network theory has really accomplished, either in its technological applications or in its efforts to explain neurobiological data. The problems that neural network theory addresses are complex, and no single model has yet "cracked" the problem of categorization, or attention, or memory. What has happened, instead, is that neural network theory has been part of a slow, steady increase in overall understanding of brain and cognitive functions. Hence, I believe that researchers in the field are justified in saying, "Don't bite my finger — look where I am pointing" (McCulloch, 1965).

Not surprisingly, the technological applications of neural networks have a head start on the biological applications. As seen in Chapter 8, artificial neural networks have achieved some preliminary successes in applications to diverse areas, such as speech recognition and synthesis; using a knowledge base to make decisions on whether to grant mortgage insurance; and classification of radar patterns. In neurobiology, network models have made some experimental predictions, particularly in vision, and have begun to suggest analyses of more central brain functions. The comfortable interplay between theory and experiment that has existed for most of this century in physics has yet to be firmly established in neurobiology or psychology. But the scientific approach to knowledge, broadly speaking, argues that mental phenomena should have *some* mechanistic basis that will eventually be understandable by human beings. As Wiener (1954, p. 263) said, the faith of scientists is that nature (including mind) is governed by ordered laws, not by the capricious decrees of a tyrant like Lewis Carroll's Red Queen. Neural network modeling provides the best methodology now available for building toward the theoretical neuropsychology of the future.

Since all current neural models are subject to modification, this book is written to give the student or other reader hands-on experience in thinking about, simulating, and ultimately designing neural networks. It begins in Chapter 2 with a historical overview of major trends and the roots of current key ideas. Later chapters are organized around cognitive tasks or structural principles or both. The structures proceed from simpler to more complex interconnected networks as the book proceeds.

2

Historical Outline

Faithfulness to the truth of history involves far more than a research ... into special facts ... The narrator must seek to imbue himself (sic) with the life and spirit of the time.

Francis Parkman, *Pioneers of France in the New World*

The abuse of truth should be as much punished as the introduction of falsehood.

Blaise Pascal, *Pensées*

2.1. DIGITAL APPROACHES

Neural network modeling as we know it today is rooted in a rich interdisciplinary history dating from the early 1940's. Much of this history is discussed in Sections 1 through 4 of Levine (1983b) and will be reviewed here. The early development of digital computers, and some perceived similarities between computers and brains, spurred interest in developing a new science called cybernetics (Wiener, 1948). In particular, the computer-brain analogy was based on the fact that neurons are all-or-none, either firing or not firing, just like binary switches in a digital computer are either on or off.

Since that time, neurophysiological data have indicated that the all-or-none outlook is oversimplified. Also, neural network models have been developed whose functional units are neuron populations rather than single neurons. In spite of these technical advances, current approaches still owe many of their formulations to pioneers from the 1940's, such as McCulloch, Pitts, Hebb, and Rashevsky.

The McCulloch-Pitts Network

This inquiry essentially began with the classical study of all-or-none neurons by McCulloch and Pitts (1943). In this article, hidden under some elaborate symbolic logic, is a demonstration that any logical function can be duplicated by some network of all-or-none neurons. That is, a neuron can be embedded into a network in such a manner as to fire selectively in response to any given spatiotemporal array of firings of other neurons in the network.

The rules governing the excitatory and inhibitory pathways in McCulloch-Pitts networks are the following:

1. All computations are carried out in discrete time intervals.
2. Each neuron[1] obeys a simple form of a *linear threshold law*: it fires whenever at least a given (threshold) number of excitatory pathways, and no inhibitory pathways, impinging on it are active from the previous time period.
3. If a neuron receives a single inhibitory signal from an active neuron, it does not fire.
4. The connections do not change as a function of experience. Thus the network deals with performance but not learning.

More general linear threshold laws are considered later in this section, in reference to the work of Rosenblatt, 1962.

An example of an all-or-none neural network is reproduced in Figure 2.1. This network was designed by McCulloch and Pitts (1943) as a minimal model of the sensation of heat obtained from holding a cold object to the skin and then removing it. The cells labeled "1" and "2" are, respectively, heat and cold receptors on the skin, whereas heat and cold are felt when cells "3" and "4" fire, respectively. Each cell has a threshold of 2, hence fires whenever it receives two excitatory (+) and no inhibitory (-) signals from other cells active at the previous time.

---

[1] In most of the models discussed in this book, network elements are called "nodes" or "units" rather than "cells" or "neurons". The exception is made for the McCulloch-Pitts network because their network is directly inspired by the all-or-none firing properties of neurons.

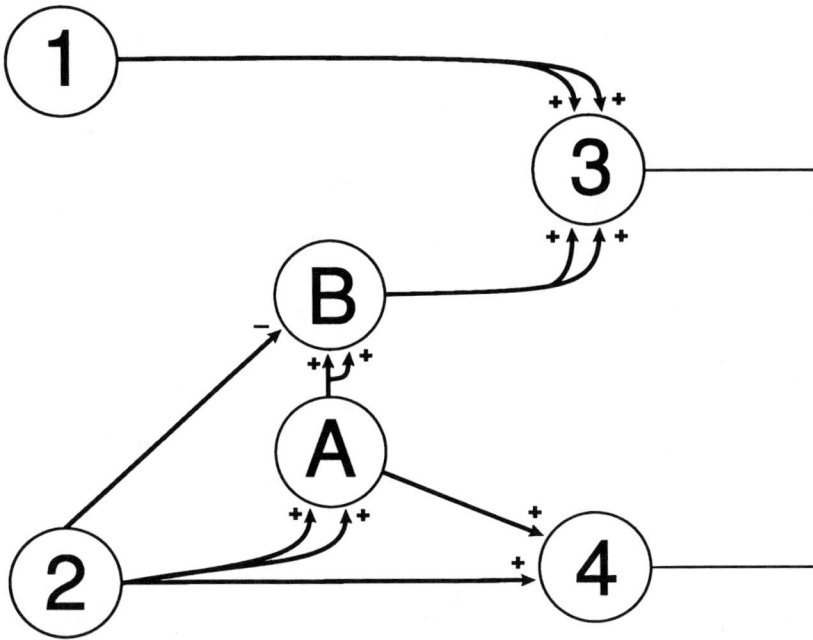

Figure 2.1. Example of an all-or-none network. Neurons labeled "1" and "2" are heat and cold receptors on skin. Heat and cold are felt when neurons "3" and "4" are active, respectively. Each neuron has threshold 2. A cold object held to the skin and then removed causes a sensation of heat. (Adapted by permission of the publisher from Levine, *Mathematical Biosciences* **66**, 1-86. Copyright 1983 by Elsevier Science Publishing Co., Inc.)

In the network of Figure 2.1, if a cold object is presented and then removed, this means that at time 1, the only cell firing is cell 2. At time 2, cell A fires because it receives two excitatory signals from cell 2. Since the cold has been removed, cell 2 does not fire again, nor do any of the other cells in the network. At time 3, cell B fires because it receives two excitatory signals from cell A. At time 4, the two excitatory signals from B to 3 cause 3 to fire, meaning that heat is felt. The time sequence of firing patterns is shown in

Table 2.1(a). In contrast, consider the same network's response to the cold object being on the skin continuously; Table 2.1(b). At time 2, cells 2 and A will both be firing. At time 3, cell B will not fire because the inhibitory signal from cell 2 prevents B's firing in response to A. Cell 4, however, will fire because it receives excitation from *both* cells 2 and A; hence, cold will be felt.

Time	Cell 1	Cell 2	Cell a	Cell b	Cell 3	Cell 4
1	No	Yes	No	No	No	No
2	No	No	Yes	No	No	No
3	No	No	No	Yes	No	No
4	No	No	No	No	Yes	No

FEEL
HOT

(a)

Time	Cell 1	Cell 2	Cell a	Cell b	Cell 3	Cell 4
1	No	Yes	No	No	No	No
2	No	Yes	Yes	No	No	No
3	No	Yes	Yes	No	No	Yes

FEEL
COLD

(b)

Table 2.1. Firings of neurons in the network of Figure 2.1 at successive time steps.

The McCulloch-Pitts model, though it uses an oversimplified formulation of neural electrical activity patterns, presages some issues that are still important in current cognitive models. For example, some of the best known modern connectionist networks contain three types of units or nodes — *input units, output units*, and *hidden units*. The input units react to particular data features from the environment (*e.g.*, "cold object on skin," "black dot in upper left corner," "loud noise to the right"). The output units generate particular organismic responses (*e.g.*, "I feel cold," "the pattern is a letter A," "walk to the right"). The hidden units (a term popularized by Rumelhart & McClelland, 1986) are neither input nor output units themselves but, via network connections, influence output units to respond to prescribed patterns of input unit firings or activities. The input-output-hidden trilogy can at times be seen as analogous to the distinction between sensory neurons, motor neurons, and all other neurons (interneurons) in the brain. At other times, though, a model neural network is designed to represent a small part of a larger behavioral process. The output may therefore not be a motor output but a particular internal state, such as a categorization or an emotion, that could be preparatory to a present or future motor response.

Note that in the McCulloch-Pitts network of Figure 2.1, there are already input units (cells 1 and 2), hidden units (cells A and B), and output units (cells 3 and 4). This distinction becomes explicit in more sophisticated linear threshold networks to be discussed below. In particular, the *perceptrons* developed by Rosenblatt (1962) contained units classified as "sensory," "associative," or "response."

Another cognitive issue raised by the "feel hot when cold is removed" network of Figure 2.1 is how to create output unit responses to given inputs that depend on the context of previous inputs. Specifically, this network responds to difference of the present input from a previous one; this may be called *temporal contrast enhancement*, by analogy with the spatial contrast enhancement (particularly observed in visual responses) which is a main topic of Chapter 4. Various forms of temporal contrast enhancement have been combined with learning in many more recent neural network models (*e.g.*, Grossberg, 1972 b, c; Sutton & Barto, 1981; Klopf, 1986; Bear, Cooper, & Ebner, 1987; Grossberg & Schmajuk, 1987). Some of these recent networks model such psychological effects as a motor act becoming rewarding when it turns off an unpleasant stimulus (relief); the withholding of an expected reward being unpleasant (frustration); and the reward value of food being enhanced if the food is unexpected (partial reinforcement acquisition effect).

McCulloch and Pitts also confronted the issue of how memory is stored. Figure 2.2(a) shows a network of the McCulloch-Pitts type in which a neuron fires if a given input (say, a light) is on for three time units in a row. A similar network can easily be constructed to respond to any fixed number of consecutive occurrences of an input. Figure 2.2(b) shows a network in which a neuron is made to fire if the light has been on at *any time in the past*. Note that the mechanism for such memory storage is a reverberatory circuit. The concept of reverberation remains central to the understanding of memory today, and some advantages and limitations of the mechanism are discussed below.

McCulloch and Pitts noted the absence of a precise sense of timing in their model (1943, p. 35): "the regenerative activity of constituent circles renders reference indefinite as to time past." To them, this makes the model useful in certain ways: "This ignorance, implicit in all our brains, is the counterpart of the abstraction which renders our knowledge useful." Yet obviously a sense of timing is necessary for some other cognitive processes. For those processes, it is necessary to include, as later models do, the possibility of changing connection strengths over time.

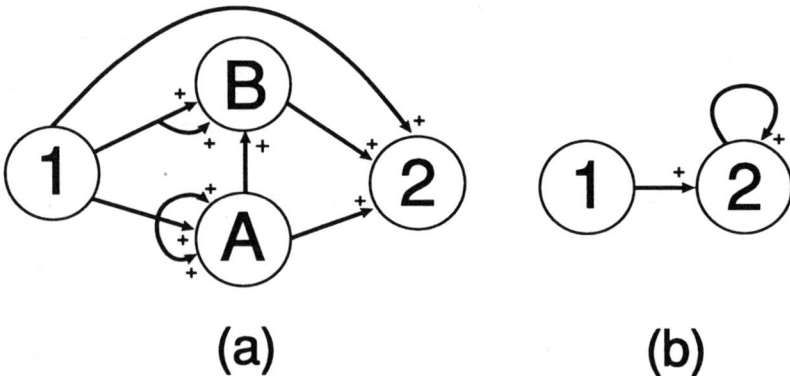

(a)   **(b)**

Figure 2.2. Two more all-or-none neural networks. In both networks, neuron "1" responds to a light being on. (a) Each neuron has threshold 3, and neuron "2" fires after the light has been on for three time units in a row. (b) Neuron "2," which has threshold 1, fires if the light has ever been on in the past.

Early Approaches to Modeling Learning: Hull and Hebb

At the same time that McCulloch and Pitts were developing a neural network formalism, psychologists were starting to consider mechanistic frameworks for studying learning and memory. This led to consideration of the issue of whether short-term memory (STM) can be distinguished from long-term memory (LTM). Hull (1943) proposed that the two memory processes involved the storage of two sets of traces. For example, consider the classic experiment of Pavlov (1927) where a bell is repeatedly paired with food until a dog salivates to the bell alone. After the experiment is stopped, conscious memory of the bell will be gone, since the dog is concentrating on other things. The memory of the *bell-food association*, however, will still be present, enabling the dog to salivate quickly on the next presentation of the bell. Hull thus distinguished between *stimulus traces* subject to rapid decay and *associative strengths* (or, in his terms, habit strengths) able to persist over a longer time period.

Hull's stimulus traces can be considered as the amounts of activity of particular nodes or functional units in a neural network. His associative strengths, then, are the strengths of connections between nodes. This suggests first that such connection strengths should change with experience, and second that they should correspond to some variable related to the *synapse*, or junction between neurons.

Hebb (1949) interpreted these memory issues with a theory that attempted to bridge psychology and neurophysiology. He declared that reverberatory feedback loops, which had been suggested as a memory mechanism by McCulloch and Pitts (1943), could be a useful mechanism for STM but not for LTM. Concerning traces arising in such reverberatory loops, Hebb (1949, p. 61) said: "Such a trace would be unstable. A reverberatory activity would be subject to the development of refractory states in the cells of the circuit in which it occurs, and external events could readily interrupt it." He was one of the first to recognize that a stable long-term memory depended on some structural change. But at the same time, he proposed (1949, p. 62) that "A reverberatory trace might cooperate with the structural change and *carry the memory until the growth change is made*" (author's italics).

Hebb went on (1949, p. 62) to describe a hypothesis for the structural change involved in long term memory: "When the axon of cell A is near enough to excite a cell B and repeatedly or persistently takes part in firing it, some growth process or metabolic change takes place in one or both cells such that A's efficiency, as one of the cells firing B, is increased." As for the nature of the structural change, he proposed that if one cell repeatedly assists in firing another, the knobs of the synapse between the cells could grow so as to increase the area of contact (see Figure 2.3).

Recent neurophysiological data have suggested that actual growth of synaptic knobs can sometimes occur (Tsukahara & Oda, 1981; Anderson, Lee, Thompson, Steinmetz, Logan, Knowlton, Thompson, & Greenough, 1989). More frequently, as seen in Chapter 3, there has been experimental support for cellular and synaptic processes that do not involve gross structural changes but that alter the effective strength of connections in other ways. Such processes can embody an associative rule such as Hebb's for changes in connection strength between cells. There has also been extensive theoretical work on alternative rules for learning of connection weights and network modeling based on these rules.

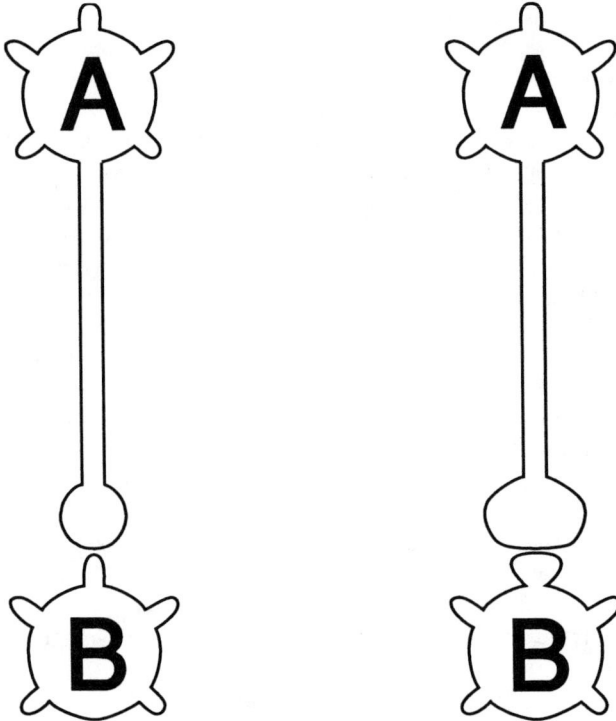

Figure 2.3. Diagram of Hebb's structural change hypothesis. The synaptic knob from *presynaptic* cell A to *postsynaptic* cell B gets larger after firing of A is repeatedly followed by firing of B. (Adapted by permission of the publisher from Levine, *Mathematical Biosciences* **66**, 1-86. Copyright 1983 by Elsevier Science Publishing Co., Inc.)

Rosenblatt's Perceptrons

In the early days of neural network modeling, considerable attention was paid to incorporating Hebb's rule for learning into a network of all-or-none neurons similar to that of McCulloch and Pitts. The modelers building adaptive networks of this variety included Rosenblatt (1962), Widrow (1962), and Selfridge (1959). In these networks, the McCulloch-Pitts form of the linear threshold law was generalized to laws whereby activities of all pathways impinging on a neuron are computed, and the neuron fires whenever some weighted sum of those activities is above a given amount.

The work of Rosenblatt was particularly influential and anticipated many of the themes of modern adaptive networks such as those of the PDP research group (*cf.* Rumelhart and McClelland, 1986). The main function he proposed for his perceptrons was to make and learn choices between different patterns of sensory stimuli.

Rosenblatt set out to study the pattern classification capabilities of networks of sensory (S), associative (A), and response units (R) with various structures of active connections between units. Figure 2.4 shows examples of perceptrons with four possible connection structure types. These types are, in order, *three-layer series-coupled* (connections one-way from S to A to R); *multilayer series-coupled* (connections from S to one level of A to another level of A to R); *cross-coupled* (like three-layer series-coupled with the addition of cross links between A units, and *back-coupled* (like series-coupled with the addition of feedback links from R to A units).

Whatever their type of coupling, Rosenblatt initially considered perceptrons with certain restrictions, which he called *elementary perceptrons*. To define an elementary perceptron, it was first necessary to define a *simple perceptron* (p. 85):

DEFINITION 22: A *simple perceptron* is any perceptron satisfying the following five conditions:

1. There is only one R-unit, with a connection from every A-unit.
2. The perceptron is series-coupled, with connections only from S-units to A-units, and from A-units to the R-unit.
3. The values of all sensory to A-unit connections are fixed (do not change with time).
4. The transmission time of every connection is either zero or equal to a fixed constant, τ.

5. All signal generating functions of S, A, and R units are of the form $u_i^*(t) = f(\alpha_i(t))$, where $\alpha_i(t)$ is the algebraic sum of all input signals arriving simultaneously at the unit u_i.

DEFINITION 23: An *elementary perceptron* is a simple perceptron with simple R- and A-units, and with transmission functions of the form $c_{ij}^*(t) = u_i^*(t-\tau)v_{ij}(t)$.

The precise form of transmission functions and signals is discussed in more detail in Section 2.3 below. Many more of Rosenblatt's mathematical definitions are also listed in that section.

Rosenblatt's book consisted of descriptions of a large number of mathematical and computer experiments on how well these different types of networks could either classify or generalize sensory patterns. The approach to modeling was described as "genotypic" rather than "monotypic." These terms were defined as follows (Rosenblatt, 1962, p. 22): "Instead of beginning ('monotypic') with a detailed description of functional requirements and designing a specific physical system to satisfy them, this approach ('genotypic') begins with a set of rules for generating a set of physical conditions, and then attempts to analyze their common functional properties."

Finally, we need to consider the learning rules for perceptrons, which Rosenblatt called the *reinforcement system*. Many of his ideas on learning were influenced by those of Hebb (1949). He distinguished two major types of reinforcement systems, *alpha* versus *gamma* systems. In the alpha system, all active connections terminating on a given active cell are changed by equal amounts, whereas inactive connections are not changed at all. In the gamma system, the total value of connection strengths is conserved, so that inactive connections are decreased while active ones are increased.

The amount of the connection change associated with reinforcement was a value δ determined by one of three *training procedures*. In a *response-controlled system*, the magnitude of δ is constant and its sign is determined by the response (that is, by the vector of R-element activities). In a *stimulus-controlled system*, the magnitude of δ is again constant but its sign is determined by the stimulus (that is, by the vector of S-element activities). In an *error-correcting* system, δ is 0 unless the response is determined elsewhere to be "incorrect." Also, reinforcement can be either *positive* or *negative*, that is, going in either the same direction as or the opposite direction to the current response.

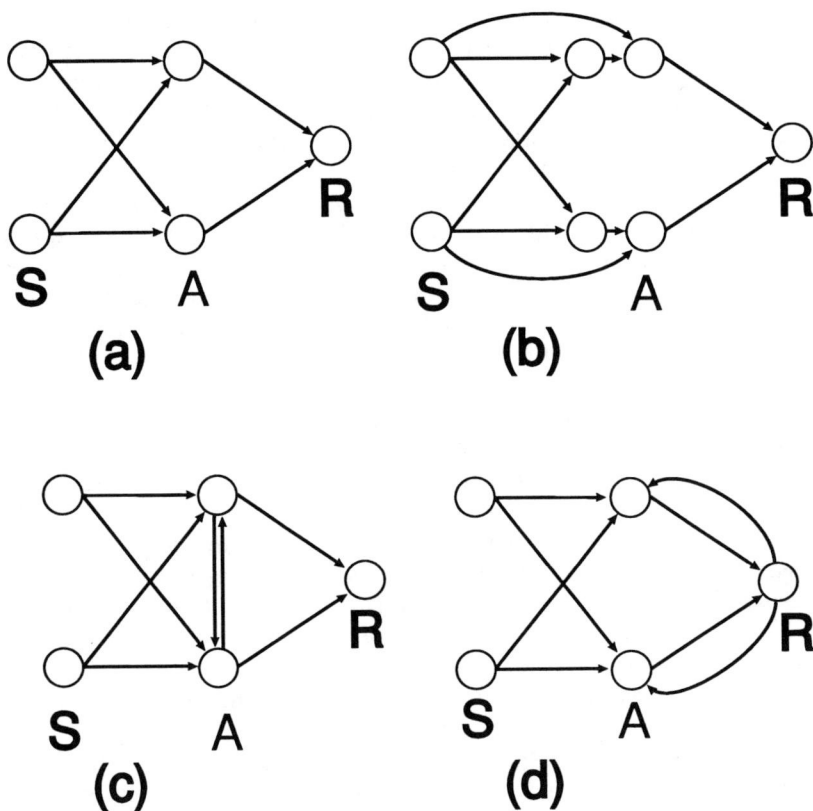

Figure 2.4. Examples of some classes of perceptrons: (a) three-layer series-coupled; (b) multilayer series-coupled; (c) cross-coupled; (d) back-coupled. (Reprinted by permission of the publisher from Levine, *Mathematical Biosciences* **66**, 1-86. Copyright 1983 by Elsevier Science Publishing Co., Inc.)

Some Experiments with Perceptrons

Rosenblatt (1962), starting with Chapter 7, ran experiments in which these different types of perceptrons were taught to discriminate classes of stimuli. A number of distinctions were found between the capabilities of perceptrons with different reinforcement rules and different training procedures, distinctions which are now mainly of historical interest. Not surprisingly, the perceptrons with error-corrective reinforcement converged faster than those with either stimulus-controlled or response-controlled reinforcement. Reinforcement rules of the error-correcting type were concurrently developed by Widrow and Hoff (1960) and are still used widely (*e.g.*, Sutton & Barto, 1981; Anderson and Murphy, 1986; Stone, 1986).

As for the distinction between alpha and gamma reinforcement, the results of the simulation experiments were equivocal. A slight advantage was found for the gamma rule if the various stimuli presented were of unequal size or frequency, while the alpha rule seemed to carry some advantage if the system included an error correction mechanism. Conservation laws similar to the gamma rule have been used in more recent neural networks. Rosenblatt found that the conservation rule made the network's responses more likely to be stable. This same property was used in later neural network models by Malsburg (1973) and Wilson (1975), who both thought this "principle of constant synaptic strengths" could be explained in terms of conservation of some chemical substance at or near synapses.

In one of Rosenblatt's major experiments (see Figure 2.5), the S-units are arranged in a rectangular grid. Connections from S- to A-units are random, whereas all A-units connect to the single R-unit. The perceptron (elementary, series-coupled) was taught to discriminate vertical from horizontal bars; variants of this experiment are given in the exercises for this chapter.

Rosenblatt found that if *all* possible vertical and horizontal bars are presented to the elementary series-coupled perceptron, and the perceptron is reinforced positively for responding to the vertical bars and negatively for responding to the horizontal, then eventually the network gives the desired response reliably to each one. However, if only some of the vertical and horizontal bars are presented and positively or negatively reinforced, the series-coupled perceptron is unable to generalize its behavior to other vertical or horizontal bars that have not been presented. What generalization the network can do is based on location rather than on any more fundamental properties of the input patterns. In models of visual pattern discrimination, issues like translation invariance (ability to recognize a given pattern regardless of where it is in the visual field) remain difficult ones today. Perhaps the best network for exhibiting this property is the Neocognitron of Fukushima (1980) (*cf.* Section 6.4 of this book).

Inability to generalize is related to another weakness of series-coupled perceptrons: their inability to separate out parts (features) of a complex pattern. This means that for a perceptron to perform categorizations, it needs an excessively large number of nodes. Minsky and Papert (1969, p. 161) remarked about a similar network that "along with its never-forgetting, it brings other elephantine characteristics." A third weakness is that these systems rely on a reinforcement signal external to the perceptron.

Further experiments and some theorems showed that generalization can be markedly improved by adding more connections to the perceptron. This can be accomplished either by interposing extra layers of associative units or by cross-coupling existing associative units. These additional connections also remove much of the perceptron's dependence on an external reinforcer. The separation of features from an overall pattern — for example, the classification of patterns on a rectangular grid ("retina") by whether there is a square in the center — proved more difficult for perceptrons. Preliminary simulations indicated that feature detection might be improved by back-coupling, that is, adding feedback from R units to A units.

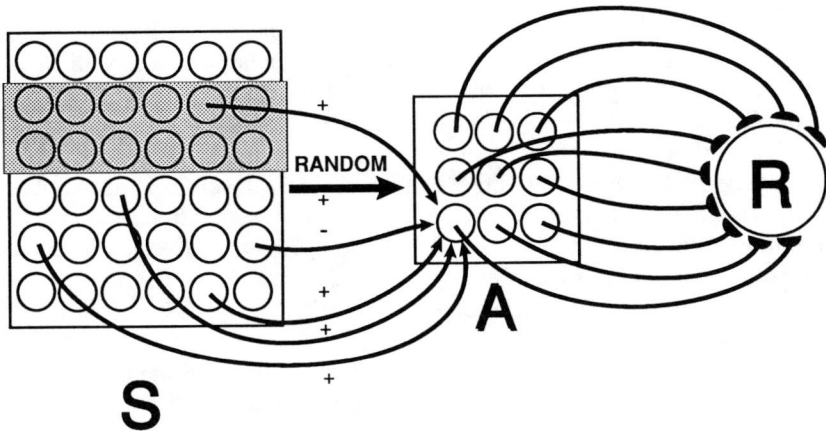

Figure 2.5. Schematic of a simplified form of one of Rosenblatt's experiments. S consists of a 6x6 grid (it was originally 20x20); each A unit receives 5 excitatory and 1 inhibitory inputs from (random) S units. A horizontal bar input is shaded.

Rosenblatt developed another idea that is a variant of back coupling — namely, back propagation of errors. This is a procedure popularized by Rumelhart, Hinton, & Williams (1986), and previously used by other modelers over the years (Werbos, 1974; LeCun, 1985; Parker, 1985). While each of these researchers made variations on this theme, the general principle was clearly stated by Rosenblatt (1962, p. 292): "The procedure to be described here is called the 'back-propagating error correction procedure' since it takes its cue from the error of the R-units, propagating corrections back towards the sensory end of the network if it fails to make a satisfactory correction quickly at the response end." In other words, if some A-to-R connection strengths need to be corrected for satisfactory response, inferences can be drawn regarding which S-to-A connections need to be changed as well. The mathematics of back propagation, and the possible biological basis for it, are discussed in Chapters 3 and 6.

The Divergence of Artificial Intelligence and Neural Modeling

From the late 1960's to the early 1980's, researchers in artificial intelligence largely abandoned neural networks of the linear threshold variety in favor of heuristic computer programs; this history was discussed in Section 3.2 of Levine (1983b). During this period, other linear threshold models contemporary with Rosenblatt's had some, although relatively minor, impact on artificial intelligence and neural modeling. Widrow (1962) developed the ADALINE (for "adaptive linear neuron"). Contrary to its author's intentions, this work was more influential among electrical engineers doing signal processing than among any group directly studying intelligent systems (Widrow, 1987). Selfridge (1959) developed the PANDEMONIUM model, which got its name from the different modules called "demons," each of them feature detectors with access to partial information from the environment. Decisions of the entire network were based on a weighted average of the decisions of the different "demons." The demon approach had some influence on some early computational models of specific brain areas such as the reticular formation (Kilmer, McCulloch, & Blum, 1969) and the hippocampus (Kilmer & Olinski, 1974). However, at that time, the detailed physiology of these brain areas was not understood well enough for such models to be widely accepted. Selfridge's work also inspired some of the abstract computational geometry of Minsky and Papert (1969).

Minsky and Papert (1969) developed their outlook in a book titled *Perceptrons*. The title was inspired by Rosenblatt's previous work, but the devices that Minsky and Papert studied are not exactly a subclass of Rosenblatt's. These machines do, however, have parts that correspond loosely to "sensory," "associative," and "response" areas. The Minsky-Papert perceptron starts with a *retina*, which is a grid consisting of small squares, each of which is at any time active or inactive ("light" or "dark"). Downstream from the retina are units that compute *partial predicates*. Each partial predicate outputs a value of 1 or 0 based on some rule depending on the activity or nonactivity of units in a given subset of the retina. The maximum size of that subset over all predicates is called the *order* of the perceptron. Finally, there is a decision-making unit that computes a linear function of those predicate outputs and responds when that linear function is above some threshold.

Minsky and Papert proved that their abstract form of the perceptron can learn any classification of patterns on its retina. However, many of the theorems stated that for a perceptron to make some geometrically important classifications, the order of the perceptron has to get arbitrarily large as the size of the retina increases. Theorems of this sort were widely interpreted as discrediting the utility of perceptron-like devices as learning machines. Now Minsky says that, in retrospect, the discrediting of perceptrons was an overreaction (Rumelhart & McClelland, 1986, Volume 1, pp. 158-159).

Moreover, some of the visual discriminations that are difficult for perceptrons are also difficult for humans. For example, consider the distinction between connected and disconnected figures, as shown in Figure 2.6. It is easy for the unaided eye-brain combination to tell that a filled-in circle is connected, whereas a pattern of two filled-in circles side by side is disconnected. But it is next to impossible for the eye and brain to tell which of the two convoluted patterns in Figure 2.6 is connected and which is disconnected without some help from finger tracing.

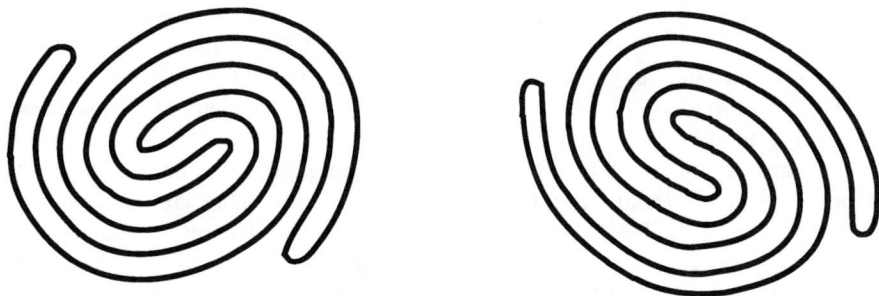

Figure 2.6. The finite-order perceptrons of Minsky and Papert (1969) cannot tell that the curve on the left is connected, whereas the curve on the right consists of two disjoint arcs. Can *you* tell that by visual inspection? (Reprinted from Minsky and Papert, 1969, with permission of MIT Press.)

The models of the PDP group, which originated about 1981 and most of which are summarized in Rumelhart and McClelland (1986), recaptured some of the threads from Rosenblatt's work. They showed that some of the distinctions that are impossible for Minsky and Papert's kind of simple perceptrons (such as between inputs that activate an odd versus an even number of retinal units) can be made by perceptrons with additional "hidden unit" layers (cross-connections) *and* nonlinear activation functions. Some of this work is discussed in Chapters 3 and 6.

2.2. CONTINUOUS AND RANDOM-NET APPROACHES

While the cybernetic revolution was stimulating discrete (digital) models of intelligent behavior, there was a concurrent proliferation of results from both experimental neurophysiology and psychology. Some of these experimental results stimulated the development of continuous (analog) neural models. We turn now to the study of continous approaches, random net approaches, and finally some partial syntheses of continuous and discrete approaches.

Rashevsky's Work

One of the pioneers in the development of continuous neural models was Rashevsky. The best exposition of his outlook was in his 1960 book, *Mathematical Biophysics*. The first edition of this book had been written in 1938 - five years before the seminal article of McCulloch and Pitts (1943). Subsequently, the evolution of his thinking had been altered by the McCulloch-Pitts article (which was published in a journal that Rashevsky himself founded and edited).

In most applications of mathematics to physical phenomena, including the biophysics of electrical current flow in single neurons, there are variables that are not all-or-none but may take on any of a range of values. Hence, such processes are typically modeled using differential equations, which are equations describing continuous changes over time in an interacting collection of physical variables. (For those desiring a "primer" in differential equations and their utility in neural network modeling, please refer to Appendix 2.) Rashevsky (1960) described how the earlier edition of his book had used differential equations to model various data in the psychophysics of perception. These data included the relation of reaction times to stimulus intensities, and the just noticeable differences among intensities.

Rashevsky went on to describe how his thinking had been influenced by the article of McCulloch and Pitts (1943), which used all-or-none neurons. He stated (Rashevsky, 1960, p. 3) that "the proper mathematical tool for representing the observed *discontinuous* interaction between neurons was not the differential equation but the Boolean Algebra or Logical Calculus." Yet it was difficult to model the observed psychophysical data using the McCulloch-Pitts postulates. This paradox was resolved with the observation that such behavioral data reflect the combined activity of very large numbers of neurons. Hence, "the discontinuous laws of interaction of individual neurons lead to a sort of average continuous effect which is described by the differential equations postulated originally" (also from p. 3 of Rashevsky, 1960).

The reconciliation effected by Rashevsky and others between continuous and discrete models is still in common use today. The description in terms of average activity is in line with the trend toward building models based on functional units or nodes that may represent large numbers of neurons (*cf.* Chapter 1). (The boundaries of "functional units" in actual mammalian brains have yet to be defined precisely. Edelman, 1987 speculates that units on the order of several thousand neurons in size encode stimulus categories of significance to the animal.)

Averaging across many neurons also allows the use of deterministic equations for unit activity even if the behavior of single neurons includes a random component. A neuron fires (*i.e.*, transmits an impulse or, more technically, an *action potential*) if its transmembrane voltage exceeds a value called the *threshold* (see Appendix 1 for details). This threshold is widely believed to vary according to some probability distribution, such as the Gaussian or normal distribution (see below). Neural models frequently average such random single-neuron effects across the functional groups of neurons that constitute network nodes; hence, the interactions between nodes become deterministic. In addition, contemporary models average random effects over short time intervals, so that the node activity variable is interpreted as representing a firing frequency rather than a voltage.

Rashevsky, however, made some simplifying assumptions about the neural averaging process, assumptions that may not always be valid. For example, he assumed that the frequency of impulses transmitted by a neuron is, on the average, a linear function of the cell's suprathreshold activity (see Figure 2.7a). That has proved to be a useful assumption for some neural models of sensory transduction, such as the model of the horseshoe crab retina developed by Hartline and Ratliff (1957). Yet averaging considerations can also lead one to consider input-output functions that are nonlinear, such as *sigmoid* functions (Figure 2.7b). As shown in Figure 2.8, if the firing threshold of an all-or-none neuron is described by a random variable with a Gaussian (normal) distribution, then the expected value of its output signal is a sigmoid function of activity.

For this reason sigmoids have become increasingly popular in recent neural models, as is seen particularly in Chapter 4. Also, there has been some physiological verification of sigmoid input-output functions at the neuron level (Rall, 1955; Kernell, 1965).

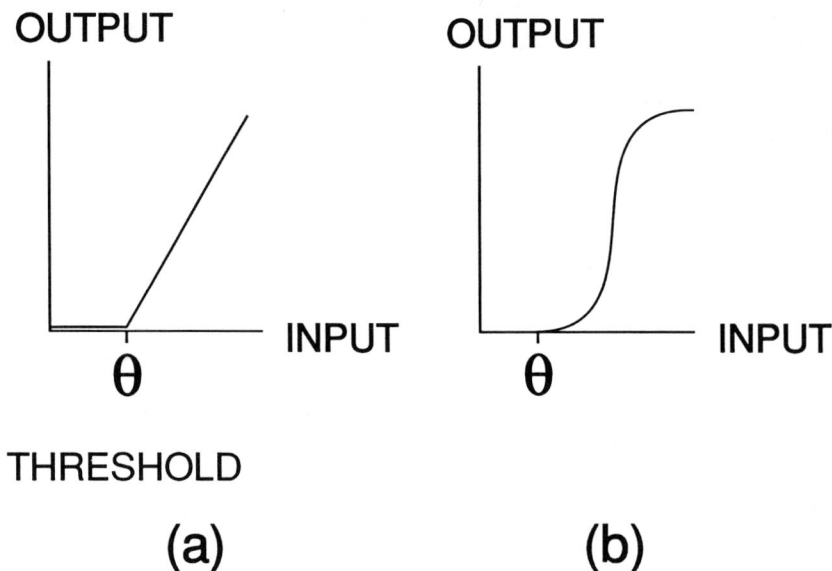

Figure 2.7. Schematic of linear (a) and sigmoid (b) functions of suprathreshold activity.

Rashevsky developed a variety of somewhat ad hoc models of different neurophysiological and psychological effects. He recognized a limitation of his models: "The vast majority of the structures in this book are highly specific, in the sense that a slight variation of the structure or a possible breakdown of a part of it renders the whole mechanism inoperative" (Rashevsky, 1960, p. 230). An example of such acute sensitivity to variation is Rashevsky's model of classical conditioning. In his conditioning model, memories are stored by the reverberation of activity within excitatory feedback loops. As seen in the above discussion of Hebb's work, this means that any momentary perturbation that inhibits the reverberation would destroy the memory.

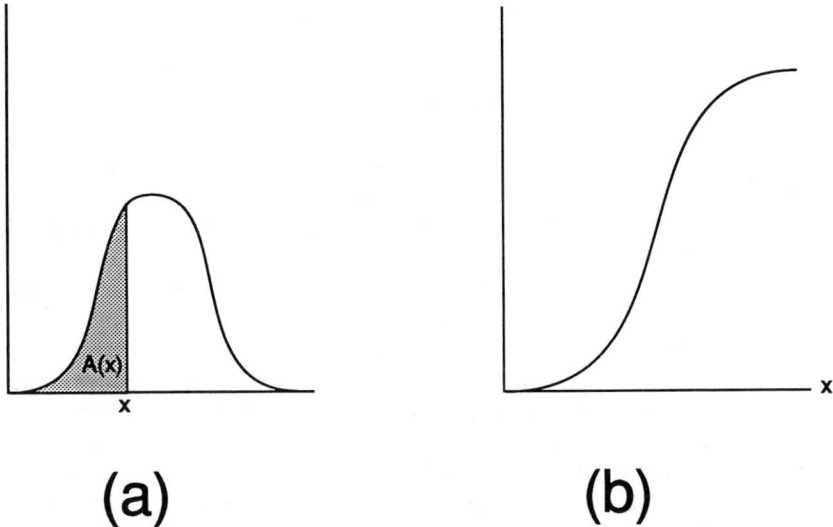

Figure 2.8. One possible biological basis for sigmoid functions: (a) Gaussian (normal) distribution of firing thresholds. If the activity (transmembrane voltage in the case of a single cell) is x, the node fires if the threshold is less than x. The probability of that happening is the area under the shaded part of the curve. (b) Schematic graph of the area in (a) as a function of x.

For all the weaknesses of his approach, Rashevsky inspired a generation of models that incorporate known neural phenomena into large networks of neurons connected more or less at random. One of these phenomena is the graded (not all-or-none) electrical potentials that occur at the dendrites of a neuron in response to all-or-none action potentials at other cells connected to it. Another is the *refractory period*, the short period of time in which a cell that has just fired (had an action potential) must remain inactive. (A historical

outline of some relevant experimental findings appears in Katz, 1966.) Some
of these models incorporated the averaging considerations described above, but
others used units that were explicitly treated as single neurons. In the
terminology of Rosenblatt (1962) (see Section 2.1 above), these random net
models tend to be genotypic rather than monotypic.

Early Random Net Models

Many of the early random net models were discussed in the last sections of
the review article by Harmon and Lewis (1968).[2] The first attempts at random
net modeling include only excitatory connections and no inhibitory ones. The
absence of inhibition in model networks, which is unrealistic from the
standpoint of known neuroanatomy, also led to unrealistic patterns of electrical
activity. The excitatory nets developed by Beurle (1956) and Ashby, Foerster,
and Walker (1962) tend, as time becomes large, to approach one of two
extremes of activity: maximal activity leading to saturation of the entire net, or
quiescence. The intermediate level of activity found in actual brains was not
modeled by these nets. Griffith (1963a, b, 1965) showed that in this random
net framework, stable submaximal activity is possible if inhibition is included.
    In the years following Griffith's articles, other modelers tried to develop
general theories for random neural networks with both excitatory and inhibitory
connections. Most of these theories were based on differential or difference
equations that include probabilistic terms. In addition, there were some neural
net models inspired by specific formalisms from other scientific fields.
Examples are models derived from statistical mechanics (Cowan, 1970) and
from nonequilibrium thermodynamics (Prigogine, 1969; Katchalsky, Rowland,
& Blumenthal, 1974; Freeman, 1975a). Application to neural network modeling
of analogies with other fields continues to this day. Some recent neural
networks, for example, have been described as arrays of two-state units. This
has led to analogies with the physics of *spin glasses*, which are structures with
an array of magnetic spins that have one of two possible values (Hopfield,
1982; Amit, Sompolinsky, & Gutfreund, 1985; Chowdhury, 1986). More
examples of physical analogies are discussed in Section 7.1.
    Yet analogies are limited by the fact that many nervous system properties
are uniquely neural, brainlike, or cognitive. Hence, the further development of
continuous and random models since the early 1970's has been influenced less

---

[2] To illustrate the field's rapid growth, Harmon and Lewis's review, unlike the later
one of Levine (1983b), covered detailed models of the single neuron as well as models
of neural networks — and in fewer pages than Levine's article!

by specific mathematical structures than by neuroanatomical, neurophysiological, and behavioral data. In particular, some data have indicated that brain connections may be random *within* certain neural populations and specific *between* these populations.

Reconciling Randomness and Specificity

The classic experiments of Lashley (1929) showed that many psychological functions, such as ability to remember specific events, are retained after extensive brain lesions. Lashley's experiments were among the first to inspire the idea, by now common, that representations of events are distributed throughout the brain rather than localized. Other experiments showed, however, that specific connections are important for other functions. Mountcastle (1957) found that the somatosensory (touch-sensitive) area of the cerebral cortex includes a well-organized topographic encoding of the body. Similarly, Hubel and Wiesel (1962, 1965) found that cells in the visual area of the cortex are organized into columns that code specific retinal positions or line orientations. (It is important to note, though, that visual and somatosensory maps are modifiable; the somatosensory maps, at least, can be altered even in adult life. If the connection to a given area of the cortex from the retinal or body area it would normally code is either cut or inactivated, the same area of cortex can learn to code a different, nearby area. Some of this evidence is summarized in Edelman, 1987).

The paradox between the Lashley data and the Hubel-Wiesel or Mountcastle data is resolved by means of a principle described in Anninos, Beek, Csermely, Harth, & Pertile (1970, p. 121) as "randomness in the small and structure in the large." This section considers some models whose equations are explicitly based on this principle. The same principle is implicit in many models discussed in later chapters. The latter models use purely deterministic equations at the population level that reflect the averaging over large ensembles of probabilistic effects at the single-cell level.

The article of Anninos *et al.* (1970) is one of a series of related articles (*e.g.*, Harth, Csermely, Beek, & Lindsay, 1970; Anninos, 1972 a, b; Wong & Harth, 1973). In this series of models, neurons are organized with random connectivities into "netlets," and netlets in turn are organized deterministically into larger nets. Evidence for such netlets was found, for example, in the organization of the somatosensory and visual areas of the cortex into functional columns (Mountcastle, 1957; Hubel & Wiesel, 1962, 1965).

Using many cell properties such as refractory periods, Anninos *et al.* (1970) derived an expression for the expected activity (defined as fractional number of neurons firing) at (discrete) time n+1 as a function of activity at time n. The crucial variable for determining long-term behavior is a parameter δ describing the number of excitatory postsynaptic potentials (here from within a netlet) needed to cause a cell to fire in the absence of inhibitory inputs. If δ is very small, netlet activity always tends to a unique positive stable steady state. If δ is very large, netlet activity always tends to 0. If δ is in a middle range, there are two stable steady states, one quiescent and one active, and a threshold exists for reaching the active state.

Anninos (1972a) pursued these principles of network organization further with simulations of multinetlet nets. He found, for example, that the dependence of activity of a single netlet in such a network on some external input can exhibit *hysteresis cycles*. That is, the effect of an input can depend on the past history of stimulation. He hinted, without giving details, that such hysteresis could be a mechanism for short-term memory.

Amari (1971, 1972, 1974) described random networks by means of differential or difference equations with two variable parameters - averaged connection weight and averaged threshold. Depending on the values of these two parameters, the network can have either a single stable steady state, many, or none. If the system has excitatory and inhibitory subnetworks, there can be oscillations of very long period. Amari's systems also modeled association of ideas, by means of connection weights.

A confluence of random net modeling with experimental data occurred in the work of Freeman (1972a, b, 1975a, b), much of which led to models of the olfactory cortex. He laid out some general principles for forming waves from pulses in large neural masses, and showed how this neural mass theory could be used to model EEG (brain wave) patterns and predict their frequencies. He has continued this general line of work to the present, with some results indicating that EEG patterns in the olfactory cortex tend to be chaotic (in the mathematical sense) in the absence of an odorant stimulus, but synchronized in the presence of an odor (Skarda & Freeman, 1987).

Some contemporary modelers are still building networks by connecting neural elements more or less at random and "seeing what happens." The mainstream, however, has shifted from random models to deterministic models that average out the random effects. A mathematical justification for this averaging process, using stochastic differential equations, was given by Geman (1979, 1980). In the deterministic approach, the networks often have particular connection patterns suggested by the cognitive task involved (such as associative learning, pattern storage, selective attention, or categorization). The next four chapters consider approaches to modeling each of those processes, and include discussion of the physiological basis for such models when it is known.

2.3. DEFINITIONS AND DETAILED RULES FOR ROSENBLATT'S PERCEPTRONS

Since Rosenblatt's perceptrons are conceptual ancestors of some current popular neural network models, it is useful to study some examples of perceptrons quantitatively. To that end we shall repeat enough of the definitions of his network concepts to be able to formulate exercises for computer simulation. We will start with the definitions of types of units and transmission functions on pp. 81 and 82 of Rosenblatt (1962)[3]:

> DEFINITION 6: A *sensory unit* (S-unit) is any transducer responding to physical energy (*e.g.*, light, sound, pressure, heat, radio signals, etc.) by emitting a signal which is some function of the input energy. The input signal at time t to an S-unit s_i from the environment, W, is symbolized by $c_{wi}^*(t)$. The signal which is generated at time t is symbolized $s_i^*(t)$.

> DEFINITION 7: A *simple S-unit* is an S-unit which generates an output signal $s_i^* = +1$ if its input signal, c_{wi}^* exceeds a given threshold, Θ_i, and 0 otherwise.

> DEFINITION 8: An *association unit* (A-unit) is a signal generating unit (typically a logical decision element) having input and output connections. An A-unit a_j responds to the sequence of previous signals c_{ij}^* received by way of input connections c_{ij}, by emitting a signal $a_j^*(t)$.

> DEFINITION 9: A *simple A-unit* is a logical decision element, which generates an output signal if the algebraic sum of its input signals, α_i, is equal or greater than a threshold quantity, $\Theta > 0$. The

---

[3] The terminology used in this section for unit activities and connection weights is quoted directly from Rosenblatt's book. Hence it differs from the usage in other chapters, which has been made as uniform as possible. Connection weights, labeled v_{ij} here, are called w_{ij} in Chapters 3-7. Node activities and signals, denoted by a variety of symbols depending on the type of unit, are usually called x_i and y_j in later chapters.

output signal a_i^* is equal to +1 if $\alpha_i \geq \Theta$ and 0 otherwise. If $a_i^* =$ +1, the unit is said to be *active*.

DEFINITION 10: A *response unit* (R-unit) is a signal generating unit having input connections, and emitting a signal which is transmitted outside the network (*i.e.*, to the environment, or external system). The emitted signal from unit r_i will be symbolized by r_i^*.

DEFINITION 11: A *simple R-unit* is an R-unit which emits the output $r^* = +1$ if the sum of its input signals is strictly positive, and $r^* = -1$ if the sum of its input signals is strictly negative. If the sum of the inputs is zero, the output can be considered to be equal to zero or indeterminate.

DEFINITION 12: *Transmission functions* of connections in a perceptron depend on two parameters: the *transmission time* of the connection, τ_{ij}, and the *coupling coefficient* or *value* of the connection, v_{ij}. The transmission function of a connection c_{ij} from u_i to u_j is of the form: $c_{ij}^*(t) = f[v_{ij}(t), u_i^*(t-\tau_{ij})]$. Values may be *fixed* or *variable* (depending on time). In the latter case, the value is a *memory function*.

The concepts relating to reinforcement are defined precisely in pp. 88-92 of Rosenblatt (1962), as follows:

DEFINITION 33: *Positive reinforcement* is a reinforcement process in which a connection from an active unit u_i which terminates on a unit u_j has a value changed by a quantity $\Delta v_{ij}(t)$ (or at a rate dv_{ij}/dt) which agrees in sign with the signal $u_j^*(t)$.

DEFINITION 34: *Negative reinforcement* is a reinforcement process in which a connection from an active unit u_i which terminates on a unit u_j has its value changed by a quantity Δv_{ij} (or at a rate dv_{ij}/dt) which is opposite in sign from $u_j^*(t)$.

(Note: the "active units" u_i in the above definitions could be either A-units or R-units.)

DEFINITION 37: *Alpha system reinforcement* is a reinforcement system in which all active connections c_{ij} which terminate on some unit u_j (*i.e.*, connections for which $u_i^*(t-\tau)$ is not equal to 0) are changed by an equal quantity $\Delta v_{ij}(t) = \delta$ or at a constant rate while reinforcement is applied, and inactive connections ($u_i^*(t-\tau) = 0$) are unchanged at time t.

DEFINITION 38: *Gamma system reinforcement* is a rule for changing the values of the input connections to some unit, whereby all active connections are first changed by an equal quantity, and the total quantity added to values of the active connections is then subtracted from the entire set of input connections, being divided equally among them. Such a system is said to be *conservative in the values*, since the total of all values can neither increase nor decrease. The change in v_{ij} is equal to

$$\Delta v_{ij}(t) = [w_{ij}(t) - \frac{\sum_i w_{ij}(t)}{N_j}]\delta,$$

where $w_{ij}(t) = 1$ if $u_i^*(t-\tau) \neq 0$
                    0 otherwise,
N_j = number of connections terminating on u_j
δ = reinforcement quantity (typically 1, -1, or 0).

DEFINITION 39: A *response-controlled reinforcement system* (R-controlled system) is a training procedure in which the magnitude of δ is constant, and the sign of δ is entirely determined by the current response, r^*, regardless of the current stimulus, S. In general, unless otherwise specified, this term implies that the reinforcement is always positive (*i.e.*, the sign of δ agrees with the sign of r^*, in a simple perceptron).

DEFINITION 40: A *stimulus-controlled reinforcement system* (S-system) is a training procedure in which the magnitude of δ is constant, and the sign of δ is determined entirely by the current stimulus, S, and a predetermined classification ...

DEFINITION 41: An *error-correcting reinforcement system* (error correction system) is a training procedure in which the magnitude of δ is 0 unless the current response of the perceptron is wrong, in which case, the sign of δ is determined by the sign of the error.

In this system, reinforcement is 0 for a correct response, and *negative* (see Definition 34) for an incorrect response ... .

EXERCISES FOR CHAPTER 2

○1. Hebb's rule for synaptic modification states that a connection strength will increase if activities of the two connected units are both high at the same time (with suitable delays, perhaps).   Give some possible advantages and disadvantages of this rule for network models of learning.

2. Design a McCulloch-Pitts network with heat and cold receptors and a cell that fires after the sequence "heat cold" or the sequence "cold heat" but nothing else. Assume that each cell takes exactly one time step to compute its output, and that cold and heat cannot be simultaneously felt at the same time step.

   Design another McCulloch-Pitts network, possibly a modification of the first one, so that the last cell fires after alternating sequences of three — "heat cold heat," "cold heat cold," "heat cold neither," or "cold heat neither" — but not after sequences that include repeats — "heat cold cold" or "cold heat heat."

3. Rosenblatt did a simulation of teaching an elementary perceptron to distinguish between 20-by-4 vertical bars and 20-by-4 horizontal bars. The following problem is based on a simplification of his experiment, so that the bars are 8-by-2 instead.

   In the simulation to be done here, S-units are arranged in an 8-by-8 grid ("retina"). There are 9 A-units and 1 R-unit. Only the A-to-R connections are modifiable, as shown in Figure 2.9.

   Each A-unit receives 8 excitatory and 2 inhibitory connections from S-units. The connection strengths v_{ij} are +1 for excitatory pathways and -1 for inhibitory pathways. For each A-unit, the sources of the connections are randomly distributed. The program listed in Figure 2.10, among others, generates a random number uniformly distributed between 0 and 1.   Then multiply RANDOM by 64, truncate to an integer, and add 1 to get an integer between 1 and 64. The last integer determines which S-unit connects to the current A-unit. (More than one connection to that unit can come from the same S-unit.)

In each run, there are 17 successive input stimuli, each a horizontal or vertical bar of width two. The connectivity within the S grid is *toroidal*: that is, the top row is considered to be adjacent to the bottom row and the leftmost column to the rightmost column. The topmost horizontal bar activates units 1 through 16; the second horizontal bar activates units 9 through 24, and so on down to the eighth and last horizontal bar, which activates the bottom row and the top row of units, that is units 57 through 64 and 1 through 8. Likewise, the leftmost vertical bar activates the left two columns of units, that is, units 1, 9, 17, 25, 33, 41, 49, 57, 2, 10, 18, 26, 34, 42, 50, and 58, up through the last vertical bar which activates units 8, 16, 24, 32, 40, 48, 56, 64, 1, 9, 17, 25, 33, 41, 49, 57. OBJECT: To teach R to respond positively to vertical, negatively to horizontal.

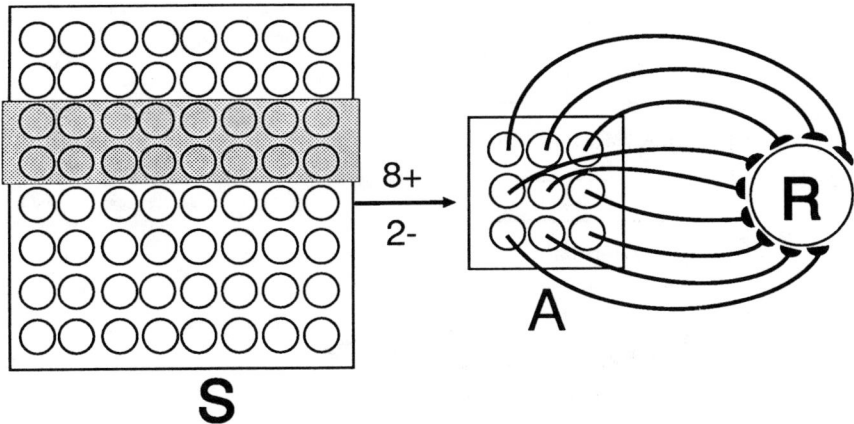

Figure 2.9. Perceptron, of similar architecture to the one shown in Figure 2.5, used in the simulations of Exercise 2.

If the ith S-unit is activated, then $s_i^*(t)=1$, otherwise $s_i^*(t)=0$. An A-unit computes $\alpha_j = \sum_i s_i^*(t) v_{ij}$, and its activity $a_j^*(t)=1$ if $\alpha_j > 2$,

$$0 \text{ if } \alpha_j \leq 2.$$

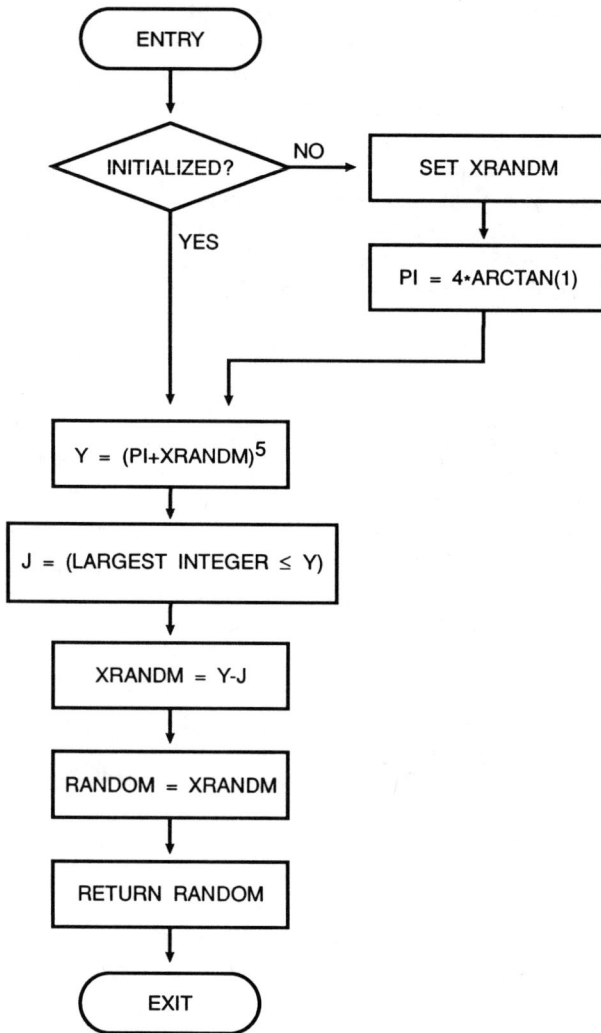

Figure 2.10. Generic program segment that, starting from any initial "seed" number between 0 and 1, generates a different random number in that interval on each pass.

At the start of each run, the A-to-R connection strengths w_j are set to values that are randomly (uniformly) distributed between -1 and 1. Error-correction and alpha reinforcement are used. That is, whenever a horizontal line is input and R (incorrectly) responds positively, any w_j that are positive while $a_j^*(t) = 1$ are reduced by an amount δ; whenever a vertical line is input and R (incorrectly) responds negatively, any w_j that are negative while $a_j^*(t) = 1$ are increased by the same value δ. R in turn responds negatively if

$$\Phi = \sum_j a_j^*(t)w_j \le 0, \text{ and positively if } \Phi > 0. \text{ Time delays can all be set to 0.}$$

Do 6 runs, 2 with each of three different δ values (2.0, .5, and .1). In the first run for each δ, all possible vertical and horizontal bars are presented. Then any bar to which the first response was incorrect is tested as the 17th stimulus. The network should, at least some of the time, learn the correct response to a bar that it has previously seen. If, even on the test, it still responds incorrectly, repeat the procedure (with the same connections and learning rate) and see if it learns it on succeeding iterations. In the second run for each δ, the top 4 horizontal and leftmost 4 vertical bars are each presented twice. Then another vertical or horizontal line not yet presented is tested as the 17th. No generalization should occur. (You could also do a variation on this problem in which the set of possible stimuli are as above, but presented in a random order until all of them are learned correctly.)

4. Do the same simulation as in Exercise 3 but with actual 20-by-4 bars, 300 associative units, 3 excitatory and 1 inhibitory connection to each A-unit, and a threshold of 2.

5. Do the same simulation as in Exercise 4 but with gamma instead of alpha reinforcement.

o6. Design a simulation in which a perceptron is trained to discriminate between two types of figures. Examples would be a square of a given size versus a triangle of a given size (that is, translates of a fixed square and a fixed triangle anywhere along the grid) or a square and a diamond. Another example would be to discriminate whether a figure does or does not contain the letter "X."

7. Do a simulation of the ADALINE network of Widrow (1962). In the ADALINE model, a set of bipolar (1 or -1) inputs is filtered through a corresponding set of adaptive weights, and the sum of the weighted inputs is then compared with a desired output. Then error-correcting reinforcement is applied.

The ADALINE equations are as follows. Let I_i, $i = 1$, ..., n represent specific inputs, and I_0 a constant ("bias") input equal to one. Let w_i, $i = 0$, ..., n represent the corresponding weights. Then the actual output, called y, is 1 if the weighted input

$$S = \sum_{i=0}^{n} I_i w_i \geq 0$$

and -1 if $S < 0$. Let y_0 be a desired output; for example, if the network is trained to learn the logical "AND" operation, n=2, and the inputs are 1 and 1, the desired output is also 1. Then at each time step, weights are updated according to the rule

$$\Delta w_i = a(y_0 - y)\frac{I_i}{n+1}$$

Weights are changed until the network has learned the desired output to every bipolar input vector.[4]

For the following simulations, the number n of non-bias inputs is 2, and the learning rate a is 1.

(a) Teach the network to learn the logical "AND" operation, which maps (1,1) to 1, (1,-1) to -1, (-1,1) to -1, and (-1,-1) to -1. Show that this can be learned in two passes through the sequence of four input vectors.

(b) Teach the network the logical "OR," which maps (1,1) to 1, (1,-1) to 1, (-1,1) to 1, and (-1,-1) to -1. Show that this can be learned in two passes through the sequence.

(c) Teach the network the logical "NAND," which always gives the sign opposite to the one given by the "AND." Show that this can be learned in three passes through the sequence.

(d) Show that the network *cannot* learn the "exclusive OR," which maps (1,1) to -1, (1,-1) to 1, (-1,1) to 1, and (-1,-1) to -1, by going through the sequence of training inputs and getting an infinite loop. (The exclusive OR can be learned by multilayer nonlinear networks, as will be seen in Chapter 6).

---

[4] Faster learning occurs in the ADALINE if y_0 is compared to the weighted input S instead of to the quantized output y. Widrow, Pierce, and Angell (1961) discuss the differences between these two error criteria; in that article, the error we use is called the *neuron error* and the other is called the *measured error*.

3

Associative Learning and Synaptic Plasticity

The present contains nothing more than the past, and what is found in the effect was already in the cause.

Henri Bergson, *L'Évolution Créatrice*

The mind is slow in unlearning what it has been long in learning.

Seneca, *Troades*

3.1. PHYSIOLOGICAL BASES FOR LEARNING

Recall from Section 2.1 the contribution of Hebb (1949) to the bridging of psychology and neurophysiology. Hebb proposed on psychological grounds the existence of synaptic modifications during learning, in the absence, then, of any physiological evidence for such modifications. Since that time, it has been experimentally demonstrated that correlated activity at the pre- and postsynaptic cells of many synapses in animal nervous systems alters the efficacy of the synapse in causing action potentials at the postsynaptic cell. Whereas the relationship of any of these cellular changes to actual storage of cognitive information remains speculative (see Brown, Chapman, Kairiss, & Keenan, 1988), these results at least provide a starting point for cellular analyses of learning.

Since Hebb's work, a wide variety of synaptic modification mechanisms has been suggested. There has been just as wide a variety of mathematical rules proposed for such modification. The term *associative learning* lacks a precise definition, but is typically used for learning rules where connection strengths tend to increase with correlated activity of the two connecting nodes (or neurons). This is sometimes referred to as *Hebbian learning*, but the term *associative* is preferred because such rules have diverged from Hebb's original conception (see the discussion in Grossberg and Levine, 1987, part II). *Associative learning* is also used to describe psychological effects in networks obeying such rules, such as classical conditioning and serial learning of lists. This section and the next discuss rules for associative learning, whereas Section 3.3 discusses rules where the change is opposite in direction to associative learning.

As discussed in Section 2.1, Hebb proposed that changes in the efficacy of synapses could take place via growth of synaptic knobs. Since such growth has not been commonly observed to take place in adult animals (though there are possible exceptions, *e.g.*, Anderson *et al.*, 1989), many alternative mechanisms have been suggested for changing synaptic efficacy. The ability for such change to take place is called synaptic *plasticity* or *modifiability*. Grossberg (1969b), for example, proposed that there could be a change in amount of usable presynaptic transmitter substance correlated with a change in postsynaptic protein synthesis. Stent (1973) proposed that there could be changes in the conformation of a receptor protein at the postsynaptic membrane. Woody, Buerger, Ungar, and Levine (1976) and Levine and Woody (1978) proposed that there could be resistance changes at the postsynaptic membrane, leading to altered electrical flows in that membrane's vicinity.

The existence of multiple mechanisms for plasticity is supported by a host of physiological results. Some of these results, in invertebrates since the mid-1960's and in vertebrates since the early 1970's, are briefly reviewed here. A much more extensive review of this body of experimental work is found in Byrne (1987).

Of the many articles that have shown some form of synaptic plasticity in experimental animals, two perhaps stand out as setting the tone for later work. The tone-setter for invertebrate studies was the article of Kandel and Tauc (1965). These investigators discovered in the sea slug *Aplysia* a mechanism called *heterosynaptic facilitation.* Heterosynaptic facilitation has been defined

(Byrne, 1987, p. 354) as "a change in synaptic efficacy (or cellular excitability) in one neuron as a result of release of a modulatory transmitter from another neuron." This work initiated a long series of cellular studies of learning in *Aplysia* and other invertebrates, which is still taking place. These studies started with nonassociative facilitation (strengthening) and habituation (weakening) of specific pathways, then went on to different kinds of associative modifications. Hawkins and Kandel (1984) review how invertebrate findings may relate to different forms of conditioning (see Chapter 5 for further discussion).

The tone-setter for mammalian studies was the article of Bliss and Lomo (1973). These investigators demonstrated in the rabbit hippocampus the phenomenon of *long-term potentiation* (LTP). LTP is defined (Byrne, 1987, p. 389) as "a persistent enhancement of synaptic efficacy generally produced as a result of delivering a brief (several second) high-frequency train (tetanus) of electrical stimuli to an afferent (incoming) pathway." This potentiation can last up to several hours in an isolated cellular preparation and several days in an intact animal. Since that time, LTP has been demonstrated elsewhere in the brain. Also, the hippocampal preparation has been studied quantitatively, and there have been recent demonstrations of associative influences on LTP (Levy, Brassel, & Moore, 1983; Lynch & Baudry, 1983; Levy, 1985; Kelso & Brown, 1986). For example, if a weak electrical stimulus to one pathway leading to a given neuron is paired with a much stronger stimulus to another pathway going to the same neuron, that same weak stimulus may subsequently become more effective in causing that cell to fire.

In summary, there now appears to be a sufficient physiological basis for many if not all of the neural network learning rules that have been suggested on cognitive grounds. We now proceed to a discussion of what these different rules are and some of the results of using them in neural networks.

3.2. RULES FOR ASSOCIATIVE LEARNING

Work on translating an associative rule for synaptic modification into explicit equations essentially began in the late 1960's, as discussed in Section 6 of Levine (1983b). Early efforts in this regard included some nonlinear models by Grossberg, and some models by Anderson and Kohonen, which started out being linear and later added some nonlinearities.

Outstars and Other Early Models of Grossberg

Grossberg's formal rules and related equations were derived from psychological considerations. The networks implementing these rules, in turn, suggested analogs of neural elements and interconnections. The most general form of these equations (Grossberg, 1969c) remains in current use in the modeling of multilevel adaptive networks with modifiable synapses between levels (*cf.* Chapter 6).

The psychological and neurophysiological implications of the theory thus derived were described in Grossberg (1969a). This article posed the question of how an organism learns to produce one sound (say B) in response to another (say A) after repeatedly hearing them in šequence. The network designed to answer that question was motivated by the following psychological postulates among others:

> 1. Language appears to be spatiotemporally discrete. That is, a sound like A is psychologically treated as an "atom" instead of being subdivided.
> 2. Such discrete symbols as occur in language are used to represent sensory experience, which is spatiotemporally continuous.
> 3. Learning changes from continuous to discrete; for example, a child learning to walk must concentrate continuously on his or her movements, whereas an adult walking has an automatic sequence of discrete steps.

The network that Grossberg used to satisfy these postulates consists of discrete elements with time-varying activities that satisfy continuous differential equations. Mathematical results about the equations, showing what is learned by particular networks, were proved in other articles (Grossberg, 1968a, b, 1969c, 1972a). For the variables defining these equations, Grossberg borrowed from Hull (1943) the notions of *stimulus trace* and *associative strength*. For each stimulus "atom" such as A, a stimulus trace $x_A(t)$ is defined that measures how active the memory for A is at any given time t. For each pair of "atoms" A and B, the associational strength $w_{AB}(t)$ measures how strongly the sequential association AB is in the network's memory at time t. In Figure 3.1, the ith stimulus trace x_i is located at the node v_i, and the association w_{ij} between the ith and j^{th} traces is located along the edge e_{ij}. Grossberg drew an analogy between the v_i and cell bodies, the e_{ij} and nerve axons, and the junctions e_{ij}-to-v_j and synapses (see Appendix 1).

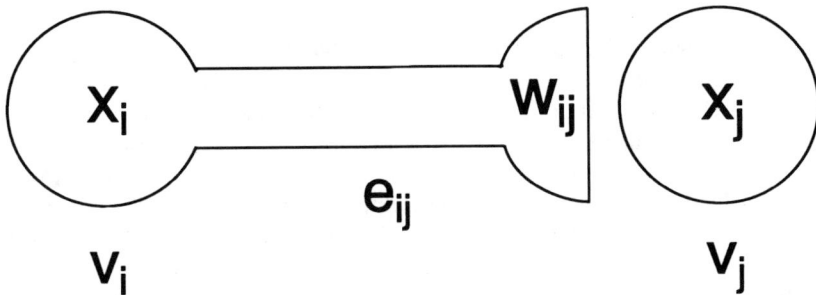

Figure 3.1. Schematic of two nodes and one modifiable connection between them, based on Grossberg's 1968 and 1969 articles.

Table 3.1 summarizes the effects that Grossberg incorporated into his differential equations. As this table shows, in the case of the sequence AB, it was desired that B should be produced if, and only if, A has been presented *and* the sequence AB is strong in memory. Similarly, the sequence AB should become stronger if, and only if, A is presented and followed by B. Hence, replacing A and B by the ith and j[th] stimuli in general, the variable x_j should increase if both x_i and w_{ij} are high, which means that the equation for the rate of change of x_j should include a (nonlinear) term like the product $x_i w_{ij}$. Likewise, the variable w_{ij} should increase if both x_i and x_j are high, which means that the equation for w_{ij} should have a term like the product $x_i x_j$. (See Appendix 2 for an introduction to the general process of incorporating neural or cognitive variables into differential equations.)

The import of Grossberg's approach to learning can be gleaned from study of a specific type of network architecture called the *outstar* (Grossberg, 1968a; see Figure 3.2). In the outstar one node v_1, called a *source*, projects to other nodes $v_2, v_3, ..., v_n$, called *sinks*. Long-term storage can be interpreted as residing in the *relative* weights of $w_{12}, ..., w_{1n}$, that is, in the functions

$$W_{1i} = \frac{w_{1i}}{\sum_{j=2}^{n} w_{1j}} \tag{3.1}$$

where "Σ" denotes summation (in this case, the sum of all weights w_{1j} of connections from source to sinks).

The outstar is affected by an input I_1 to the source node v_1, and a pattern (*vector*) of inputs I_2, ..., I_n to the sink nodes v_2, ..., v_n. (Grossberg sometimes interpreted the source input as a conditioned stimulus, and the vector of sink inputs as an unconditioned stimulus; see the discussion of conditioning models in Chapter 5.) The activity x_1 of v_1 tends to increase if the input I_1 is present, and decay back toward a baseline (interpreted as 0) in the absence of input. As illustrated in Table 3.1(a), the activity x_i of each v_i also tends to increase if both x_1 (activity of v_1) and w_{1i} (associative strength between v_1 and v_i) are significant, and decay otherwise. Finally, as illustrated in Table 3.1(b), w_{1i} tends to increase if both x_1 and x_i are significant, and decay otherwise.

A is presented	AB has been learned	B is expected
Yes	Yes	Yes
Yes	No	No
No	Yes	No
No	No	No

(a)

A is presented at a given time	B is presented a short time later	AB is learned
Yes	Yes	Yes
Yes	No	No
No	Yes	No
No	No	No

(b)

Table 3.1. Effects incorporated into Grossberg's differential equations.

If x_1 is interpreted as encoding A and x_i as encoding B, the decay of w_{1i} implies that the association between A and B is weakened while the network is not actively hearing A. This property is contrary to intuition: it is to be expected, rather, that the association will decay when A is presented without being followed by B, but remain constant when A is not presented at all. In a modified form of the outstar equations, used in much of Grossberg's later work, w_{1i} decreases only if x_1 is large and x_i is small. This implies that an association such as AB remains intact if neither A nor B is presented, but is weakened if A is presented and not followed by B.

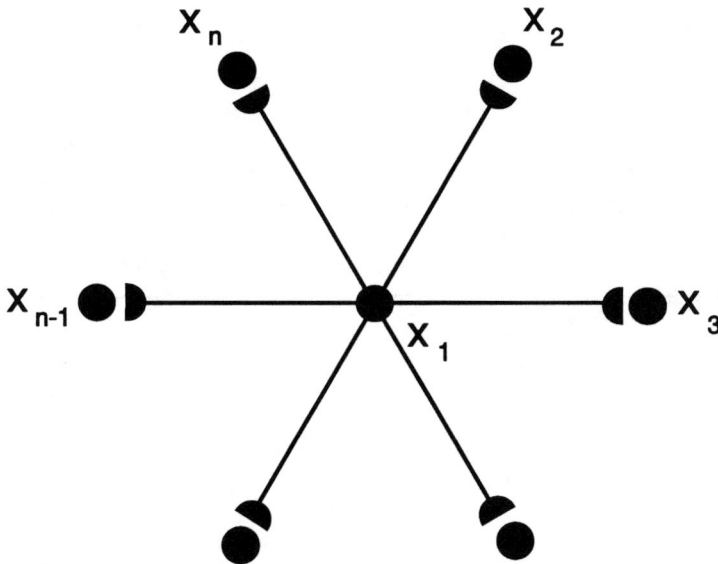

Figure 3.2. Outstar architecture. (Adapted by permission of the publisher from Levine, *Mathematical Biosciences* **66**, 1-86. Copyright 1983 by Elsevier Science Publishing Co., Inc.)

The outstar equations are given in Section 3.5. Before discussing the major results about those equations, let us briefly consider the implications of this theory for memory. The stimulus traces x_1 and x_i can be considered as analogs of *short-term memory* (often abbreviated STM) while the associative strengths w_{1i} are analogous to *long-term memory* (often abbreviated LTM). It is desired that STM traces should decay quickly after inputs cease, while LTM traces should be relatively stable. Hence, the decay rate for w_{1i} is set much smaller than the decay rates for x_1 and x_i.

Grossberg (1968a) studied the *asymptotic behavior*, that is, behavior as time increases, of the outstar equations given in Section 3.5. He showed that for many classes of inputs, the functions $W_{1i}(t)$ defined by (3.1) and the analogous functions

$$X_i = \frac{x_i}{\displaystyle\sum_{j=2}^{n} x_j} \tag{3.2},$$

the relative stimulus traces, both approach limits as these equations evolve in time. Moreover, for each j, the limiting values of X_j and W_{1j} are equal. Thus, the same distribution of weights is coded both in the relative stimulus traces and the relative associational strengths at the outstar sinks.

Simulation exercises at the end of this chapter illustrate how the outstar's limiting behavior may relate to different classes of inputs. A particularly important case is where the inputs to the sink nodes form what Grossberg called a *spatial pattern*, that is, where the relative proportions of inputs to the different sink nodes are unchanged over time (see Figure 3.3). The mathematical definition of a spatial pattern input is that

$$\textit{For } j > 1, \; I_j(t) = \theta_j I(t), \; \sum_{j=2}^{n} \theta_j = 1 \tag{3.3}.$$

In the expression (3.3), the values θ_j represent relative pattern weights, since $\theta_j(t) = (I_j / I)$, while I, which equals the sum of all the I_j, represents the total input to all sink nodes of the outstar. In this case, under suitable conditions on the inputs $I_1(t)$ and $I(t)$, the input pattern weights were shown to be stored in long-term memory, that is,

$$\lim_{t \to \infty} W_{1j}(t) = \lim_{t \to \infty} X_j(t) = \theta_j, \; j = 2, \ldots, n \tag{3.4}.$$

The conditions under which (3.4) holds, which are listed in Section 3.5 below, can be interpreted as meaning that inputs to both the source and the sink are presented "often enough" for large times. After learning, the learned spatial pattern can be reproduced by activating the source node with an input I_1.

Other articles (*e.g.*, Grossberg, 1968b, 1969c, 1972a) extended the above results to equations describing networks with learning laws similar to the outstar's but different in architecture. In some cases, learning theorems for spatial patterns, analogous to (3.2), can be proved. In other cases, spatial patterns can be shown to be learned for some parameter values and forgotten (leading to asymptotically uniform synaptic weights) for other parameter values. Such variability of dynamics occurs, for example, in the *complete graph with loops*, where every node projects to every node including itself, and the *complete graph without loops*, where every node projects to every other node.

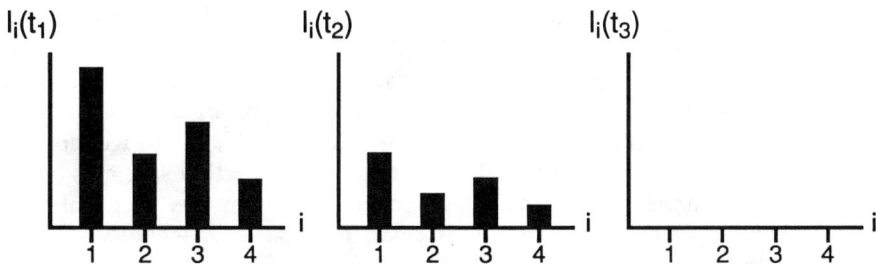

Figure 3.3. Example of a *spatial pattern* input, in which absolute inputs may change over time but always remain in the relative proportions θ_i. (Adapted by permission of the publisher from Levine, *Mathematical Biosciences* **66**, 1-86. Copyright 1983 by Elsevier Science Publishing Co., Inc.)

Some new psychological properties, separate from the original postulates, emerged from these mathematical results. These properties are:

1. The more often the network hears B following A, the more likely it is to say B after hearing A.
2. An isolated network can remember without practicing overtly.
3. Memory can sometimes improve spontaneously on recall trials.
4. If the network has learned the association AB, it can be changed to AC by sufficient presentation of the new sequence, but the change takes a long time if the association AB has become very strong.

Grossberg made use of these associative pattern learning results in models of more complex cognitive processes, which are discussed in Chapters 5 and 6. In the next section, we see that Grossberg also developed a class of modifiable synapses where the associative tendency is counteracted by habituative tendencies. The explanation for habituation is that repeated pairing of pre- and postsynaptic activities increases *production* of chemical transmitter at the synapse, but also increases *release* of transmitter. Hence, net available transmitter (production minus release) can either increase or decrease with correlated activities. Section 3.3 includes discussion of some psychological consequences of net transmitter decreasing with pairing.

Anderson's Connection Matrices

Other early approaches to associative learning in neural networks include the work of Anderson (*e.g.*, 1968, 1970, 1972, 1973) and Kohonen (*e.g.*, 1977; see also Kohonen, Lehtio, Rovamo, Hyvarinen, Bry, & Vainio, 1977). In contrast to Grossberg's mathematical formalism, which is largely nonlinear and continuous, the formalism of Anderson's early work (and some of Kohonen's) is linear and discrete, much of it using matrices of connection strengths.

The linear model of learning in Anderson (1972) employs vector algebra and probability theory. This model is based on models of memory storage, retrieval, and recognition due to Anderson (1968, 1970). In all these articles, a memory trace is described as a vector

$$\vec{x} = (x_1, x_2, ..., x_n)$$

each of whose components is the activity of a single element of the network. Hence, these memory traces are similar to the stimulus traces of Hull (1943) and Grossberg (1969a). If the elements are neurons, this activity is interpreted as instantaneous firing frequency; if the elements are populations of neurons, activity is interpreted as average firing frequency. All of the memory traces present in those elements are summed into a total storage vector

$$\vec{s} = \sum_{k=1}^{K} \vec{x}_k \qquad .$$

The model is discrete in time, although synaptic efficacy can take on any of a continuum of values.

The emphasis in Anderson's articles was on developing a simple model that would capture some of the basic properties of memory without resorting to much physiological detail. In Anderson (1968), the problems considered were:

> 1. *Recognition*: If an input pattern (vector of activities) is given, how can one tell whether a trace similar to that pattern is already stored as part of the storage vector?
> 2. *Retrieval*: Once recognition has occurred, how can the stored trace be reconstructed?
> 3. *Association*: Once retrieval has occurred, how can other stored traces be found which are also similar to the input pattern?

Retrieval poses a particular challenge in this model because it involves recovering one term of a vector sum, which is subject to error. The techniques described for solving the recognition and retrieval problems uses various kinds of filters designed so as to minimize the probability of error — that is, to maximize a certain signal-to-noise ratio.

The mathematical theory of these linear filters was discussed further in Anderson (1970). The input trace is used to construct a *matched filter* whose output is the dot product of the stored array with the input trace, that is,

$$V = \vec{s} \cdot \vec{x} = \sum_{k=1}^{N} s_k x_k$$

where N is the number of nodes in the network. If $\vec{s}$ is considered to equal the input $\vec{x}$ added to a noise component $\vec{n}$, the signal-to-noise ratio is defined as

$$\frac{\left(\sum x_k^2\right)^2}{\left(\sum n_k x_k\right)^2_{avg}}$$

the average being taken over all possible storage vectors. Using probability theory, the signal-to-noise ratio was shown to be close to N/K, with N the number of nodes and K the number of traces. The derivation of the signal-to-noise ratio is sketched in Section 3.5 below. Thus increasing the number of nodes makes the network more reliable, as von Neumann (1951) had discovered in a much earlier model. Also, increasing the number of traces makes recognition more difficult, which is to be expected because of mutual interference. The signal-to-noise ratio achieved is the maximum possible for a linear filter, but can be improved by using a suitable nonlinear filter (derived from a spatial cross-correlation vector).

The problem of association was discussed in Anderson (1968), and its relationship with a theory of synaptic connection weights was described in Anderson (1972). In these articles, a model for association was proposed which involves two sets of nodes, α and β. The following assumptions were made for simplicity: (1) both α and β have the same number N of nodes; (2) there is another number M such that M nodes of α project to every neuron of α, and every neuron of α projects to M nodes of β. There was assumed to be an input trace $\vec{x} = (x_1, ..., x_n)$ at α, a trace $\vec{y} = (y_1, y_2, ..., y_n)$ at β, and values a_{ij} for the efficacy of the synapse from the ith element of α to the jth element of β. It was assumed that $\vec{y}$ should be as close as possible to $A\vec{x}$, where A is the matrix of connection weights a_{ij} (see Figure 3.4). Under this assumption, the optimal matrix A (from the standpoint of signal-to-noise ratio) is obtained by

$$a_{ij} = cx_i y_j \qquad\qquad (3.5)$$

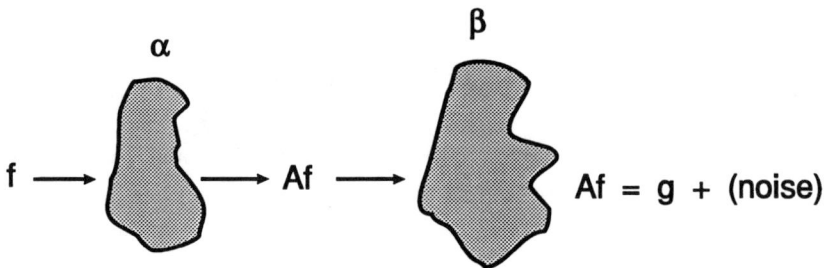

Figure 3.4. Scheme for associating distinct memory traces $\vec{x}$ and $\vec{y}$, with α and β two groups of nodes, and A a matrix. (Adapted by permission of the publisher from Levine, *Mathematical Biosciences* **66**, 1-86. Copyright 1983 by Elsevier Science Publishing Co., Inc.)

for some constant c. (The matrix defined by (3.5), for c = 1, is called the *outer product* of the vectors $\vec{x}$ and $\vec{y}$).

Equation (3.5) illustrates that optimality considerations lead to a rule similar to the Hebb rule for connection weights. Nass and Cooper (1975) extended Anderson's analysis to develop a network where the matrix of connection weights start with arbitrary values and converge to the optimal weights; algorithms of this sort are still in frequent use. This optimality analysis could be extended to some cases of multiple input traces and multiple output traces. If there are K input traces $\vec{x}_k = (x_{k1}, x_{k2}, ..., x_{kn})$, $1 \leq k \leq K$, and K output traces $\vec{y}_k = (y_{k1}, y_{k2}, ..., y_{kn})$, all with equal "power"

$$P = \sum_{i=1}^{N} x_{ki}^2 \, ,$$

then (3.5) generalizes to

$$a_{ij} = \sum_{k=1}^{K} c x_{ki} y_{kj} \qquad (3.6).$$

Equation (3.6) indicates that synaptic weights in this model are calculated instantaneously from the inputs. The dynamics of the system over time were not developed, and decay terms (as in the outstar equations) were not included. Thus, in this model, a trace can only be forgotten if it is interfered with by the memory of another stimulus whose trace involves the same neural elements.

The extension of this work to include temporal dynamics with decay terms was suggested in Anderson (1973) and done explicitly in Nass and Cooper (1975). The 1973 article dealt with the modeling of some psychological data on the learning of lists. In these data, a subject's reaction time to an item is related to the probability of occurrence of the item. This result was modeled by letting the kth trace coefficient (analogous to the factor c in (3.5) and (3.6)) be $c_k = 1 + \Gamma p_k$, where Γ is a weighting constant and p_k is the probability of presentation of the kth item. Anderson (1973, p. 429) explained his model as follows: "The probabilistic term due to the immediate past history could easily arise in the following way. Each time the item is tested or rehearsed, the stored trace is slightly strengthened. This can easily occur since the trace has just been present in the correct form. If an exponential decay of trace strength in short-term memory is assumed, the probability dependent term will appear to be of this form." He thus suggested, without proposing a formula, that the memory trace of an item decays exponentially while the item is not being presented, and increases while the item is being presented.

Nass and Cooper (1975) showed that an optimal matrix of the form (3.6) arises as the limit of a matrix that is modified at time t according to the rule

$a_{ij}(t+1) = \Gamma a_{ij}(t) + \mu x_i y_j$, with Γ and μ positive constants, $\Gamma < 1$ but close to 1. (The decay rate of associative strengths is represented by $1-\Gamma$.)

These models of Anderson, Cooper, and their co-workers presaged later models of some multilevel processes, which are discussed in Chapter 6. These processes include feature detection in the visual cortex (which was already considered in the Nass-Cooper article) and categorization of pattern classes. In this later work, the linear models are extended to include thresholds that bound the amount of activity at each node.

Kohonen's Work

Anderson's theories of associative memory and category learning are similar to theories of the same phenomena by Kohonen (1977) and Kohonen *et al.* (1977). Kohonen *et al.* (1977) used the terms "associative memory" and "associative learning" for two distinct yet interrelated sets of cognitive phenomena. One set of phenomena involves memory or learning of associations that develop, for example, by classical conditioning or serial learning of lists. The other set involves recollection of a total pattern if part of the pattern is perceived (Kohonen *et al.*, 1977, p. 1065):

> The term associative memory is here restricted to denote a process in which a signal pattern is recalled upon the basis of a fraction of it (the key). The signal pattern is composed of the activities in a set of axons, acting as the input to a neural network during a short period of time. These axons may originate within one modality, with the signal pattern thus representing one sensory stimulus, or they may originate within different parts of the central nervous system, and thus represent the simultaneous activity in these structures. Consequently, for instance, both the recall of a visual image from its fraction, and a paired association in the classical conditioning, can be regarded as different aspects in the functioning of the associative memory.

He continued to develop this distinction in later work (Kohonen, 1984) where he referred to the above two aspects of associative memory as *autoassociative* and *heteroassociative*; this later work is discussed in Section 3.4 below.

Kohonen *et al.* (1977) and Kohonen (1977, Chapter 1) illustrated the recognition process by means of simulations on recognition of human faces. We shall now discuss the algorithm for these simulations, as an example of Kohonen's general method. The nodes in his model were assumed to be analogs of pyramidal cells in the cerebral cortex (the largest type of cortical neuron), or else of columns of cortical neurons.

Kohonen *et al.* (1977, p. 1069) assumed that each unit modulates the activities of other units in proportion to its activity, and that a nonspecific background activity is present. The variables defining the network are output firing frequencies y_i, which depend on input firing frequencies x_i. (Kohonen *et al.* referred to these variables as *spiking* frequencies. *Spike* is a synonym for action potential, based on the characteristic shape of the action potential as a function of time; see Appendix 1.) If the direct connectivity between input and output is denoted by w_i, the long-range connectivity from unit j onto unit i by w_{ij}, and the background activity by y^*_b, the input-output transformations can be represented as a system of linear equations

$$y_i = w_i x_i + (\sum_j w_{ij} y_j) + y^*_b \qquad (3.7).$$

The values w_{ij} of the connection weights in Equation (3.7) can be positive for excitatory connections, negative for inhibitory ones, and 0 for units that do not project to the given unit.

In some earlier examples of Kohonen's work, the connectivities w_{ij} were assumed to be constant, and, by methods similar to Anderson's, the optimal weights were found for recall of specific patterns. But Kohonen also incorporated memory effects, modeled by a Hebb-like associative law of the form $dw_{ij}/dt = ay_i(y_j-y_b)$, where y_b is another measure of baseline activity. The equations on which Kohonen's simulations were based, to be given in Section 3.5 below, combine (3.7) with such a learning law.

The actual simulations of Kohonen et al. (1977) used an approximation to these equations combined with a preprocessing (sharpening) of the pattern using lateral inhibition. Lateral inhibition, which is the main topic of Chapter 4, is widely used in neural networks to enhance contrast between different locations within a pattern. Kohonen and his co-workers achieved some success in recovering a recognizable face from a blurred or partially missing image. Like Anderson, they found that the larger the number of stored images, the more difficult it is to get a sharp recollection of each image. It was found that inclusion of lateral inhibition markedly sharpens the recollected image.

Kohonen et al. (1977) also commented that the model could be improved by incorporation of *temporal* as well as spatial contrast enhancement (p. 1075): "The temporal differentiation of signals exerts an effect of improvement, similar to that of lateral inhibition, since the most relevant information is usually associated with changes in state." A previous article (Kohonen & Oja, 1976) had used temporal difference effects to construct a "novelty filter," which selectively enhances those parts of a pattern which have not previously been present. Learning laws based on various forms of temporal difference are discussed extensively in the next section.

3.3. LEARNING RULES RELATED TO CHANGES IN NODE ACTIVITIES

In the associative rules discussed in the last section, connection weights are modified by correlated presynaptic and postsynaptic activities. By contrast to rules based on correlated node activities, many other modelers have suggested learning rules in which the crucial variable is *change* in postsynaptic (and also, in some cases, presynaptic) activity. The error-correcting rules of Widrow and Hoff (1960) and Rosenblatt (1962) (*cf.* Section 2.1) fit into this general category. More recent modelers employing similar rules, for a variety of reasons, include Klopf (1979, 1982, 1988), Sutton and Barto (1981, 1991), Kosko (1986b), and Rumelhart *et al.* (1986), among others.

Klopf's Hedonistic Neurons and the Sutton-Barto Learning Rule

Klopf (1982) proposed that a synapse is increased in efficacy if its activity is followed by a net increase in the depolarization (positive stimulation) received by the postsynaptic cell. In other words, he proposed that depolarization acts as *positive reinforcement* for neurons. Klopf's theory was based on an analogy between single neurons and whole brains, both of them being treated as goal-seeking devices. This is the reason for the words "hedonistic neuron" in the title of his book.

The importance of activity change, as opposed to activity itself, was also highlighted in a psychological theory, developed by Rescorla and Wagner (1972), based on the results of many classical conditioning experiments. These experiments had indicated that associative learning of a conditioned stimulus can be greatly influenced by the background stimuli present during both training and recall trials. The basis of Rescorla and Wagner's theory (1972, p. 75) was that "organisms only learn when events violate expectations. Certain expectations are built up about the events following a stimulus complex: expectations initiated by the complex and its component stimuli are then only modified when consequent events disagree with the composite expectation."

Sutton and Barto (1981) set out to explain classical conditioning with a theory that included elements of both the Rescorla-Wagner and Klopf theories. Their conditioning model includes n stimulus traces $x_i(t)$, an output signal $y(t)$, and n synaptic weights $w_i(t)$, as shown in Figure 3.5. These weights are considered to denote associations between conditioned stimuli (CS's) and a primary reinforcer or unconditioned stimulus (US or UCS).

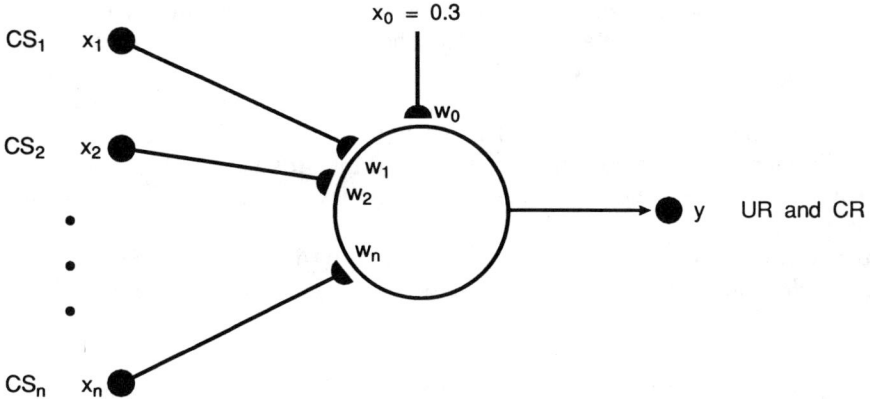

Figure 3.5. Adaptive element analog of classical conditioning. This network has n modified conditioned stimulus (CS) pathways, and a pathway with fixed weight w_0 for the unconditioned stimulus (US). The node y represents the unconditioned and conditioned responses (UR and CR). (From Sutton & Barto, *Psychological Review* **88**, 135-170, 1981. Copyright 1981 by the American Psychological Association. Adapted by permission.)

Sutton and Barto proposed that in addition to traces $x_i(t)$ which denote the duration and intensity of given CS's, there are additional traces, called $\bar{x}_i(t)$, that are separate from the stimuli and longer lasting. These traces were termed *eligibility traces* because they indicate when a particular synapse is eligible for modification. The existence of these two separate sets of traces had been previously proposed by Klopf (1972). Possible cellular mechanisms were suggested for eligibility traces, involving calcium ions and cyclic nucleotides. Finally, the current amount of reinforcement, $y(t)$, was compared with the weighted average $\bar{y}(t)$ of values of y over some time interval preceding t. These variables are governed by Equations (3.23) below.

The two innovations in Sutton and Barto's model — including eligibility traces and making learning depend on the comparison term $y(t) - \bar{y}(t)$, rather than simply the postsynaptic activity $y(t)$ — were motivated by experimental results on timing in classical conditioning. In particular, the model can explain the fact that in many conditioning paradigms, the optimal interstimulus interval (time interval between the two stimuli to be associated) is greater than 0; this explanation is further developed in Chapter 5. Sutton and Barto's network can also simulate other contextual effects in classical conditioning, such as the blocking of formation of associations to a new stimulus if another stimulus that has already been conditioned is simultaneously present.

The synaptic learning law involving change in postsynaptic activity is not the only possible way to simulate timing effects or blocking in classical conditioning. The same data were simulated by Grossberg and Levine (1987) using a form of the earlier Grossberg learning law (see Section 3.2) combined with competitive attentional effects in a larger network. Yet while the behavioral data studied by Sutton and Barto did not lead to a unique physiological model, they inspired a class of models that is now being actively studied by a large number of researchers.

Error Correction and Back Propagation

Sutton and Barto (1981) noted some formal analogies between their learning rule, where present is compared to previous reinforcement, and other rules where actual is compared to desired reinforcement. Rules of the latter sort were derived from considerations both in psychology (Rescorla & Wagner, 1972) and engineering (Widrow & Hoff, 1960). Such error-correcting rules fall into the category of *supervised learning*; that is, they rely on a "teacher" to tell whether particular neural responses are "right" or "wrong." The best known of these supervised learning rules is the *generalized delta rule* used in the back propagation algorithm.

The essentials of the back propagation algorithm were developed by Werbos (1974), as a procedure for optimizing the predictive ability of mathematical models.   LeCun (1985) and Parker (1985) discovered this procedure independently, and Rumelhart *et al.* (1986) placed it in a widely studied connectionist framework. It is often applied to discrimination or classification of sensory input patterns (*cf.* Chapter 6). The principle is as follows. The network is feedforward with three layers, composed of input units, hidden units, and output units (*cf.* Section 2.1).   A particular pattern of output responses to particular input patterns is desired. To the extent that the actual response to the current input deviates from the desired response, the weights of connections from hidden to output units change. Then those weight changes propagate backward to cause changes in weights from input to hidden units that will reduce the error in the future. The hidden units thereby come to respond to, hence encode, specific patterns of input activities (the "internal representations" in the title of the article by Rumelhart *et al.*, 1986).

The original delta rule (similar to that of Widrow & Hoff, 1960 and many others) was formulated by Rumelhart *et al.* as a rule for changing weights following presentation of a given pattern, labeled by the index p. If w_{ij} is the weight from the ith input unit to the jth output unit, then

$$\Delta w_{ij} = \eta \delta_{pj} i_{pi} \qquad (3.8a)$$

where η is a learning rate constant, δ_{pj} is a measure of desired change in the jth component of the output response, and i_{pi} is the value of the ith component of the input pattern. In the simplest network with only input and output units and no hidden units, the desired change in the jth output component is

$$\delta_{pj} = t_{pj} - y_{pj} \qquad (3.8b)$$

where y_{pj} is j^{th} component of the actual output response and t_{pj} is the jth component of the desired or "target" response. The work of Rumelhart *et al.* and their predecessors consisted of generalizing (3.8b) to networks with hidden units. We now give a condensed description of the learning law arising from this generalization.

Rumelhart and McClelland (1986, Vol. I, Chapter 2) had previously demonstrated that there is no advantage to including hidden units if their *activation functions* (outputs as a function of total signal received) are linear. They therefore posited activation functions that are nondecreasing and differentiable but nonlinear. Typically, they used sigmoid functions (see Figure 2.7).

Let f be a sigmoid function, and let f′ be its rate of change or derivative (see Appendix 2). Let the net signal received by the jth unit in any given layer of the network be

$$net_{pj} = \sum_i w_{ij} y_{pi} \qquad (3.9),$$

a linear sum of the outputs y_{pi} from the previous layer weighted by the connections w_{ij}. If j is a hidden unit so that i is an input unit, y_{pi} equals the input component i_{pi}. If j is an output unit so that i is a hidden unit, then y_{pi} equals f(net$_{pi}$), the activation function f applied to the ith net signal net$_{pi}$.

Let L = 0, ..., N indicate the Lth layer of the network, where L = 0 represents the input layer and L = N (N = 2 in most applications) represents the output layer. The rule of Rumelhart *et al.* (1986) for back propagation of errors, which generalizes (3.8b), is

$$\delta_{pj} = f'(net_{pj})[t_{pj} - y_{pj}] \qquad (3.10a)$$

if L = N and

$$\delta_{pj} = f'(net_{pj})\sum_k \delta_{pk}w_{jk} \qquad (3.10b)$$

if L $\neq$ N, where the jth node in layer L can either be an output or a hidden unit, and net$_{pj}$ is defined by (3.9). The sum in (3.10b) is over all units k in the next layer downstream, that is, in layer L+1. The changes in weights of connections to the jth node are in turn calculated from the errors using a generalization of (3.8a), namely

$$\Delta w_{ij} = \eta\, \delta_{pj}\, y_{pi} \qquad (3.11).$$

The derivation of (3.10b) is given in Section 3.5 below.

The back propagation algorithm is an example of a long-established class of mathematical methods known as *steepest descent* (*cf.* Duda & Hart, 1973, and Section 6.5 of this book for details). That is, an expression is found for the total error in the network's response (based on the desired or target response), and the weight changes that cause the sharpest possible decrease in this error measure are computed. Heuristically, (3.10b) says that weight changes are greatest at connections from node activities sending signals net$_{pj}$ whose values are on the sharpest rising slope of the sigmoid function f. This means that those values are in the intermediate range; that is, the units sending those signals are furthest from an established "yes-or-no" response to the input.

Back propagation of synaptic weights occurs because the changes in input-to-hidden weights are computed via (3.8) from the δ_{pj} values, which in turn are computed via (3.10b) from the changes in hidden-to-output weights. Hence, this scheme allows for *credit assignment*, that is, deciding which connections at an earlier level in the network to alter if the responses of later stages are inappropriate (see also Barto & Anandan, 1985). In Chapter 6 we discuss the convergence properties of the back propagation algorithm and compare its success with that of other pattern classification schemes.

The Differential Hebbian Idea

Learning rules including changes in postsynaptic activity, as in the Sutton-Barto model (and also, sometimes, changes in presynaptic activity), are sometimes called *differential Hebbian* rules (Kosko, 1986b). This term is used to contrast with the term *Hebbian* for rules including a simple cross-correlation of pre- and postsynaptic activities (see the discussion in Section 3.1 above).

Klopf (1988), building on Sutton and Barto's work, developed a differential Hebbian learning model which he called the *drive-reinforcement model*. In Klopf's model, the synaptic efficacy changes as a function of changes in both presynaptic and postsynaptic activities. But in order to account for the positive optimal interstimulus interval in classical conditioning, the change in postsynaptic activity is delayed in time. Also, the change in efficacy of a given synapse is made proportional to the current efficacy. The purpose of this latter rule is to account for the initial positive acceleration in the S-shaped acquisition curves observed in animal learning (see Figure 3.6). Hence, his basic learning rule is

$$w_i(t+1) = w_i(t) + \Delta y(t-1) \sum_{j=1}^{\tau} c_j \left| w_i(t-j) \right| \Delta x_i(t-j-1) \qquad (3.12)$$

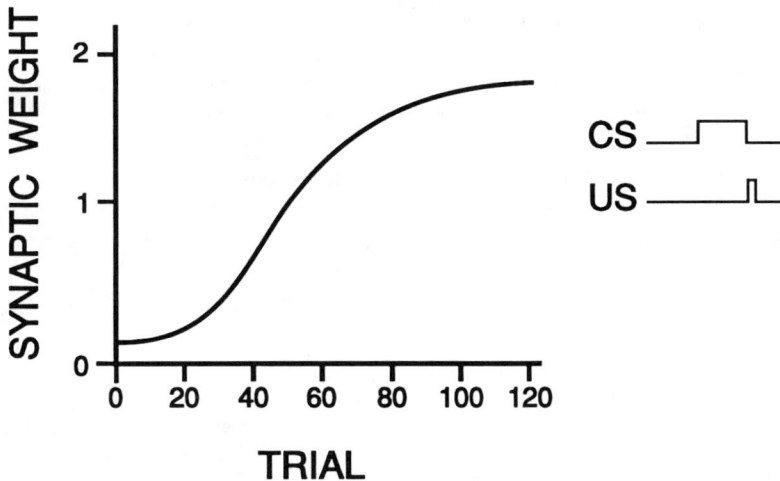

Figure 3.6. Synaptic weight between CS and US representations in a simulated classical conditioning experiment using the drive-reinforcement learning rule. The model yields an S-shaped (sigmoid) acquisition curve, consistent with some animal learning data. The CS is shut off at the time of US onset (*delay conditioning*).

where for any given time T, $\Delta x_i(T) = x_i(T+1)-x_i(T)$ and $\Delta y(T) = y(T+1)-y(T)$; the $c_{j's}$ denote weighting constants for the influences of different past times of presynaptic input activity; and "$| \, |$" denotes the absolute value.

Klopf was led to the learning rule of Equation (3.12) by consideration of the animal learning literature, and noted that Kosko (1986b) had independently been led to a rather similar learning law by philosophical and mathematical considerations. Using this law, Klopf was able to simulate a wide variety of classical conditioning data including effects of stimulus duration, partial reinforcement, and compound stimuli. Chapter 5 considers these conditioning data in more detail, comparing Klopf's models of such phenomena, and those of Barto and his co-workers, with the models of Grossberg and Levine (1987) and Grossberg and Schmajuk (1987). One essential difference between these two sets of models is worth noting now. The Klopf and Barto approaches rely on complex learning laws that are suggested to represent processes at the neuronal level. Indeed, Klopf (pp. 85-86) states:

> the model offers a way of defining drives and reinforcers at a neuronal level such that a neurobiological basis is suggested for animal learning. In the theoretical context that the neuronal model provides, I will suggest that *drives*, in their most general sense, are simply *signal levels in the nervous system* and *reinforcers*, in their most general sense, are simply *changes in signal levels*.

The Grossberg approach, by contrast, relies on simpler learning laws (associative cross-correlation combined with exponential decay) combined with certain characteristic network-level interactions which are discussed next.

Gated Dipole Theory

One of the network interactions used in Grossberg's models of classical conditioning is competition, or lateral inhibition; this is a common feature in the networks of other neural modelers, and is the main topic of Chapter 4. Competition is particularly important to models of attentional effects such as blocking. The other type of network interaction that occurs repeatedly in models of the Grossberg group is known as *opponent processing*, and is the basis for an architecture called the *gated dipole*. The gated dipole theory, like the differential Hebbian theory, was motivated by an effort to compare current values of stimulus or reinforcement variables with recent past values of the same variables.

Gated dipoles were introduced by Grossberg (1972b, c) to answer the following question about reinforcement. Suppose an animal receiving steady electric shock presses a lever which turns off the shock. Later, in the same context, the animal's tendency to press the lever is increased. How can a motor response associated with the *absence* of negative reinforcement (shock) become itself positively reinforcing? Absence of shock is, clearly, not rewarding per se: if you walk to the back right corner of a room and do not get shocked, that corner of the room does not become more attractive for you. Hence, zero shock must become (transiently) rewarding by contrast with the ongoing shock level.

Figure 3.7 shows a schematic gated dipole, which obeys Equations (3.26) below. The synapses w_1 and w_2, marked with squares, have a chemical transmitter that tends to be depleted with activity, as indicated by the $-y_i w_i$ terms in the differential equations for those w_i values. Other terms in those equations denote new transmitter production. The amount produced is greatest when the transmitter is much less than its maximum.

In Figure 3.7, the input J represents shock, for example. The input I is a nonspecific arousal to both channels y_1-to-x_1-to-x_3 and y_2-to-x_2-to-x_4. While shock is on, the left channel receives more input than the right channel, hence transmitter is more depleted at w_1 than at w_2. But the greater input overcomes the more depleted transmitter, so left channel activity x_1 exceeds right channel activity x_2. This leads, by feedforward competition between channels, to net positive activity from the left channel output node x_3. For a short time after shock is removed, both channels receive equal inputs I but the right channel is less depleted of transmitter than the left channel. Hence, right channel activity x_2 now exceeds x_1, until the depleted transmitter recovers. Again, competition leads to net positive activity from the right channel output node x_4. Whichever channel has greater activity either excites or inhibits x_5, thereby enhancing or suppressing a particular motor response.

The network is called a gated dipole because it has two channels that are opposite ("negative" and "positive") and that "gate" signals based on the amount of available transmitter. Characteristic output of one gated dipole is graphed in Figure 3.8. This graph illustrates the "rebound" in x_4 activity after the cessation of x_3 activity. The mathematical relationships between J, I, and the other system parameters needed for such a rebound to occur are studied in detail in Grossberg (1972c).

The idea of opponent processing is an old one in vision. For example, the retina contains pairs of receptors for opponent colors, such as green and red, and one of the two opponent colors is transiently perceived after removal of the other one. Concurrently with Grossberg's work, Solomon and Corbit (1974) developed an opponent processing theory of motivation, arguing that significant events elicit both an initial reaction and a subsequent counter-reaction.

Grossberg and Levine (1987, p. 5027) compared the gated dipole model for measuring temporal differences with the differential Hebbian model. They argued that the dipole model is better at reproducing two important psychological effects. In the context of shock avoidance data, one effect is that

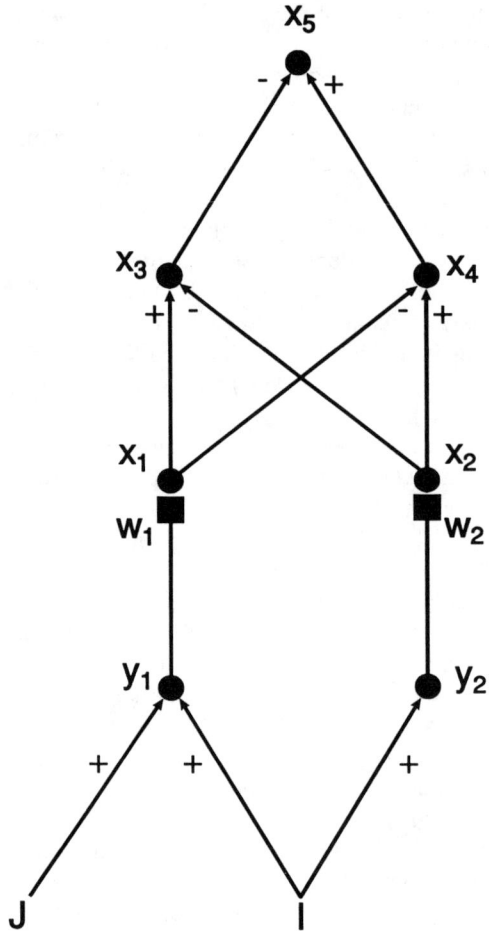

Figure 3.7. Schematic gated dipole network. J is a significant input (in the example of Grossberg, 1972b, electric shock) while I is nonspecific arousal. Synapses w_1 and w_2 undergo depletion. After J is shut off, $w_1 < w_2$ (transiently), so $x_1 < x_2$. By competition, x_4 is activated, enhancing a motor output suppressed by J.

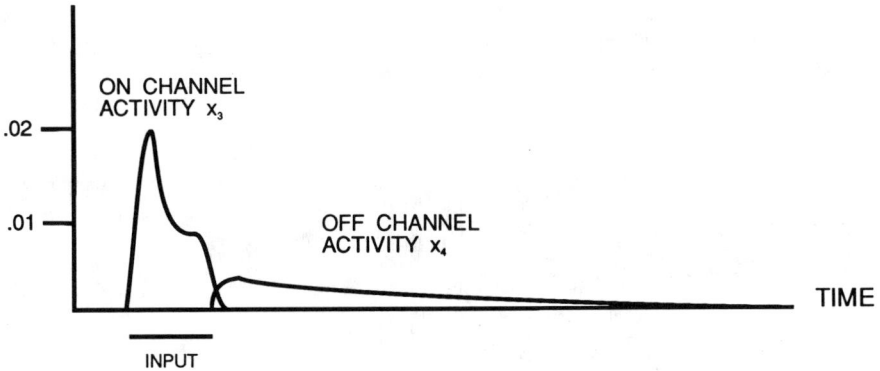

Figure 3.8. Typical time course of the channel outputs of a gated dipole. (Adapted with permission from Levine & Prueitt, *Neural Networks* **2**, 103-116. Copyright 1989 Pergamon Press.)

the amount of reinforcement from escaping shock is sensitive not only to the shock's intensity but also to its duration. The other effect is that the amount of reinforcement depends on the overall arousal level of the network (or organism).

If the two channels in Figure 3.7 are reversed in sign so that the channel receiving input is the "positive" one, the network provides an explanation for frustration when a positively reinforcing event either is terminated, or does not arrive when expected. The rebounds between positive and negative also explain the partial reinforcement acquisition effect (PRAE), whereby a motor response learned by an animal under intermittent reinforcement is more stable than the same response learned under continuous reinforcement (*e.g.*, Gray & Smith, 1969). According to the gated dipole theory (or the differential Hebbian theory), a reward is enhanced by comparison with an expected lack of reward.

The dipole idea can also be extended from the reinforcement domain to the sensory domain, with "on cells" and "off cells" responding to presence or absence of specific sensory stimuli (Grossberg, 1980). On cells and off cells for different stimuli are joined into a "dipole field." Transient rebounds in such a dipole field were used in Grossberg (1980) to model various visual phenomena such as color-dependent tilt after-effects. Also, gated dipoles have been applied to the modeling of motor systems (*cf*. Chapter 7). Grossberg and Kuperstein (1986) and Bullock and Grossberg (1988) used dipoles to simulate the actions of neuron populations innervating agonist-antagonist muscle pairs.

The on-cells and off-cells in the gated dipole are reminiscent of the novelty filter developed in Kohonen and Oja (1976) for selectively enhancing those parts of a pattern that have not been seen before. An example of such a novelty filter, used in visual pattern recognition, is shown in Figure 3.9.

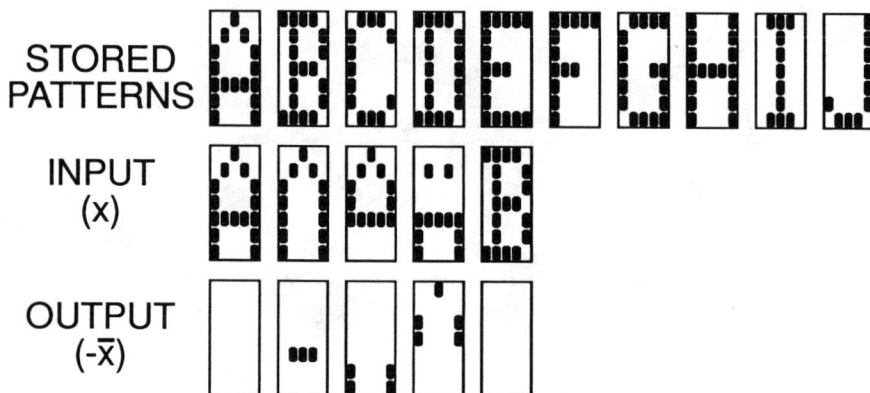

Figure 3.9. A demonstration of the novelty filter. The output selectively enhances those parts of the stored patterns that are absent from the current input pattern. (Reprinted from Kohonen & Oja, 1976, with permission of Springer-Verlag.)

To simplify the discussion in this section and the previous sections of this chapter, we have treated the stimuli that are being associated as if they activate single nodes. Other modeling considerations arise when we look instead at the learning of associations between activity patterns that span large numbers of nodes. Building on some ideas previously introduced in Section 3.2, the next section presents aspects of associative learning of patterns in some influential recent neural network models.

3.4. ASSOCIATIVE LEARNING OF PATTERNS

Recall from the discussion in Section 3.2 that a sensory pattern can be encoded as a *distribution* or *vector* of activities across different nodes; this theme is taken up again in Chapter 4. Early neural network models yielded some insights into how a single node can learn a particular pattern (*e.g.*,

Grossberg, 1968a, b) and how a network can learn to respond with one given pattern to another given pattern (*e.g.*, Anderson, 1970, 1972). In recent years, more sophisticated versions of this type of associative learning have played a large role in applications of neural networks to pattern recognition.

There are likely to be differences between a single node learning an association with a pattern of other node activities, as occurs, for example, in the outstar (see Figure 3.3), and the same node learning an association with a single other node. One possible difference is illustrated in Figure 3.10. If the association to be learned is simply between single node activities, a law (such as that of Hebb, 1949) whereby contiguous presentations increase connection strength seems to make sense. But if a pattern is to be learned, what is important is that the distribution of learned synaptic weights comes to approximate the distribution of original pattern intensities. Under these conditions, contiguous presentation can sometimes lead to increases in some connection strengths and decreases in others.

In Section 3.1 we discussed the early work of Anderson (1972) on association between vector patterns. Work along these lines has been developed further, with both mathematical theory and implementation, by several investigators, most notably Kohonen (1984) and Kosko (1987a, 1987c, 1988). (A discussion of associations between patterns also appears in Rumelhart & McClelland, 1986, Vol. 1, pp. 33-40).

Kohonen's Recent Work: Autoassociation and Heteroassociation

Kohonen (1984, p. 162) defined:

> two types of transformation operations, the *autoassociative recall*, whereby an incomplete key pattern is replenished into a complete (stored) version, and the *heteroassociative recall* which selectively produces an output pattern y_k in response to an input pattern x_k; in the latter case the paired associates x_k and y_k can be selected freely, independently of each other. This operation is a generalization of the simple stimulus-response (S-R) process.

Autoassociative theory was implemented, for example, in the restoration of human faces from blurred or partially missing images; this process is illustrated in Kohonen, Reuhkala, Makisara, & Vainio (1976), Kohonen *et al.* (1977) and Chapter 1 of Kohonen (1977). Examples of some of these simulations are shown in Figure 3.11. Detailed neural network connections for this process are best presented in Kohonen et al. (1977); the simulations in that article are based on his differential equation for associative learning, which are given in Section 3.5 below.

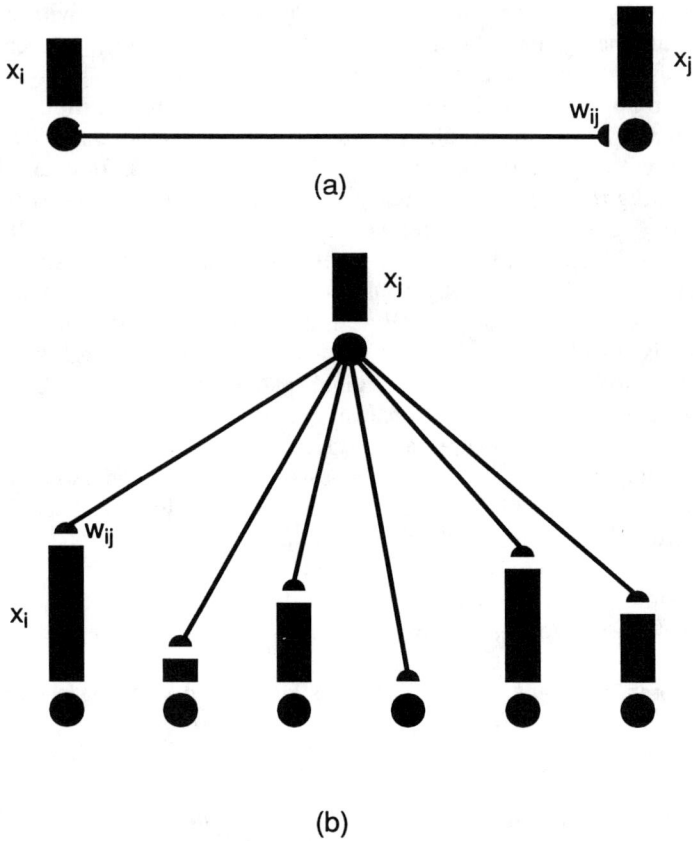

Figure 3.10. (a) Hebbian associative learning at single nodes. Correlation between node activities x_i and x_j always increases the LTM trace, or synaptic strength, w_{ij}. (b) Non-Hebbian associative learning of patterns. Correlation of a spatial pattern of node activities x_i with a single node activity x_j enables the LTM traces w_{ij} either to increase or decrease to match the spatial pattern. (Adapted from Grossberg & Levine, 1987, with permission of the Optical Society of America.)

Kohonen's simulations employed an approximation to this associative learning equation combined with a preprocessing (sharpening) of the pattern using lateral inhibition. In the primary patterns x_i, each "pattern element" x_i

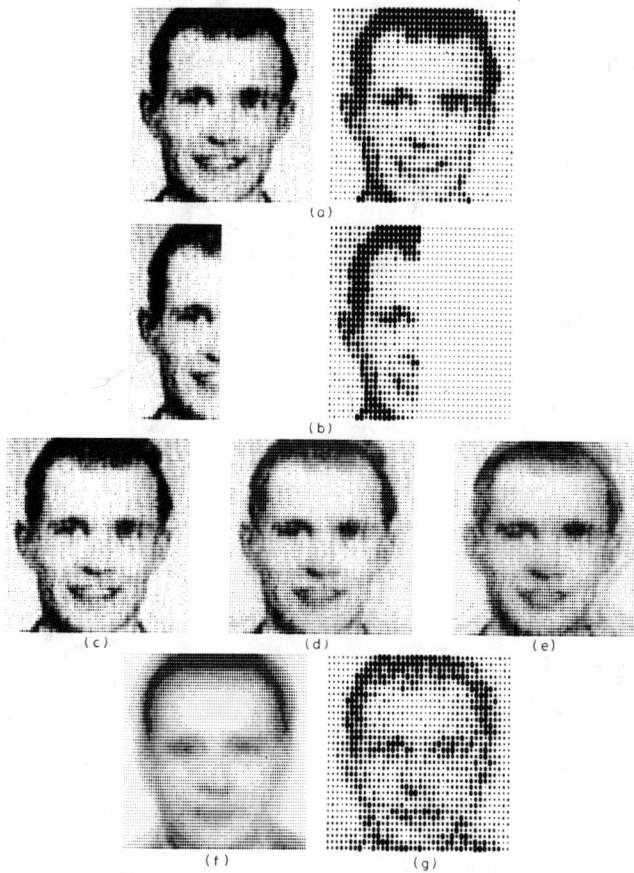

Figure 3.11. Face recall by an autoassociative memory network. (a) A stored face, as represented in a network with 5120 and 1280 nodes respectively. (b) Key images tested in recall. (c), (d), (e) Recollections from a 5120-unit network with 16, 160, and 500 stored images respectively. Note that quality gets worse as there are more images. (f) Recollection from a 5120-node network with 16 stored images and no lateral inhibition. (g) Recollection from a 1280-node network with lateral inhibition. (Reprinted from Kohonen *et al.*, 1977, with permission of Academic Press.)

was replaced by a numerical value x'_i which is a weighted sum of itself and its neighboring elements, where the weighting factors λ_{ia}, for a given i, add up to 0. It was further assumed that λ_{ii} is positive for each i; examples of weighting factor distributions are given in Figure 3.12.

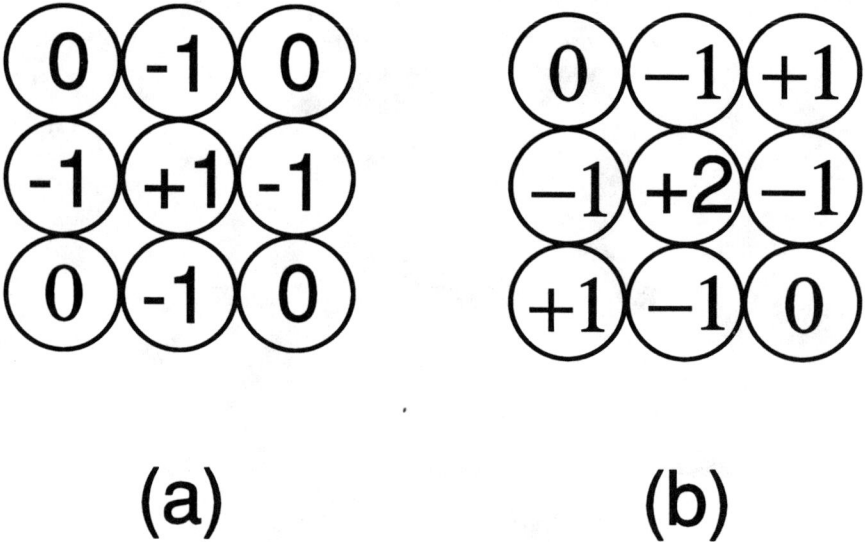

(a)   **(b)**

Figure 3.12. Two types of weighting parameters λ_{ia} used in numerical simulation of lateral inhibition. A new value for excitation at each point was obtained from the activity pattern by forming a weighted sum of the activities of the neighboring nodes, using the indicated numbers as weights. (Reprinted from Kohonen *et al.*, 1977, with permission of Academic Press.)

The heteroassociative theory was developed in Kohonen (1984), Chapter 6. As Kohonen stated, his approach to heteroassociation is less well developed at the level of practical implementation than either the autoassociation developed above, or the novelty filter discussed in the last section. But his general theory of optimal linear mapping for paired associations encompasses the autoassociative as well as the heteroassociative case. For if the set of pattern pairs to be encoded is ($\vec{x}_k$, $\vec{y}_k$), k = 1, ..., n, then $\vec{x}_k$ can be thought of as a key for recovering a desired pattern, and $\vec{y}_k$ as the pattern to be recovered. In that framework, the autoassociative case occurs if $\vec{x}_k$ is a subpattern of $\vec{y}_k$ with some of the components missing (as is true, for example, in Figure 3.11 where $\vec{x}_k$ is part of the face and $\vec{y}_k$ the whole face.

Kosko's Bidirectional Associative Memory

The theory of associations between pattern pairs was further advanced by the development in Kosko (1987a, 1987c, 1988) of the *bidirectional associative memory* or BAM[1]. Kosko developed a dynamical system of differential equations for a general heteroassociative link between collections of nodes as shown in Figure 3.13. In the case where the activity pattern vectors are binary (consisting of 1's and 0's) or *bipolar* (consisting of 1's and -1's), Kosko (1988) proved that the states of the system converge to a stable equilibrium value denoting the pairing of patterns. (For those unfamiliar with the mathematical notion of equilibrium, it is discussed in Section 4.2 and again in Appendix 2.) In the autoassociative case, where the a_i and b_i represent the same patterns, Kosko showed that his system is a generalization of the models of Hopfield (1982, 1984) and a special case of the system for which Cohen and Grossberg (1983) proved a convergence theorem (see Chapter 4).

All these proofs have in common that the convergence to a steady state is based on a symmetry assumption in the connection weights: if w_{ij} denotes the strength of the connection from x_i to y_j in Figure 3.13, it is also the strength of the connection from y_j to x_i. This is true both for the non-adaptive case where w_{ij} are constant, and for the adaptive case, studied in Kosko (1987a), where w_{ij} vary over time according to an associative learning rule with decay.

But strict symmetry of connections is not realistic for all forms of associative learning. Asymmetric associative learning occurs, for example, in Pavlovian conditioning. As the conditioning models to be discussed in Section 5.2 make clear, what is typically learned is that the CS *precedes* the US, in fact, predicts it.

Grossberg (1969d, 1970a) had previously generalized the learning of spatial patterns in the outstar (see Section 3.2) to the learning of *spatiotemporal patterns*, that is, time sequences of spatial patterns, in a network called the *outstar avalanche* (Figure 3.14). The avalanche consists of a collection of outstars that share the same sink nodes but whose sources are connected in series. Each source has learned, via synaptic weights, a different pattern of sink node activities. Hence, the spatial patterns determined by these weights are

---

[1] The order of the dates in Kosko's three articles seems to be an accident of journal scheduling, since the most basic of these articles is the one dated 1988!

activated in sequence. In this manner, for example, a ritualistic sequence of movements or musical notes can be learned.

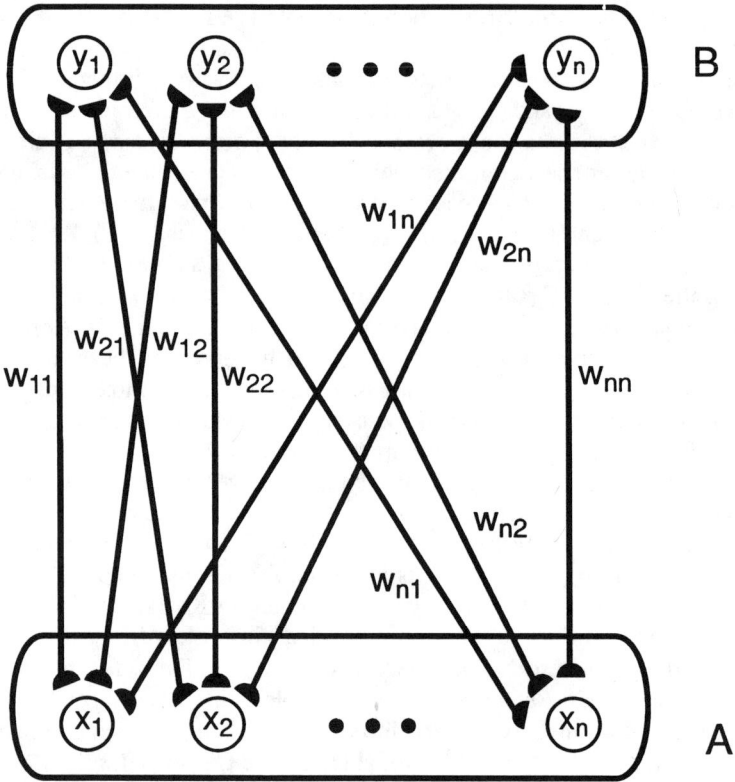

Figure 3.13. Bidirectional associative memory. x_i and y_j are nodes whose activity levels are either binary (1 or 0) or bipolar (1 or -1). w_{ij} are their connection weights, which are usually symmetric ($w_{ij}=w_{ji}$). Patterns are vectors of the x_i and y_j activity levels; the w_{ij} determine what pairs of patterns at A and B the network will learn to associate. (Adapted from Kosko, 1988, copyright © 1988 IEEE; reprinted by permission.)

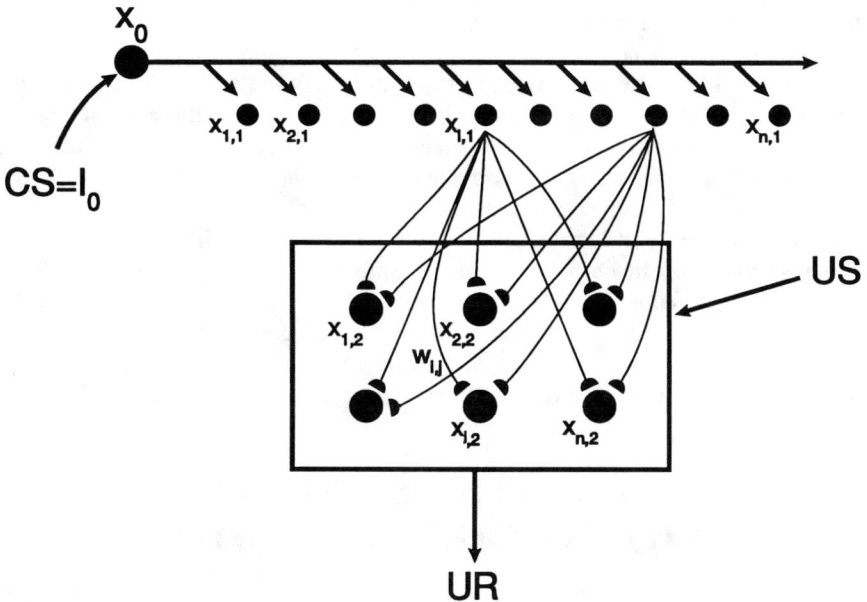

Figure 3.14. Outstar avalanche. A CS input excites an axon starting from x_0 with collaterals at $x_{1,1}$, $x_{2,1}$, ..., $x_{n,1}$, which are activated sequentially. Each $x_{i,1}$ is the source of an outstar with sink $x_{j,2}$ (inside a rectangle), and has learned a pattern of sink node activities that is encoded in the weights w_{ij}. The space-time pattern consists of the sequence of w_{ij} distributions (each a spatial pattern) as i runs from 1 to n. (Adapted from Grossberg, 1969d, with permission of the Department of Mathematics, University of Indiana.)

Just as Kosko (1988) interpreted his BAM's as reciprocal outstars, he generalized the BAM's to nonsymmetric dynamical systems known as TAM's (for *temporal associative memories*) that can be interpreted as outstar avalanches. One example of a TAM is a network that can learn a repeating temporal pattern, say (A_1, A_2, A_3, A_1). In that case, where the symmetry assumption is no longer valid, the system can be shown to converge not to an equilibrium but to an oscillatory solution or limit cycle (see Section 4.2 for a discussion of these mathematical terms).

Kosko (1987c) added competitive, or lateral inhibitory, interactions (*cf.* Chapter 4) within a level (a_i or b_j) to the adaptive BAM's. He showed that the system still converges to an equilibrium, and the equilibrium behavior approximates that of the adaptive resonance theory (ART) network of Carpenter and Grossberg (1987a). ART is a network designed particularly for pattern classification, and is discussed extensively in Chapter 6.

The multiplicity of learning laws in the neural network literature reflects an immense variation in both biological capabilities and cognitive tasks. Yet in spite of this variability, a few basic types of laws are widely repeated. A similar repeatability is seen in the laws for neural competition discussed in the next chapter. These two sets of laws provide most of the "building blocks" needed for the models of larger-scale processes discussed in Chapters 5, 6, and 7.

3.5. EQUATIONS AND SOME PHYSIOLOGICAL DETAILS

Neurophysiological Principles

The article of Byrne (1987) ended with a statement of four general principles that are emerging as points of agreement among neurobiologists. They are:

1. *Plasticity involves changes in existing neural circuits.* This means that cellular correlates of associative learning typically do not, at least in adult animals, involve growth of new synaptic connections but rather changes in the efficacy of existing connections. (There may be exceptions: Tsukahara and Oda, 1981 found that classical conditioning caused new postsynaptic electrical potentials to appear in the red nucleus (an area of the midbrain), and conjectured that new synaptic connections had been formed. Anderson et al., 1989 found growth of synaptic knobs in the cerebellum during conditioning of an eyeblink response. This might indicate either creation of links between neurons that were not previously connected, or addition of synapses between neurons that were already connected.)

2. *Plasticity is not localized to one site or type of neuron.* Evidence for modifiable synapses has been found at motor neurons in some experiments (for example, eyeblink conditioning in the cat) and at

sensory neurons in other experiments (for example, heart-rate conditioning in the pigeon).

3. *Plasticity involves second messenger systems.* Second messengers are particular chemical substances that are important in neuronal biochemistry. More detailed references on this general subject are given in Appendix 1. Briefly (*cf.* Figure 3.15), action potentials involve characteristic patterns in the transport across nerve membranes of potassium, sodium, and chloride ions. Transmission of impulses across a synapse is mediated by a chemical transmitter that affects the "channels" carrying those ions across the postsynaptic membrane. There are over twenty known neurotransmitters; some of the most common are acetylcholine, norepinephrine, serotonin, gamma-amino butyric acid (GABA), and dopamine. Transmitter production and release are in turn affected by second messengers. There seem to be only a few such second messengers — cyclic AMP, cyclic GMP, and the calcium ion.

4. *Plasticity at one site involves multiple synergistic processes.* For example, in the sea slug, *Aplysia*, there can be coordinated effects on the release of a chemical transmitter and on a postsynaptic potassium channel affected by that same transmitter.

Equations for Grossberg's Outstar

A form of the outstar equations is as follows. As discussed in Section 3.2 above, the activity x_1 of the source node is affected positively by the source node input I_i, and negatively by exponential decay back to a baseline rate (interpreted as 0). Recalling that the rate of change of x_1 as a function of time can be described by its derivative, dx_1/dt, this leads to a differential equation of the form

$$\frac{dx_1}{dt} = -ax_1(t) + I_1(t) \tag{3.13}$$

where a is a positive constant (the decay rate). The activities x_i of the sink nodes, $i = 2, ..., n$, are affected by inputs and decay in the same manner as is x_1. Each sink node is also affected by source node activity, weighted by the strength w_{1i} of the synapse from the source node. Hence

$$\frac{dx_i}{dt} = -ax_i(t) + bx_1(t-\tau)w_{1i}(t) + I_i(t), \tag{3.14}$$

$$i = 2, ..., n$$

where b is another positive constant and τ is a transmission time delay.

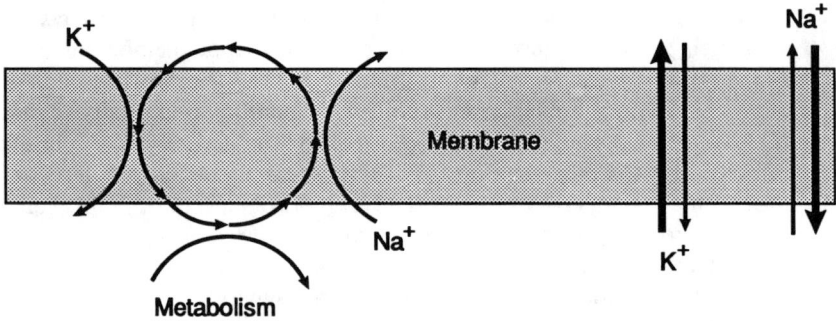

Figure 3.15. Schematic diagram of ionic mechanisms for resting and action potentials of the nerve membrane. On the left, ionic "pumps" help to preserve the concentration difference of sodium (Na^+) and potassium (K^+) ions inside and outside the cell membrane during rest. This concentration difference changes during the action potential, as shown on the right. (Adapted from Thompson, 1967, with permission of Harper and Row Publishers.)

In one version of the theory, the synaptic weights (long-term memory traces) w_{1i} at the source-to-sink synapses undergo a passive decay which is counteracted by correlated activities of x_1 (with a time delay) and x_i. Hence

$$\frac{dw_{1i}}{dt} = -cw_{1i} + ex_1(t-\tau)x_i \tag{3.15}.$$

Synaptic weights, since they encode long-term memory, are assumed to decay much more slowly than potentials; hence c << a. In some examples of this theory, a threshold term is subtracted from x_1 in Equations (3.14) and (3.15).

A modification of the outstar equations is suggested by learning considerations and used in much of Grossberg's later work on different network architectures. The modification is to make the synaptic weights decay only if the source node is activated and not followed by sink node activation. This is achieved by replacing (3.15) (if τ is set to 0) with

$$\frac{dw_{1i}}{dt} = x_1(-cw_{1i}+ex_i) \tag{3.16}$$

so that w_{1i} remains unchanged while $x_1 = 0$ but decreases while $x_1 > 0$ and $x_i = 0$.

Grossberg (1968a) studied the large time (asymptotic) behavior of a general system of equations which includes both (3.13) - (3.15) and (3.13), (3.14), (3.16) as subcases. This general system is

$$\frac{dx_i}{dt} = a(t)x_i(t)+b(t)w_i(t)+I_i(t)$$
$$\frac{dw_i}{dt} = c(t)w_i(t)+d(t)x_i(t) \tag{3.17}$$

with the restrictions that a(t), b(t), c(t), and d(t) are continuous functions, and that b(t) and d(t) are nonnegative. (The function $x_1(t)$ is incorporated into b(t) and d(t).)

Recall, from Section 3.2 above, Grossberg's definition of a spatial pattern input as a vector of inputs $I_i(t)$ such that

$$\textit{For } j>1, \; I_j(t) = \theta_j I(t), \; \sum_{j=2}^{n} \theta_j = 1 \tag{3.3}.$$

Recall also his definition of the *relative* node activities

$$X_i = \frac{x_i}{\sum_{j=2}^{n} x_j} \tag{3.2}$$

and the relative synaptic weights

$$W_i = \frac{w_i}{\sum_{j=2}^{n} w_j} \tag{3.1.}$$

(In (3.1), the weights were doubly subscripted as w_{1i}, but the "1" is dropped for the more general system. Also, the sums were previously taken from 2 to n instead of 1 to n, when the source node x_1 was included in the equations.)

The outstar learning theorem says that for a network obeying equations (3.17a, b), the relative synaptic weights defined by (3.1) converge to the relative weights θ_i of the input pattern, under certain technical conditions on the inputs. These conditions are such as to guarantee that inputs are presented to both the source and the sinks for arbitrary large times. Mathematically, in the case of an outstar obeying (3.13) - (3.15), this means that there exist two positive constants r and t_0 such that for all times $t \geq t_0$,

$$\int_{\tau}^{t} e^{-a(t-v)} I_1(v)\, dv \geq r \quad and \quad \int_{\tau}^{t} e^{-a(t-v)} I(v)\, dv \geq r \quad .$$

The basic method of proof of the outstar learning theorem involves transforming (3.17) into a system of equations in the relative activities and relative weights. The system derived from (3.17), (3.1), (3.2), and (3.3) is

$$\frac{dX_i}{dt} = A(t)(W_i - X_i) + B(t)(\theta_i - X_i)$$
$$\frac{dW_i}{dt} = C(t)(X_i - W_i) \tag{3.18}$$

where $A(t)$, $B(t)$, and $C(t)$ are nonnegative functions (specifically, if $x(t) = \sum_{j=1}^{n} x_j(t)$ and $w(t) = \sum_{j=1}^{n} w_j(t)$, then $A(t) = b(t)w(t)/x(t)$, $B(t) = I(t)/x(t)$, and $C(t) = d(t)x(t)/w(t)$). Equation (3.18b) shows that as t increases, each $W_i(t)$ moves closer to $X_i(t)$, whereas (3.18a) shows that $X_i(t)$ moves closer to $\theta_i(t)$. Hence, the relative node activities converge to the relative input pattern weights and then bring the relative synaptic weights toward themselves. This ultimately causes the relative synaptic weights also to converge to the θ_i.

Derivation of the Signal-to-noise Ratio for Anderson's
Linear Filter

As discussed in Section 3.2 above, Anderson (1970) considered the retrieval of a single vector pattern $\vec{x}$ from a stored trace $\vec{s} = \vec{x} + \vec{n}$, where $\vec{n}$ is regarded as noise. Putting the trace with a *matched linear filter*, that is, taking its dot product with $\vec{x}$, yields an output of

$$V = \vec{s} \cdot \vec{x} = \vec{x} \cdot \vec{x} + \vec{n} \cdot \vec{x} \qquad (3.19)$$

If N is the number of nodes in the network, therefore of components in each trace, the *mean* of a trace $\vec{x}$ is defined as the average of its components $x_1, ..., x_n$. For simplicity, and also to prevent biases in favor of one stored trace over others, Anderson assumed that all stored traces had the same mean m. Hence, if K is the number of traces other than the one to be retrieved, and if $\vec{I}$ is defined to be the vector $(1, 1, ..., 1)$, then $\vec{x}$ and $\vec{n}$ can be expressed as $\vec{x} = \vec{x}_0 + m\vec{I}$, $\vec{n} = \vec{n}_0 + mK\vec{I}$ for vectors $\vec{x}_0$ and $\vec{n}_0$ with mean 0. If P_0 is defined as $\vec{x}_0 \cdot \vec{x}_0$ (the *power* of the trace $\vec{x}_0$), then from (3.18), the signal component of the output V is

$$\begin{aligned} \vec{x} \cdot \vec{x} &= (\vec{x}_0 + m\vec{I}) \cdot (\vec{x}_0 + m\vec{I}) \\ &= \vec{x}_0 \cdot \vec{x}_0 + 2m(\vec{I} \cdot \vec{x}_0) + m^2(\vec{I} \cdot \vec{I}) = P_0 + Nm^2 \end{aligned} \qquad (3.20a)$$

and the noise component is

$$\begin{aligned} \vec{n} \cdot \vec{x} = (\vec{n}_0 + mK\vec{I}) \cdot (\vec{x}_0 + m\vec{I}) &= \vec{n}_0 \cdot \vec{x}_0 + mK(\vec{I} \cdot \vec{x}_0) + m^2 K(\vec{I} \cdot \vec{I}) \\ &= \vec{n}_0 \cdot \vec{x}_0 + nKm^2 \end{aligned} \qquad (3.20b).$$

The actual signal-to-noise ratio is calculated as the ratio of the *squared* signal (to make all terms positive) to the average squared noise over all possible stored traces. This is $(S/N)_0 = (\vec{x}_0 \cdot \vec{x}_0)^2 / [\vec{n} \cdot \vec{x}]^2_{avg}$, which by (3.20a,b) is equal to

$$\frac{(P_0 + Nm^2)^2}{[(\vec{n}_0 \cdot \vec{x}_0)^2]_{avg} + N^2 K^2 m^4}$$

Since $\vec{n}_0 \cdot \vec{x}_0$ has mean 0, the average value of its square is equal to the variance of $\vec{n}_0 \cdot \vec{x}_0$. Letting n_{0i} and x_{0i} denote components of $\vec{n}_0$ and $\vec{x}_0$, that variance is

$$var\left(\sum_{i=1}^{N} n_{0i} x_{0i}\right) = \sum_{i=1}^{N} x_{0i}^2 \, var(n_{0i}) \tag{3.21}.$$

Each n_{0i} is the sum of K random variables with mean 0 and equal variance; if it is assumed that those traces have, on the average, the same power as $\vec{x}_0$, each of those variances becomes P_0/N, yielding a total variance of KP_0/N for each n_{0i}. This, combined with (3.21) and the definition of P_0, yields KP_0^2/N for the average value of $\vec{n}_0 \cdot \vec{x}_0$. Hence, the signal-to-noise ratio $(S/N)_0$ equals

$$\frac{\left(P_0 + Nm^2\right)^2}{\left[K\frac{P_0^2}{N} + N^2 K^2 m^4\right]} \tag{3.22}.$$

By (3.22), if $m = 0$, this ratio equals N/K, the number of nodes divided by the number of traces (see the discussion in (3.22)). It can also be shown, using elementary calculus, that the maximum value of $(S/N)_0$ occurs for $m^2 = P_0/(N^2 K)$ and equals $N/K (1 + (1/NK))$, which is not appreciably larger than N/K.

Equations for Sutton and Barto's Learning Network

Recall the network of Sutton and Barto (1981) shown in Figure 3.5, including n stimulus traces $x_i(t)$, an output signal $y(t)$, and n synaptic weights $w_i(t)$ representing connections between x_i and y. The conditioning model is based on the existence of two additional sets of variables. One of these sets of variables consists of the nonstimulating or eligibility traces $\bar{x}_i(t)$ for each sensory stimulus. The value of $\bar{x}_i(t)$ is assumed to be large when the x_i-to-y synapse is "eligible" for modification. The other is the ongoing activity level $\bar{y}(t)$ of the output node, which represents a weighted average of its past activities.

All these effects (eligibilities, weighted averages, and delta rule for synaptic modification) are incorporated in the following equations for the changes in these variables from time t to time t + 1:

$$\bar{x}_i(t+1) = \alpha \bar{x}_i(t) + x_i(t)$$
$$\bar{y}(t+1) = \beta \bar{y}(t) + (1-\beta)y(t)$$
$$w_i(t+1) = w_i(t) + c(y(t) - \bar{y}(t))\bar{x}_i(t) \tag{3.23}$$

$$y(t) = f\left[\sum_{j=1}^{n} (w_j(t)x_j(t)) + z(t)\right]$$

where α and β are constants between 0 and 1; f is a sigmoid function (*cf.* 2.7); c is a positive constant determining the rate of learning; and $z(t)$ is the trace denoting the intensity of the US.

Derivation of Rumelhart, Hinton, and Williams' Back Propagation Algorithm

As discussed in Section 3.3, Rumelhart *et al.* (1986) assumed that that unit j (whether hidden or output) receives a signal equal to the linear sum of the outputs y_{pi} from the previous layer weighted by the connections w_{ij}. (If j is a hidden unit so that i is an input unit, y_{pi} equals the input component i $_{pi}$. If j is an output unit so that i is a hidden unit, then y_{pi} equals the activation function f applied to the ith net signal net_{pi}.) Hence, the signal it receives is

$$net_{pj} = \sum_i w_{ij}y_{pi} \tag{3.9}.$$

If f is the activation function[2], then the output of unit j is

$$y_{pj} = f(net_{pj}) = f(\sum_i w_{ij}y_{pi}) \tag{3.24}.$$

Now let the measure of the total error in the pth output pattern be

$$E_p = \frac{1}{2}\sum_j (t_{pj} - y_{pj})^2 \tag{3.25}.$$

---

[2] Rumelhart *et al.* (1986) demonstrated this rule with a separate activation function f_j for each node index j. Since this does not affect the demonstration herein, and since most of their actual simulations used the same activation function for all nodes, we are using a single f for simplicity.

Then if (3.8b) holds, the response change δ_{pj} is simply the negative derivative of the total error E_p with respect to y_{pj}; in other words, it is a measure of how much the jth unit contributes to the incorrectness of the response.

In the case where there are hidden units and nonlinear activation functions, it is desired therefore to compute δ_{pj} by taking the derivative of E_p, from (3.24), with respect to the jth signal net_{pj}. Using the expressions (3.24) and (3.25) and the chain rule for derivatives (see Appendix 2), this gives us the transformed learning rule

$$\delta_{pj} = f'(net_{pj})[t_{pj} - y_{pj}] \qquad (3.10a),$$

if the jth unit is an output unit. If the jth unit is instead a hidden unit, then again using the chain rule, we obtain from (3.24) through (3.26) ("∂" denoting partial derivative) that

$$\delta_{pj} = -\frac{\partial E_p}{\partial net_{pj}} = -f'(net_{pj})\left[\frac{\partial E_p}{\partial y_{pj}}\right] \qquad .$$

If k is the generic index of output units that receive projections from hidden unit j, we obtain, again by the chain rule and (3.10a), a value for the above expression in brackets, namely

$$\frac{\partial E_p}{\partial y_{pj}} = \sum_k \left[\frac{\partial E_p}{\partial net_{pk}}\right]\left[\frac{\partial net_{pk}}{\partial y_{pj}}\right] = \sum_k \left[\frac{\partial E_p}{\partial net_{pk}}\right]w_{jk} = -\sum_k \delta_{pk}w_{jk} \qquad .$$

Combining the above two expressions, we obtain finally that if unit j is a hidden unit,

$$\delta_{pj} = f_j'(net_{pj})\sum_k \delta_{pk}w_{jk} \qquad (3.10b).$$

Gated Dipole Equations Due to Grossberg

The general equations for the gated dipole of Figure 3.8 are given in Grossberg (1972c). The simplified form of those equations, with thresholds and time delays set to 0, is

$$\frac{dy_1}{dt} = -ay_1 + I + J \qquad \frac{dy_2}{dt} = -ay_2 + I$$

$$\frac{dw_1}{dt} = b(c-w_1) - ey_1w_1 \qquad \frac{dw_2}{dt} = b(c-w_2) - ey_2w_2$$

$$\frac{dx_1}{dt} = -fx_1 + gy_1w_1 \qquad \frac{dx_2}{dt} = -fx_2 + gy_2w_2 \qquad (3.26)$$

$$\frac{dx_3}{dt} = -hx_3 + k(x_1 - x_2) \qquad \frac{dx_4}{dt} = -hx_4 + k(x_2 - x_1)$$

$$\frac{dx_5}{dt} = -mx_5 + (x_3 - x_4)$$

where a, b, c, e, f, g, h, k, and m are all positive constants. Equations (3.26) reflect a symmetry between the "positive" and "negative" channels. Both have the same set of activity decay rates (a, f, and h), the same transmitter depletion rate (e), the same transmitter recovery rate (b), the same maximum amount of depletable transmitter (c), and the same coefficients for signal transmission between levels (g and k).

Equations (3.26) were written under the assumption that the "negative" channel is the one receiving the phasic input J, as in the case of relief from electric shock. If instead the "positive" channel receives the phasic input, then the term $I + J$ in (3.26a) is replaced by I, and the term I in (3.26b) is replaced by $I + J$.

In a variant of (3.26) (see Exercise 4 below), the signals received by x_3, x_4, and x_5 from lower levels are constrained to be positive or zero. Hence, each of the terms $x_1 - x_2$ in (3.26g), $x_2 - x_1$ in (3.26h), and $x_3 - x_4$ in (3.26i) is replaced by 0 if it becomes negative.

Kosko's Bidirectional Associative Memory (BAM)

In the simplest form of the bidirectional associative memory, described in Kosko (1988), there are two *fields* or collections of nodes, A and B. The aggregate activation of the ith node in A is denoted by x_i and the activation of the jth node in B by y_j. These variables can either be binary (taking on the values 0 or 1), bipolar (taking on the values 1 or -1), or analog (taking on any of a continuous range of values). In the continuous case, the x_i and y_j are governed by the system of equations

$$\frac{dx_i}{dt} = -x_i + \sum_j f(y_j)w_{ij} + I_i$$

$$\frac{dy_j}{dt} = -y_j + \sum_i f(x_i)w_{ij} + J_j$$

(3.27)

where f is a sigmoid function, the w_{ij} denote the (symmetric) interfield connection weights, and I_i and J_j denote the (constant) inputs to the ith and j^{th} cells. In the adaptive version of the BAM, as described in Kosko (1987a), the w_{ij} obey associative learning equations of the form

$$\frac{dw_{ij}}{dt} = -w_{ij} + f(x_i)f(y_j)$$

(3.28).

Kosko (1987c) extended the ideas of (3.27) - (3.28) to include competition. For the competitive BAM, in addition to connections between A and B described by coefficients w_{ij}, there are interactions within A, described by coefficients r_{ij}, and within B, described by coefficients s_{ij}. The competitive nature of these connections is ensured by the rules $r_{ii} > 0$, $s_{ii} > 0$, and for i and j unequal, $r_{ij} = r_{ji} < 0$ and $s_{ij} = s_{ji} < 0$. Equations (3.27) are then replaced by equations of the form

$$\frac{dx_i}{dt} = -x_i + \sum_j S(y_j)w_{ij} + \sum_k S(x_k)r_{ik} + I_i$$

$$\frac{dy_j}{dt} = -y_j + \sum_i S(x_i)w_{ij} + \sum_k S(y_k)s_{ik} + J_j$$

The coefficients r_{ik} and s_{ik} can either be constant or obey learning equations similar to (3.28). In either case, Kosko showed, the system converges to a solution corresponding to a set of associations between patterns in A and patterns in B.

Kohonen's Autoassociative Maps

Of the articles by Kohonen and his colleagues on autoassociative maps, the one in which the neural network connections were best developed was Kohonen *et al.* (1977). The simulations in that article were based on the difference equations that combine a linear input-output transformation with a Hebb-like associative learning law. Recall from Section 3.2 above that the output firing frequencies y_i depend on input spike frequencies x_i, direct input-output connectivities y_i, and inter-unit connectivities w_{ij} in a manner shown by the linear equations

$$y_i = w_i x_i + (\sum_j w_{ij} y_j) + y^*_b \qquad (3.7).$$

The w_{ij} in turn are assumed to obey the associative learning law $dw_{ij}/dt = ay_i(y_j - y_b)$, where y_b, like y^*_b, is a measure of baseline activity. In actual simulations, this differential equation is approximated by a stepwise solution of the form

$$w_{ij}(t) = \alpha(\Delta t)\sum_{k=1}^{m} y_i(t_k)[y_j(t_k)-y_b] + w_{ij}(0)$$

where Δt is the step size, m the number of steps from time 0 to time t, and t_k the time at the end of the kth step. This difference equation is in turn substituted into (3.7) to yield

$$y_i(t) = x^*_i(t) + \alpha(\Delta t)\sum_j \sum_k y_i(t_k)[y_j(t_k)-y_b]y_j(t)$$
$$+ \sum_j w_{ij}(0)y_j(t) + y^*_b \qquad (3.29)$$

where $x^*_i(t)$ aenotes the ith *effective input excitation* $w_i x_i(t)$, and the values t_k denote all times previous to t.

The simulations of face recognition in Kohonen *et al.* (1977) combined an approximation to Equation (3.29) with a preprocessing of the pattern using lateral inhibition. In the primary patterns x_i at time t_k, each "pattern element" x_i is replaced by a numerical value x^*_i which is a weighted sum of itself and its neighboring elements. Hence, the equation $x^*_i = w_i x_i$ used above is replaced by

$$x^*_i = \sum_a \lambda_{ia} x_a \qquad (3.30)$$

where the weighting factors λ_{ia}, for a given i, add up to 0. It was further assumed that λ_{ii} is positive; examples of weighting factor distributions are given in Figure 3.12. Then the recollections of the patterns, weighted by past experience, are given by

$$\hat{x}_i(t) = \sum_{k=1}^{m} \Gamma_i(t,t_k)x_i(t_k) \qquad (3.31a)$$

with the Γ_i defined for each given unit i and time t_k by

$$\Gamma_i(t,t_k) = \sum_j x^*_j(t)x^*_j(t_k) \qquad (3.31b).$$

The sum in (3.31b) (which is a correlation between the pattern at time t and the pattern at a previous time t_k) is taken over those units j that are assumed to have connections with unit i.

EXERCISES FOR CHAPTER 3

o1. Can you think of a way, using the framework of the 1970 and 1972 Anderson articles, to model selective attention to one trace rather than another (say, because of motivational significance)? If you believe that is difficult to do in Anderson's framework, why?

2. Consider a Grossberg outstar whose source node has activity x_1 and whose 4 sink nodes have activities x_2, x_3, x_4, and x_5. Let w_2, w_3, w_4, and w_5 be the corresponding synaptic strengths as shown in Figure 3.16.
The equations defining the network are

$$\frac{dx_1}{dt} = -5x_1 + I_1$$

$$\frac{dx_i}{dt} = -5x_i + x_1 w_i + I_i, \; i=2,3,4,5$$

$$\frac{dw_i}{dt} = x_1(- .1w_i + x_i), \; i=2,3,4,5$$

(Time delays have been set to 0 for ease of implementation. Other parameters have been set to obey a boundedness criterion — see Grossberg, 1970b). The inputs I_i, i = 1 to 5, are defined below, differently for two subcases.
Solve these equations numerically using the simple Euler method (see Appendix 2) or some other differential equation solving algorithm. The two cases are:

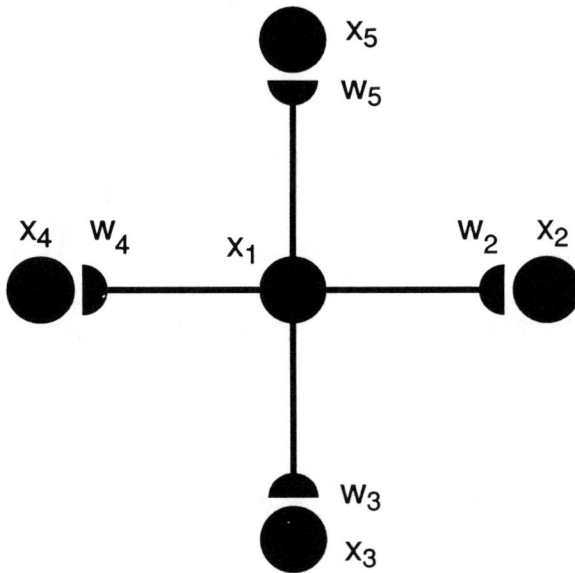

Figure 3.16. Outstar network used in the simulation of Exercise 2.

(a) Set $I_1 = 2$ for two steps on every tenth time step, starting with the first, and 0 on all other time steps. I_2, I_3, I_4, I_5 form a spatial pattern $I_i = \theta_i I$, where $\theta_2 = .4$, $\theta_3 = .3$, $\theta_4 = .2$, $\theta_5 = .1$; $I = 2$ for the two time steps directly *after* those times when $I_1 = 2$, and 0 on other time steps, as shown in Figure 3.17. (Hence $I_2 = .8$, $I_3 = .6$, $I_4 = .4$, and $I_5 = .2$ when they are not zero.)

The starting value of x_1 is 0; x_i, i = 2 to 5 start at positive numbers of your own choosing but *not* proportional to .4, .3., .2., .1, and so do w_i.

Run up to 10000 time steps.

Define $x = x_2 + x_3 + x_4 + x_5$, $w = w_2 + w_3 + w_4 + w_5$, and for each i, i = 2,3,4,5, define

$$X_i = \frac{x_i}{x}, \quad W_i = \frac{w_i}{w}$$

Graph (for each i, i=2,3,4,5) X_i, W_i, and θ_i on the same axes, showing values at every 100th time step. (Graphing may be done either by hand or by

ASSOCIATIVE LEARNING

computer. By the outstar learning theorem, these three variables should get closer together as time increases.)

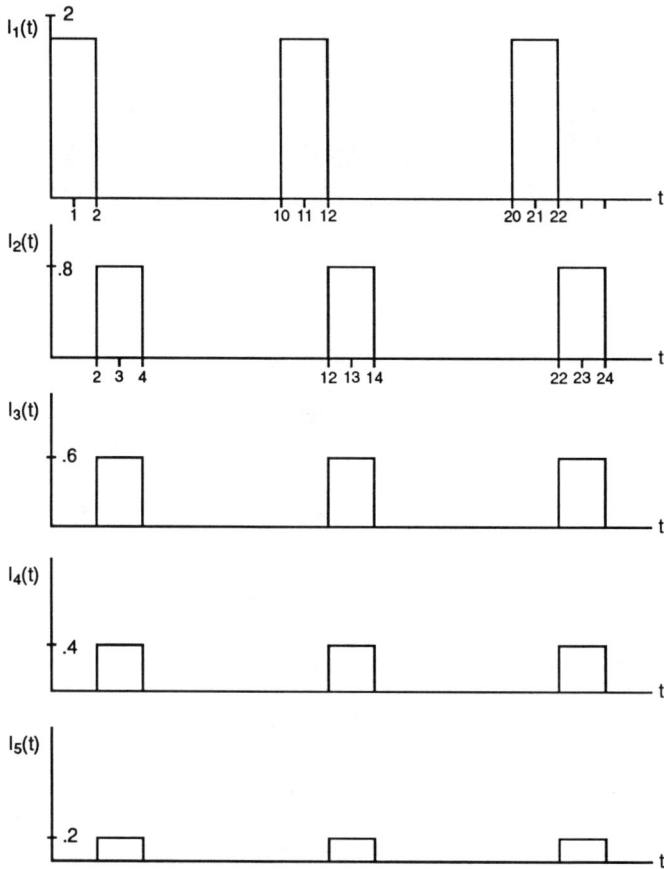

Figure 3.17. A spatial pattern input to the outstar shown in Figure 3.16.

(b) Do the same as in (a) except that I_2, I_3, I_4, and I_5 are a mixture of two spatial patterns. After times where I_1 is nonzero, the other I_i become nonzero for one time step, but alternate on different presentations between

$I_2 = .8, I_3 = .6, I_4 = .4, I_5 = .2$
$I_2 = .5, I_3 = .5, I_4 = .3, I_5 = .7.$

(In Part (b), the X_i should be graphed at times directly *after* the pattern presentation times in order to observe their oscillations.)

3. (a) Do the simulation in Exercise 2(a) for a variety of different initial conditions. Show that convergence is fastest when the initial values of X_i and W_i are closest to θ_i.

(b) From the result of Part (a), an outstar that has come close to learning one spatial pattern will be slow to learn another, vastly different spatial pattern (*cf.* the quote from Seneca at the start of this chapter). Confirm this by running an outstar simulation with the two patterns from Exercise 2(b) presented in succession, each for 5000 time steps.

4. Do a series of simulations of a slight modification of Grossberg's gated dipole equations, (3.26). The modification is that in the equations for x_3 and x_4, the quantities in parentheses ($x_1 - x_2$ or $x_2 - x_1$) are replaced by 0 whenever they are negative. Use the following parameter values: the decay rates a, f, h, and m are all 5; c, e, and k are 1; b = .5; g = 10. Set the "shock" J to 2 units, and keep it on for a length of time that varies between runs. Study the maximum over time of the rebound x_4 as a function of

(a) intensity of arousal I
(b) time duration of shock J.

Make tables of the results.

○5. Consider some phenomenon from experimental psychology that involves response to a change in stimulation. One example is *extinction*: a response learned by classical conditioning is weakened if the conditioned stimulus is presented and not followed by reinforcement. Another would be *conditioned inhibition*: first a stimulus CS_1 is associated with a US and thereby conditioned to a response followed by a second stage where a combination of two stimuli CS_1 and CS_2 is associated with absence of the US. As a consequence, CS_2 subsequently leads to suppression of the same response when it is associated with other stimuli.

Whichever psychological phenomenon you choose, give a network interpretation of it using

(a) the differential Hebbian model
(b) the gated dipole model.

Does this suggest advantages or disadvantages of either model?

6. Run the following simulations of a bidirectional associative memory or BAM, from Kosko (1987d). In this version, Equations (3.27) and (3.28) are replaced by an algorithm which combines difference equations and linear threshold mappings. There are 6 nodes at the x_i level of Figure 3.14, and 4 nodes at the y_j level. The network may be simultaneously taught to associate several pairs of binary vector patterns.

The BAM algorithm used is as follows:

Step 1: For all i, j, reset w_{ij}, a_i, and b_j to 0.
Step 2: Get the binary inputs into the A and B arrays for an association to be learned.
Step 3: (a) For each i, let $x_i = 2a_i - 1$. (Hence, the x_i vector is bipolar, taking on values of 1 or -1).
   (b) For each j, let $y_j = 2b_j - 1$.
   (c) For each pair i, j, let $w_{ij} = w_{ij} + x_i y_j$.
Step 4: If there is another association to learn, return to Step 2.
Step 5: Input binary A and B vectors to be run on the network.
Step 6: Run the A-to-B iteration of the network. For each j,
   (a) The new $b_j = 1$ if $\sum_i a_i w_{ij} > 0$;
   (b) The new $b_j = 0$ if $\sum_i a_i w_{ij} < 0$;
   (c) The new b_j equals the current value of b_j if $\sum_i a_i w_{ij} = 0$.
Step 7: Run the B-to-A iteration of the network. For each i,
   (a) The new $a_i = 1$ if $\sum_j b_j w_{ij} > 0$;
   (b) The new $a_i = 0$ if $\sum_j b_j w_{ij} < 0$;
   (c) The new a_i equals the current value of a_i if $\sum_j b_j w_{ij} = 0$.
Step 8: Repeat steps 6 and 7 until there are no changes in the A and B vectors.

(a) Teach the network to associate the patterns A^1 = (1,0,1,0,1,0) and B^1 = (1,1,0,0), and the patterns A^2 = (1,1,1,0,0,0) and B^2 = (1,0,1,0).

(b) After the weights from (a) are established, input A^1 = (1,0,1,0,1,0) to the x_i level and B^3 = (0,0,0,0) to the y_j level, and show that the network converges to the pair (A^1,B^1).

(c) With the same weights, input A^3 = (1,0,1,0,0,0) and B^3 = (0,0,0,0) and see what the network converges to. What does this say about possible steady states of the network?

(d) Add to the associations (a) one pattern pair at a time, and study how the number of time steps to convergence increases. If the number of associations increases beyond 4, the minimum of the numbers of nodes in the two levels, the network may in fact be unable to learn all the associations simultaneously.

7. Kosko's simplified algorithm is very similar to the algorithm for Kohonen's *correlation matrix memory* (see p. 183 of the 1988 edition of Kohonen, 1984). For this algorithm, if the vector pairs $(\vec{x}_k,\vec{y}_k)$, k =1, ..., n, are to be associated, an optimal matrix **W** is chosen for that purpose, and

$$W = \sum_{k=1}^{n} \vec{y}_k \vec{x}_k^T$$

where T denotes the transpose.
     For example, if x_k = (1,0,0), and y_k = (1,1,0), then

$$\vec{y}_k \vec{x}_k^T = \begin{bmatrix} 1 \\ 1 \\ 0 \end{bmatrix} [1 \ 0 \ 0] = \begin{bmatrix} 1 & 0 & 0 \\ 1 & 0 & 0 \\ 0 & 0 & 0 \end{bmatrix}$$

using standard matrix multiplication. If the x_k that are encoded are *orthogonal*, that is, the dot product of any two of them is 0, then $W\vec{x}_k = \vec{y}_k$ for each k. The response of the system to any pattern is obtained by multiplying the vector encoding that pattern by the matrix **W**.
     If $x_k = y_k$ for each k, the correlation matrix memory is called *autoassociative*; otherwise it is called *heteroassociative*. Now consider the following set of binary vectors:

$$\vec{A} = (1,0,0,0,0) \qquad \vec{E} = (1,0,1,0,0)$$
$$\vec{B} = (0,1,0,0,0) \qquad \vec{F} = (1,0,0,0,1)$$
$$\vec{C} = (0,0,1,0,0) \qquad \vec{G} = (1,1,1,0,0)$$
$$\vec{D} = (0,0,0,1,0)$$

(Note that the vectors $\vec{E}$, $\vec{F}$, and $\vec{G}$ can each be considered as noisy versions of one of the vectors $\vec{A}$, $\vec{B}$, $\vec{C}$, and $\vec{D}$ or of some sum of these vectors. Note also that $\vec{A}$, $\vec{B}$, $\vec{C}$, and $\vec{D}$ are orthogonal.)

(a) Simulate an autoassociative, correlation matrix memory for the vectors $\vec{A}$, $\vec{B}$, $\vec{C}$, and $\vec{D}$. Then list and discuss the response of the system to each of the above seven patterns.

(b) Do the same as (a) with the vector $\vec{E}$ added to the correlation matrix memory.

(c) Simulate a heteroassociative correlation matrix memory that associates each of the vectors $\vec{A}$, $\vec{B}$, $\vec{C}$, and $\vec{D}$ to their bitwise complements, $\vec{P}$, $\vec{Q}$, $\vec{R}$, and $\vec{S}$ respectively:

$$\vec{P} = (0,1,1,1,1)$$
$$\vec{Q} = (1,0,1,1,1)$$
$$\vec{R} = (1,1,0,1,1)$$
$$\vec{S} = (1,1,1,0,1)$$

Then list and discuss the response of the system to $\vec{P}$, $\vec{Q}$, $\vec{R}$, and $\vec{S}$.

4

Competition, Lateral Inhibition, and Short-term Memory

Victory at all costs, victory in spite of all terror, victory however long and hard the road may be; for without victory there is no survival.

Winston Churchill (Speech, May 13, 1940)

O for a life of Sensations rather than Thoughts!

John Keats (*Letter to Benjamin Bailey*)

Inhibitory connections in neural networks serve a variety of purposes. In our discussion of random nets (Section 2.2), we noted that inhibition can facilitate the stabilization of network activity levels. Also in our discussion of network principles (Chapter 1), we noted that inhibition can provide a mechanism for making choices. These choices might be, for example, between input patterns for short-term memory storage, between categories for classification of a single input pattern, or between drives for activation. It must be added, though, that the choices are not always all-or-none.

Both the stabilization and choice properties have been achieved in neural networks using mechanisms suggested by sensory (particularly visual) anatomy and physiology. We shall now give some history of the ideas behind those mechanisms.

4.1. EARLY STUDIES AND GENERAL THEMES (CONTRAST ENHANCEMENT, COMPETITION, NORMALIZATION)

The systematic study of visual perception was advanced in the middle to late 19th century by the noted physicists Helmholtz and Mach. In particular, both of these scientists observed that edges or contours between light and dark portions of a scene tend to be enhanced relative to the light or dark interiors of the scene. They explained this phenomenon by means of networks of retinal cells, each excited by light within a central area and inhibited by light within a surrounding area. Receptive fields with that structure were later found experimentally, in the compound eye of the horseshoe crab *Limulus* (Hartline & Ratliff, 1957) and in the vertebrate retina (Kuffler, 1953), as shown in Figures 4.1 and 4.2. This kind of structure is variously referred to as *lateral inhibition* or *on-center off-surround organization*.

Figure 4.3 schematizes two types of lateral inhibitory architectures used in pattern processing models: *nonrecurrent* or feedforward, and *recurrent* or feedback inhibition. The principle of lateral inhibition generalizes to networks where nodes *both* excite and inhibit each other, but inhibition operates over a greater distance than excitation, as shown in Figure 4.4.

In this chapter, we explore the functions of lateral inhibition in the transformation and short-term storage of patterns in model neural networks. There is some experimental evidence that in actual mammalian nervous systems, the lateral inhibitory principle is operative at central as well as peripheral areas. Szentagothai (1967 a, b, 1975), Stefanis (1969) and many others have found that the largest neurons in the cerebral cortex, which are called *pyramidal cells*, typically excite smaller neurons called interneurons, which in turn project to other pyramidal cells. While the projections to the other cells have not been verified to be inhibitory, they are of a structural type that is usually believed to be inhibitory. Similar kinds of interactions between large and small cells occur in subcortical areas such as the hippocampus

(a)

(b)

(c)

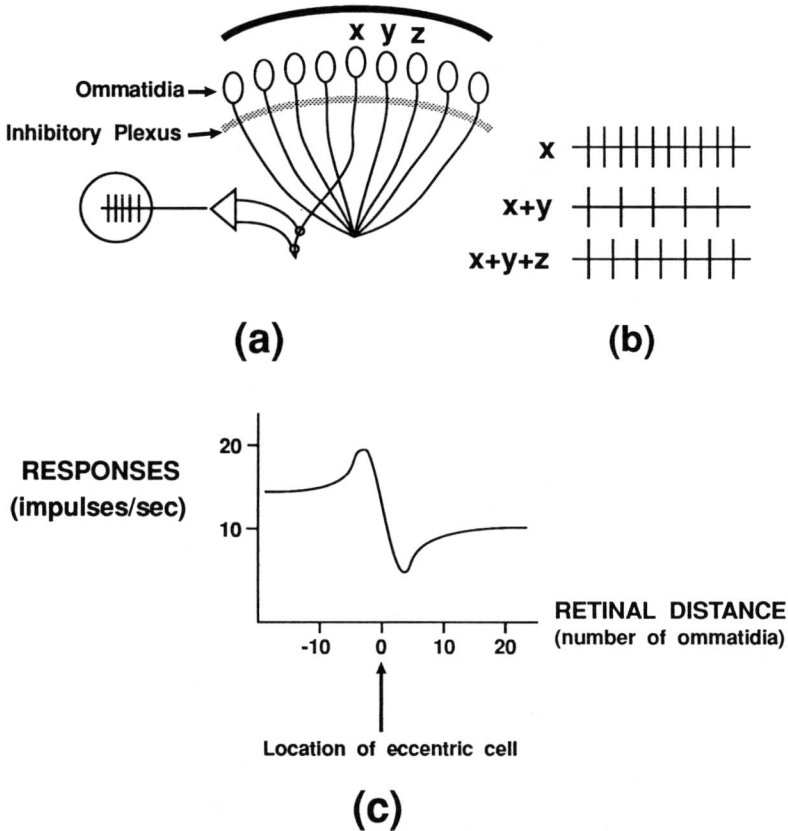

Figure 4.1. Data on the compound eye of *Limulus*. (a) Sketch of the *Limulus* eye, made up of *ommatidia* ("little eyes"). The axons are those of *eccentric cells*, from which the electrical recordings in (b) were taken. (b) Recordings from eccentric cell x during illumination first of x, then of x and y, and finally of x, y, and z. Vertical lines denote action potentials. y (next to x) inhibits x, reducing its firing rate. z inhibits y, thereby disinhibiting x. (c) Firing frequency of a single eccentric cell in response to a sharp dim-to-bright edge being moved across the array of ommatidia. Lateral inhibition accentuates firing close to the edge. (Adapted from THE RETINA: AN APPROACHABLE PART OF THE BRAIN by John E. Dowling, Cambridge, Mass.: The Belknap Press of Harvard University Press, Copyright © 1987 by John E. Dowling; reprinted by permission of the publishers.)

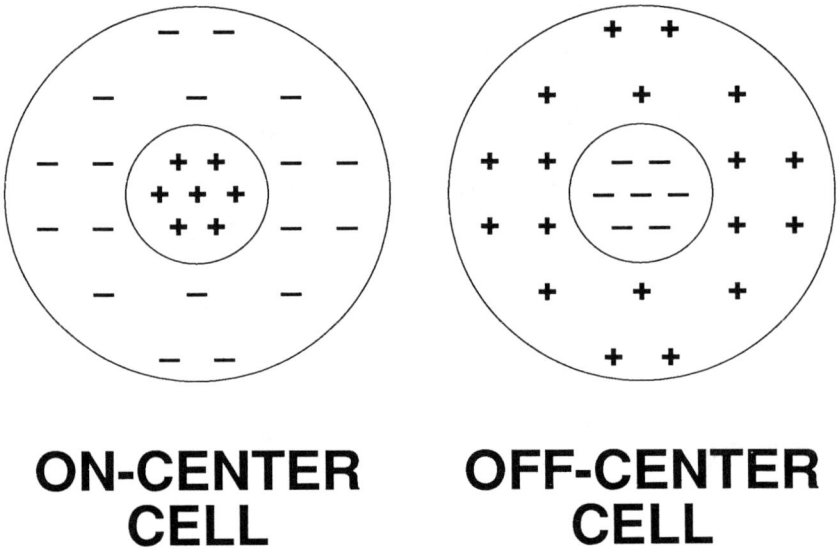

ON-CENTER CELL   OFF-CENTER CELL

Figure 4.2. Idealized receptive field maps of the two types of retinal ganglion cells (the cells that send retinal signals along the optic nerve to the brain). *On-center* cells are excited by light within a circle (*center*) on the receptive field and inhibited by light within an outlying annulus (*surround*); the reverse is true for *off-center* cells. "+" denotes increase in the cell's firing rate when a given area is illuminated, "-" decrease. (Adapted from THE RETINA: AN APPROACHABLE PART OF THE BRAIN by John E. Dowling, Cambridge, Mass.: The Belknap Press of Harvard University Press, Copyright © 1987 by John E. Dowling; reprinted by permission of the publishers.)

(Andersen, Gross, Lomo, & Sveen, 1969) and cerebellum (Eccles, Ito, & Szentagothai, 1967).

In the models discussed in this chapter, the functional units or nodes are frequently collections of neurons sharing some common response properties. This idea, previously suggested in Section 2.2, has a partial physiological basis in the organization of the visual cortex (Hubel & Wiesel, 1962, 1965) and somatosensory cortex (Mountcastle, 1957) into columns of cells with the same preferred stimuli. Moreover, columns that are close together also tend to have

preferred stimuli that are close together.   Evidence for such columnar
organization has also appeared, more recently, in multimodality association
areas of the cortex (Rosenkilde, Bauer, & Fuster, 1981; Fuster, Bauer, & Jervey,
1982; Goldman-Rakic, 1984).   Because the evidence from vision is the most
compelling so far, many of the models in this chapter are inspired by visual
data, though the modeling principles they incorporate may be more broadly
applicable.

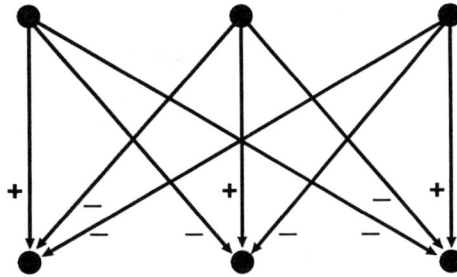

(a)

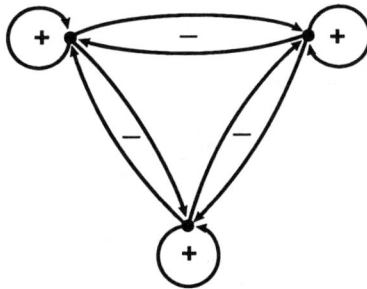

(b)

Figure 4.3. Examples of two kinds of lateral inhibitory networks:
(a) nonrecurrent (feedforward); (b) recurrent (feedback).
(Reprinted by permission of the publisher from Levine,
Mathematical Biosciences **66**, 1-86. Copyright 1983 by Elsevier
Science Publishing Co., Inc.)

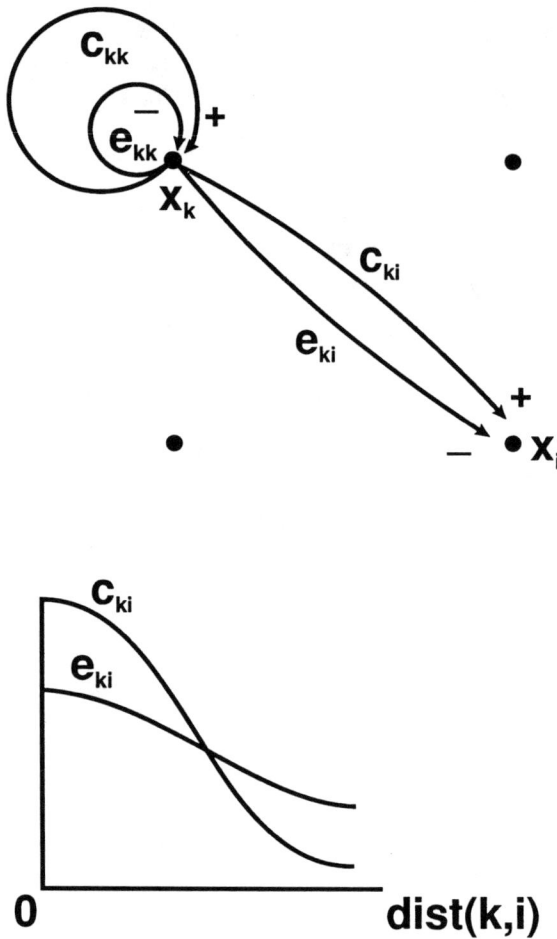

Figure 4.4. Generalization of the recurrent lateral inhibition in Figure 4.3(b). Each node x_k sends *both* excitation and inhibition to itself and to all other node x_i. Excitatory interaction strengths c_{ki} and inhibitory interaction strengths e_{ki} both decrease with distance, reflecting fewer synaptic connections as distance increases, but c_{ki} decreases more rapidly, as seen in the graph.

Some general themes about the functions of lateral inhibition emerged in early modeling studies from the late 1960's and early 1970's. Many of the lateral inhibitory networks studied at that time did not include learning, but were later embedded in multilevel architectures that included learnable connection weights between levels (see, in particular, Chapter 6).

Hartline and Ratliff's Work, and Other Early Visual Models

Hartline and Ratliff (1957) modeled inhibition in the horseshoe crab eye by means of a pair of simultaneous algebraic equations for two mutually inhibiting receptors, as follows:

$$x_1 = (I_1 - k_{12}(x_2 - \theta_2)^+)^+$$
$$x_2 = (I_2 - k_{21}(x_1 - \theta_1)^+)^+$$

Here x_i denotes the impulse frequency in the axon of cell i, and I_i denotes the excitation of cell i by an external stimulus. θ_2 and θ_1 are threshold frequencies that each cell has to exceed before it can exert inhibition, and k_{12} and k_{21} are coefficients of inhibitory action. Finally, for any real number x, the quantity x^+ denotes x if x is positive, and 0 if x is negative or 0; for example, if $x_2 \leq \theta_2$, then $(x_2 - \theta_2)^+ = 0$, whereas if $x_2 > \theta_2$, then $(x_2 - \theta_2)^+ = x_2 - \theta_2$.

The linear Hartline-Ratliff equations proved effective in the modeling of a variety of experimental data, and several other early lateral inhibitory models used extensions of these equations. But other effects, many of them nonlinear, had to be introduced to model some additional complexities of vertebrate vision. For example, Sperling and Sondhi (1968) developed a lateral inhibitory model of effects in the mammalian retina in order to explain certain data on luminance and flicker detection. Their model includes both feedback and feedforward stages. In the feedback stage, as shown in Figure 4.5, the jth node is excited by the $j-1^{st}$ node, for $j > 1$, and inhibited by feedback from the nth node.

The type of inhibition exerted by the feedback stage in Sperling and Sondhi's model is *shunting* rather than *subtractive*. In subtractive inhibition, the incoming signal is linearly weighted, and an amount proportional to that signal is subtracted from the activity (or firing frequency) of the receiving node. In shunting inhibition, the amount subtracted is also proportional to the activity of the receiving node. Thus the inhibiting node acts as if it *divides* the receiving node's activity by a given amount, that is, as if it "shunts" a given fraction of the node's activity onto another, parallel pathway.

In addition to shunting (multiplicative) inhibition, recent lateral inhibitory models often include *shunting excitation*, whose strength is proportional to the difference of a node's activity from its maximum possible level. This is in contrast to *additive* excitation, the opposite of subtractive inhibition, which simply adds an amount proportional to excitatory signal to the activity of a receiving node. Shunting interactions in neural networks have been suggested by experimental results on the effects of a presynaptic neuron on the conductances of various ions across the postsynaptic membrane (*cf.* Hodgkin, 1964; Grossberg, 1973; Appendix 1 of this book). Additional evidence for shunting interactions in actual neurons has been summarized in Blomfield (1974) and Freeman (1983).

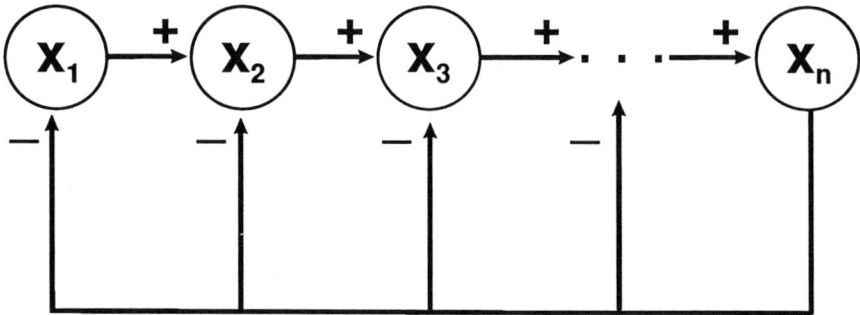

Figure 4.5. Schematic feedback connections in the flicker detection model of Sperling and Sondhi (1968). (Their actual diagrams used an electrical analogy with resistors and capacitors.)

Sperling and Sondhi (1968) described the effect of shunting inhibition as "reducing dynamic range." In other words, while sensory inputs can be arbitrarily intense, the response of network nodes to these inputs has an upper limit. But while lateral inhibition can reduce distinctions between input intensities at extreme ranges, it can enhance such distinctions at intermediate ranges. Variants of the latter effect have been called "contour enhancement" (Ratliff, 1965; Grossberg, 1973); "input-pattern sharpening" (Morishita & Yajima, 1972); and "contrast enhancement" (Ellias & Grossberg, 1975; Grossberg and Levine, 1975). The latter term is the one we shall use below.

Contrast enhancement is an outgrowth of decision or competition between inputs. Competition can be biased in favor of either more intense or less intense inputs by nonlinear interactions. As we shall see in multilevel networks

(Chapters 5 and 6), competition can also be biased in favor of motivationally significant inputs.

Also, similar competitive mechanisms can exist at many levels in the brain. While different sensory inputs compete for storage in short-term memory, for example, different drives or gross modes of behavior can also compete for activation, as in a model by Kilmer *et al.* (1969) of the midbrain reticular formation. Sections 4.2 and 4.3 of this chapter concentrate mainly on the dynamics of competition between inputs, particularly at the level of sensory areas of the cerebral cortex. This includes a sketch of the growing body of mathematical results on competitive neural systems. In Section 4.4 we return to competition at other cognitive levels: between drives, between categories, and between behavior sequences.

Besides contrast enhancement, another common property of lateral inhibitory networks is *pattern normalization* (see Figure 4.6). Normalization (for example, Grossberg, 1970a) means that a pattern of node activities x_1, x_2, ..., x_n at one ("input") level is replaced by activities y_1, y_2, ..., y_n at the next ("output") level that are proportional to the x_i's but independent of the total intensity $\sum_i x_i$. Normalization has often been used in lateral inhibitory networks to keep the total network activity bounded. This concept is reminiscent of Sperling and Sondhi's reduction of dynamic range.

Nonrecurrent Versus Recurrent Lateral Inhibition

In early models involving lateral inhibition, nonrecurrent (feedforward) and recurrent (feedback) inhibition were preferred for different purposes and used to model different processes. Recall from Chapter 2 that recurrent (reverberating) loops have been used in neural models since the 1940's to extend the duration of a stimulus trace. Grossberg (1970a, 1972d), modeling pattern discrimination in the retina, chose nonrecurrent rather than recurrent lateral inhibition in order to shorten the duration of pattern representations. This was done because the retina is designed to encode a fairly accurate representation of ongoing visual events. The visual cortex, by contrast, is designed to encode both present events and memories of recent past ones. Hence, for modeling of pattern processing at the cortical level, it is important to keep patterns active in sensory memory for longer periods. Therefore recurrent lateral inhibition tends to be preferred in cortical models (for example, Wilson & Cowan, 1972; Grossberg, 1973). Differences between actual cellular architecture in the retina and the cortex generally reflect this functional difference.

An example of the "retinal" level of modeling is shown in Figure 4.7, adapted from Grossberg (1970a, 1972d). Two stages of nonrecurrent inhibition are constructed so that a particular node fires in response to one and only one *space-time pattern*, that is, to one time-varying input distribution. Aspects of this network's anatomy are reminiscent of particular layers of the vertebrate retina (*cf.* Figure 9.13 in Appendix 1).

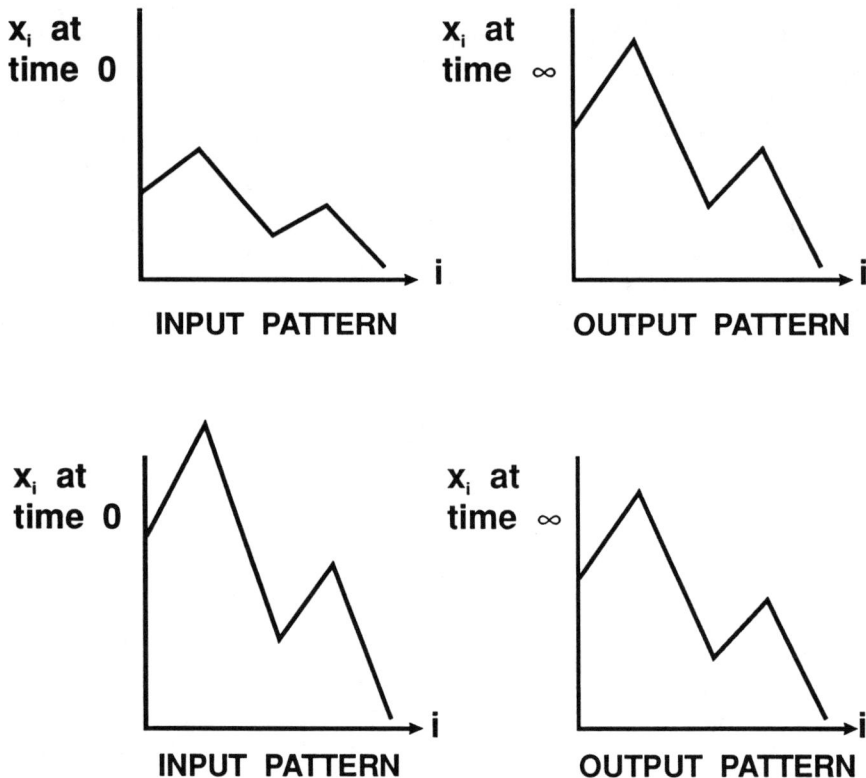

Figure 4.6. Example of pattern normalization. Output pattern has same *relative* activities as input pattern, but is independent of *absolute* input activities. (Reprinted by permission of the publisher from Levine, *Mathematical Biosciences* **66**, 1-86. Copyright 1983 by Elsevier Science Publishing Co., Inc.)

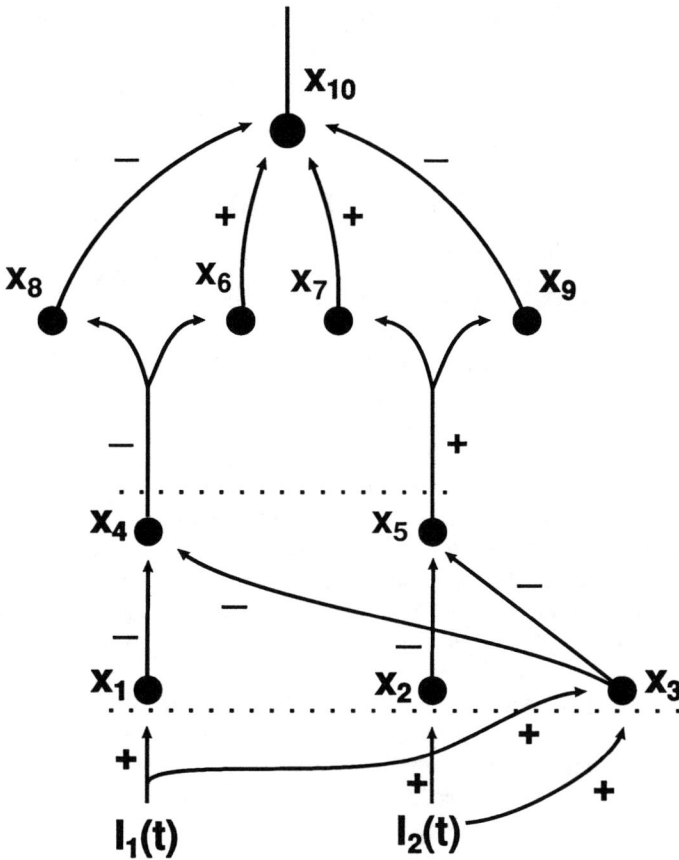

Figure 4.7. Network to recognize a single space-time pattern (consisting of inputs $I_i(t)$, two of which are shown). Subnetwork x_1 through x_5 does *low-band filtering* (filtering out patterns with activity levels lower than those in the desired pattern) and pattern normalization (see Figure 4.6). Subnetwork x_4 through x_{10} does *high-band filtering* (filtering out patterns with activities higher than those in the desired pattern). x_i correspond roughly to retinal cell layers: x_1 and x_2 to receptors, x_3 to horizontal cells, x_4 and x_5 to bipolar cells, x_6 through x_9 to amacrine cells, x_{10} to ganglion cells. (Adapted from Grossberg, 1970, with permission of Academic Press; see that article for details.)

In the next section, we concentrate on the "cortical" level of modeling. In particular, we consider the use of networks with recurrent lateral inhibition (and, sometimes, lateral excitation) to model short-term storage of sensory patterns.

4.2. LATERAL INHIBITION AND EXCITATION BETWEEN SENSORY REPRESENTATIONS

Short-term memory in recurrent lateral inhibitory networks has been modeled since the early 1970's, particularly by Wilson and Cowan, Grossberg, Amari, and their colleagues. Typically, an input pattern is regarded as the initial state of a mathematical *dynamical system*, which can roughly be defined as the movement through time of the solutions of a system of differential equations for interacting variables (see Appendix 2 for further discussion). This solution is described by a vector composed of the values of all the variables in the system at any given time (*cf.* the discussions in Section 3.2). The equations describe the transformation of this pattern and its storage in short-term memory; the stored pattern is then regarded as a limiting vector to which the system converges as time gets large.

Lateral inhibitory architectures, as stated in Section 4.1 above, tend to enhance contrasts between pattern intensities. The inhibitory connections mean that larger activities tend to suppress smaller ones; thus, after a certain time, some subcollection of nodes becomes, and remains, dominant. As a consequence, dynamical systems defined by such networks often, but not always, converge to an *equilibrium* state (also called a *steady* state). An equilibrium is a state where the system interactions are in "balance," so that once the system reaches that state, it will not be perturbed from it. (See Appendix 2 for a more precise mathematical definition of an equilibrium.) The study of pattern transformations by analysis of the equilibrium behavior of a dynamical system is common in other recent neural network models (*e.g.*, Hopfield, 1982; Geman & Geman, 1984; Hopfield & Tank, 1985, 1986; Golden, 1986; White, 1987). System vectors do not always converge to an equilibrium, however. The next section includes examples of networks where the state vector converges instead to a *limit cycle*, that is, to a periodic (oscillatory) solution, as shown in Figure 4.8.

Wilson and Cowan's Work

One of the first published mathematical studies of a neural network emulating cortical lateral inhibition was done by Wilson and Cowan (1972). This work was further elaborated in Wilson and Cowan (1973) (where the lateral inhibitory principle is most explicit) and Ermentrout and Cowan (1980).

The network of Wilson and Cowan (1972) is composed of neuron populations for which intrapopulation connections are random and interpopulation connections deterministic. This is reminiscent of the notion of "randomness in the small and structure in the large" (Anninos *et al.*, 1970) discussed in Section 2.2. The distributions of different cell thresholds are averaged into sigmoid functions (see Figure 2.7b) describing input-output relations at the node (*i.e.*, population) level.

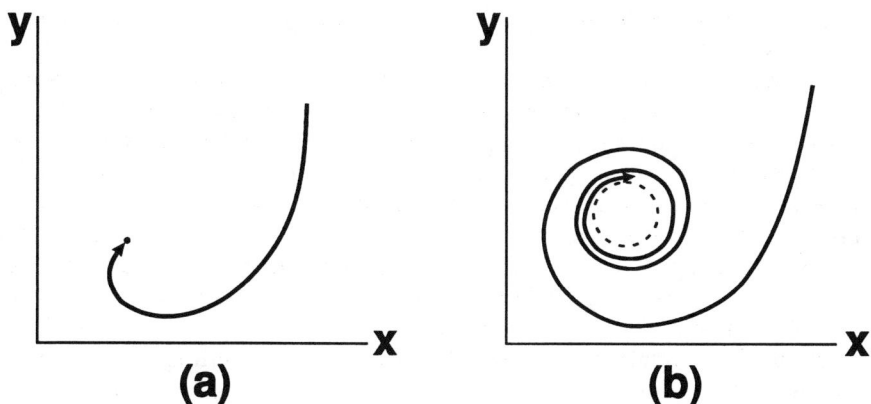

Figure 4.8. If a dynamical system has two variables (say x and y in these graphs), the changes in these variables over time can be shown by a curve. Arrows denote direction of flow as time increases. The system can either approach an *equilibrium point* (part a) or else oscillate toward a limiting curve or *limit cycle* (the dotted curve in part (b)).

The model of Wilson and Cowan (1972) includes excitation as well as inhibition between nodes. The network is described in terms of the two functions $x_E(t)$ and $x_I(t)$, which denote the proportions of excitatory and inhibitory cells, respectively, firing per unit time at time t. The change over time of the excitatory activity $x_E(t)$ reflects a combination of three influences (incorporated in Equations (4.8) below

(a) decay back to a baseline activity level (assumed to be zero, for simplicity);
(b) a signal that linearly combines excitation from excitatory cells, inhibition from inhibitory cells, and excitatory inputs from outside the network, and then is transformed by a sigmoid signal function;
(c) refractory periods of the excitatory cells themselves (see Section 2.2), so that only cells that have not fired within a recent time interval can be excited by the signal described in (b).

The inhibitory node is subject to the same set of influences, though the interaction coefficients and sigmoid functions are different from those for the excitatory node.

Mathematically, the effect of refractory periods on the excitatory node x_E is described by a term that decreases linearly with excitatory signal strength, that is, a term of the form $a_E - b_E x_E$ for some positive numbers a_E and b_E. As a consequence, any given input signal has the strongest effect on an inactive node, and no effect on a node already at a saturation point of activity (namely a_E/b_E). This term is multiplied by the strength of the incoming signal. Similarly, the signal to the inhibitory node is multiplied by a decreasing linear factor $a_I - b_I x_I$. Grossberg (1973) noted that the factors $a_E - b_E x_E$ and $a_I - b_I x_I$ are equivalent to terms for shunting (multiplicative) excitation in passive membrane equations for a single neuron (see the discussion of Sperling and Sondhi in Section 4.1); we return to this point in the next subsection.

A second article by Wilson and Cowan (1973) described a two-dimensional network for representing an area of the cerebral cortex or thalamus. (The thalamus is an area of the brain just below the cortex, much of it connected to the cortex in a one-to-one fashion and providing a "relay" to the cortex from lower brain areas; cf. Appendix 1). This network has properties similar to the network of Wilson and Cowan (1972) with the addition of distance-dependent interactions. The two variables $x_E(t)$ and $x_I(t)$ are replaced by variables $x_E(s,t)$ and $x_I(s,t)$ that depend on distance as well as time. Excitation tends to fall off more sharply with distance than does inhibition (cf. Figure 4.4; see Sholl, 1956 for anatomical justification).

Hence, different positions in the visual field, or different line orientations, can be represented at different cortical or thalamic locations. The network variables represent averaged activities of excitatory and inhibitory neurons at these locations.

The distance-dependent networks of Wilson and Cowan (1973) have the same range of limiting behavior as those of their earlier article, despite greater mathematical complexity. The large-time dynamics of Wilson and Cowan's equations include the possibility of *hysteresis*, whereby if the amount of external stimulation is changed, the dynamics are dependent on the past history of stimulation. These equations can also, for some parameters, exhibit limit cycles (see Figure 4.8). Wilson and Cowan saw limit cycles as possible analogs of the reverberatory loops between the cerebral cortex and the thalamus. These loops have often been suggested as a physiological substrate for short-term memory (*e.g.*, Andersen & Eccles, 1962).

The network of Wilson and Cowan (1973) reproduced various phenomena of visual psychophysics. This included characteristic responses to different spatial frequencies; *metacontrast*, or perceptual masking of a brief stimulus by a second, subsequent, stimulus presented elsewhere in the visual field; and a hysteresis phenomenon found in stereopsis. (Stereopsis, or three-dimensional binocular vision, is discussed further in Section 4.3.) Ermentrout and Cowan (1980), studying a more abstract version of Wilson and Cowan's network, proved the existence of periodic solutions, and discovered that these solutions had properties in common with some simple visual hallucinations.

Work of Grossberg and Colleagues

Another series of articles on lateral inhibition between cortical populations was initiated by Grossberg (1973), who used shunting interactions but combined excitatory and inhibitory influences in a different manner than did Wilson and Cowan. Later articles in this series include Grossberg and Levine (1975), Ellias and Grossberg (1975), Levine and Grossberg (1976), Levine (1975), Grossberg (1978a), Cohen and Grossberg (1983), and Cohen (1988).

The implementation of shunting recurrent lateral inhibition in Grossberg (1973) was motivated by some heuristics relating to shunting nonrecurrent inhibition. The need for inhibition in his model arose in turn from consideration of a shunting network without inhibition, as defined by differential equations. Those equations are

$$\frac{dx_i}{dt} = -Ax_i + (B-x_i)I_i \qquad (4.1)$$

for the activity x_i of the ith node as a function of time, where I_i are outside inputs, B is the maximum possible activity, and A is a decay rate. (The left-hand side of a differential equation for a variable denotes the variable's rate of change, while the right-hand side describes the interactions causing it to change; see Appendix 2 for more details.) Equation (4.1) says that shunting (multiplicative) excitation, proportional to the difference of x_i from its maximum activity B, is supplied by outside inputs, whereas shunting inhibition, proportional to x_i itself, is supplied only by spontaneous decay.

But if the inputs I_i to the nodes defined by (4.1) form a spatial pattern (see Figure 3.4), the shunting term B-x_i in (4.1) causes a distortion of relative pattern weights. (This assertion is justified in Exercise 1 of this chapter.) The distortion is removed (see also Exercise 1) by making the ith input not only excite the ith node but inhibit all other nodes (see Figure 4.9). Equations (4.1) are then modified so that shunting inhibition is from a combination of spontaneous decay and inputs to other nodes:

$$\frac{dx_i}{dt} = -Ax_i + (B-x_i)I_i - x_i \sum_{k \neq i} I_k \qquad (4.2)$$

where the "Σ" is standard notation for a sum (in this case, taken over all inputs exciting nodes other than the ith).

In many versions of Grossberg's recurrent network, the same nodes are both excitatory and inhibitory. (The exception is Ellias & Grossberg, 1975). Shunting inhibition (proportional, in general to x_i minus its minimum activity C_i) and excitation (proportional to the maximum i^{th} node activity B_i minus x_i) are now supplied not by outside inputs but by the node itself and other network nodes. Hence, (4.2) is replaced by

$$\frac{dx_i}{dt} = -Ax_i + (B_i-x_i)\sum_{k=1}^{n} f(x_k)c_{ik}$$
$$- (x_i-C_i)\sum_{k=1}^{n} f(x_k)e_{ik} + I_i \qquad (4.3).$$

The expression (4.3), while appearing as a single equation, actually represents a system of equations for the activities of arbitrarily many nodes with identical connection properties. In these equations, f is a *signal function* reflecting input-output transformations at the single-cell level. The function f might or might not be sigmoid (a point to which we return later) but must be increasing with x_k. The positive constants c_{ik} and e_{ik} are nonmodifiable excitatory and inhibitory interaction coefficients[1], and $I_i(t)$ is the input to the ith node.

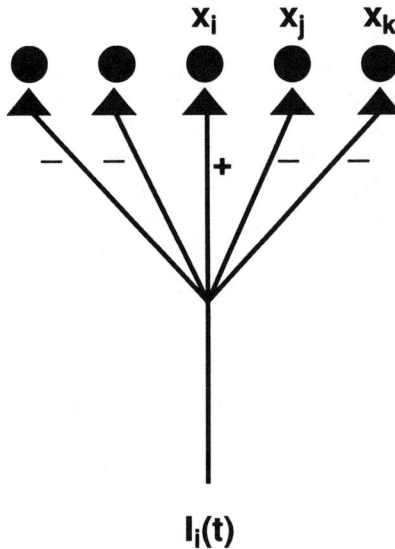

Figure 4.9. Nonrecurrent on-center off-surround interactions. (Adapted from Grossberg, 1976a, with permission of Springer-Verlag.)

Equations (4.3) are similar to Wilson and Cowan's except that in (4.3), the signal function is computed separately for the excitatory and inhibitory inputs. Therefore, excitatory and inhibitory inputs combine not linearly as in Wilson and Cowan's network, but nonlinearly via shunting terms. Grossberg (1973) interpreted these shunting interactions as separate excitatory and inhibitory "gain control," allowing for automatic tuning of network sensitivity in response to fluctuating inputs. Also, if these equations reflect averaging on the node level

------

[1] In this book, connection weights are usually called w_{ik}, but when the same network contains more than one type of connection (*cf.* also the work of Malsburg, 1973, discussed in Ch. 6), different letters may be used for weights.

of the single-neuron equations, B_i and C_i can be interpreted as equilibrium voltages for sodium and potassium ions respectively (see Appendix 1 for discussion of the physiological significance of sodium and potassium). In this interpretation, the factors under the summation signs are conductances of those ions across the neuron membrane.

Grossberg (1973) and succeeding articles include mathematical proofs that as time gets large, the variables x_i in (4.3) always converge to steady-state values (limits) for broad classes of functions f. These steady-state values can either be zero for all nodes, or nonzero for one or more nodes. The nonzero limiting node activities, and their relative sizes, were interpreted as reflecting the network's choice as to what parts, if any, of a pattern are to be stored in short-term memory. Periodic or chaotic patterns of the system variables are thereby prevented (see Appendix 2 for discussion of these alternatives in mathematical dynamical systems).

Computer simulations of Grossberg's network show that the system reaches activities close to its steady-state values quickly (in a few minutes if decay rates were chosen to reflect neuronal time constants). Hence, this long-time behavior actually approximates short-time sensory pattern transformations such as occur in short-term memory.

In Grossberg (1973), the dynamics of Equations (4.3) were studied with all maximum activities B_i equal. The values x_i at time 0 reflect the input pattern, and pattern transformations after time 0 reflect the recurrent interactions. The connectivity of the network is pure on-center off-surround; that is, the excitatory coefficients c_{ik} are set to 1 if i=k and 0 otherwise, with the reverse true for the inhibitory coefficients e_{ik}.

In the network of Grossberg (1973), the steady state approached by the system depends on the function f; in particular, it depends on whether f grows linearly with x_i, faster than linearly, slower than linearly, or in a sigmoid fashion (*i.e.*, faster than linearly for small x_i and slower for large x_i; see Figure 4.10). If f is linear in x_i, the output values, $x_i(\infty)$ (x_i at the limiting or "infinite" time) are proportional to the input values $x_i(0)$; in the case of a visual pattern, for example, relative reflectances are conserved. Such proportionality might initially appear to be a good property for a sensory memory system. But Grossberg argued that perfect proportional storage is undesirable because it means that insignificant network noise is stored along with significant signal traces. A better outcome, he stated, is contrast enhancement with noise suppression; that occurs if f is a suitable sigmoid function. For such functions, $x_i(\infty)$ is proportional to $x_i(0)$ if $x_i(0)$ is above a threshold value and equal to 0 if $x_i(0)$ is below that threshold.

Grossberg and Levine (1975) generalized many of these results to (4.3) with unequal maximum activities B_i, representing biases in the competition between nodes for storage of their preferred sensory features. Possible sources of such biases are: (1) some stimuli occur more often than others during development, causing unequal growth of relevant feature detectors (*cf.* Blakemore & Cooper, 1970, or Hirsch & Spinelli, 1970); (2) some stimuli are

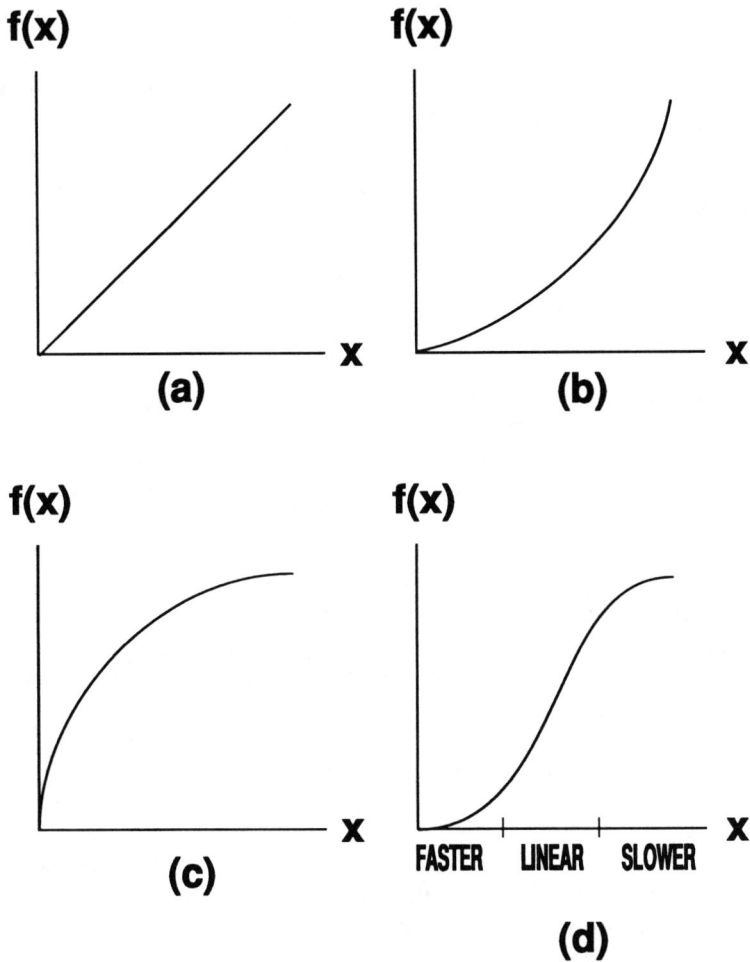

Figure 4.10. Schematic graphs of four types of signal functions: (a) linear; (b) faster-than-linear; (c) slower-than-linear; (d) sigmoid.

attended to more than others during adult life; or (3) neuron populations exhibit random inequalities of cell distribution. The results of Grossberg (1973) on contrast enhancement with noise suppression remain true within each *subfield* (defined as the subnetwork of all nodes with a given value of B_i). A choice is made *between* subfields, and typically the activities of all but a small number of subfields are suppressed as t becomes large. The subfields chosen are the winners of a "tug-of-war" between those with largest inputs and those favored by network biases (*cf.* Figure 4.11).

Later articles (Grossberg, 1978a; Cohen & Grossberg, 1983) further generalized these theorems on global existence of limits. To date, the most general theorem of this sort is Cohen and Grossberg's. They proved the existence of limits for Equations (4.3) in the case of self-excitation with distance-dependent inhibition. Self-excitation is defined as the situation where $c_{ik}=1$ when i=k and 0 otherwise; distance-dependent inhibition is defined as the situation where e_{ik} values are arbitrary as long as $e_{ik}=e_{ki}$.

Work of Amari and Colleagues

One of the pioneers in mathematical study of random neural networks, as noted in Section 2.2, was Amari. Like Wilson and Cowan's work, Amari's work evolved from consideration of randomly connected networks to study of networks with connections exhibiting lateral inhibition and lateral excitation. Amari (1977a) and Amari and Arbib (1977) studied networks of the latter type, based on nonlinear distance-dependent interactions that are additive rather than shunting. This work became the basis for a theory of categorization developed by Amari and Takeuchi (1978); see Section 6.1. (Hirsch, 1990 showed that the Amari-Takeuchi system is equivalent to that of Hopfield, 1984.)

Amari (1977a) studied neural populations arranged in a "field" in the sense used in physics. That is, unlike the network described in Equations (4.3) that consist of a finite number of distinct nodes, his network is mathematically described by an activity variable that depends continuously on both time and location. The equations describing the dynamics of this variable use separate excitatory and inhibitory weighting functions, with inhibition decreasing less sharply with distance than excitation as in Figure 4.4(b). He found that the system typically (but not always) approaches an equilibrium state in which some part of the field remains active in short-term memory. Depending on various system parameters, the part that remains active could either be the entire field, a wide range, or a narrow range.

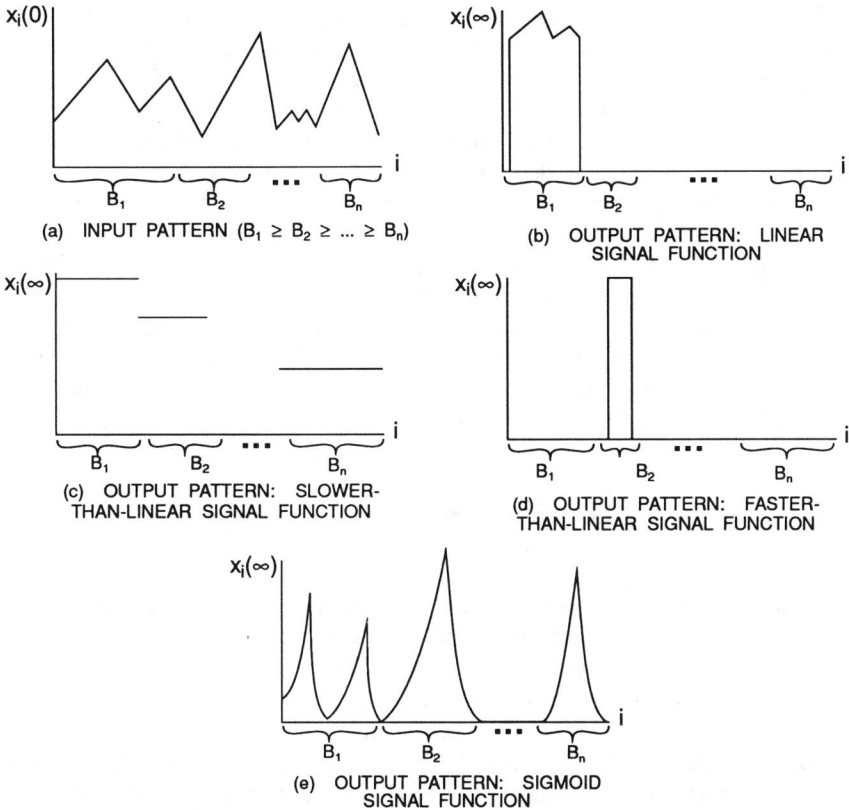

Figure 4.11. Pattern storage with unequal B_i. (a) Input pattern at time 0. (b) Output pattern at time ∞ if the function f of (4.3) is linear. Inputs to nodes with the largest B_i (called B_1) are stored in proportion to input activities $x_i(0)$, and suppress others. (c) Output pattern for f slower than linear. The pattern becomes uniform within each subfield. Activities are largest for nodes with largest B_i, regardless of $x_i(0)$. (d) Output pattern for f faster than linear. Only inputs to nodes with one B_i value (not always the largest) survive, and only those with largest $x_i(0)$ in that subfield. (e) Output pattern for f a suitable sigmoid. Inputs with $x_i(0)$ below a threshold are suppressed. Inputs with $x_i(0)$ above the threshold, at least in some subfields, are enhanced. (Adapted from Grossberg & Levine, 1975, with permission of Academic Press.)

Amari and Arbib (1977) extended the earlier model to a neural field with two dimensions. The design of that model was motivated by previous work of Dev (1975) on modeling binocular vision, which is discussed more fully in Section 4.3. Briefly, some results on binocular vision were based on detection of disparity between the positions of an image on the right and left retinae. The Amari-Arbib model includes inhibition between detectors of different disparitiesat the same position, and excitation between detectors of the same disparity at different positions.

Hence, the neural field developed by Amari and Arbib consisted of nodes indexed by the two dimensions of disparity and distance. The resulting network was termed *competitive-cooperative*, a term that has come into common usage in the field (see, for example, Amari & Arbib, 1982). In general, if lateral inhibition is interpreted as competition between different percepts for encoding in short-term memory, then by the same token, lateral excitation can be interpreted as *cooperation* between related or compatible percepts.

Energy Functions in the Cohen-Grossberg and Hopfield-Tank Models

The theorem of Cohen and Grossberg (1983) relies on a common construction from mathematical dynamical systems theory, namely, a *Lyapunov function* or "energy" function. A Lyapunov function is some function of the system's state variables whose value decreases as the system's state changes over time. An equilibrium point of the dynamical system corresponds to a local minimum of the energy function (see Figure 4.12). In dynamical systems derived from physics, this function frequently represents an actual energy; in dynamical systems for neural networks, the energy function is more abstract.

The Lyapunov function for the Cohen-Grossberg system is discussed in Section 4.5. We now discuss the Lyapunov or energy function for the related but simpler system of equations introduced in Hopfield (1982). The Hopfield networks do not always include lateral inhibition, but are discussed in this section because they deal with issues of short-term pattern storage and obey equations formally similar to Cohen and Grossberg's. Hopfield's 1982 article and more recent articles by Hopfield and Tank are discussed again in Chapter 7, in the context of optimization problems, which have formed the major application of their work.

In the simplest form of the network of Hopfield (1982), the ith node has two possible states: $x_i = 0$ ("not firing") or $x_i = 1$ ("firing at the maximum rate"). Hence, the instantaneous state of the system is a binary vector whose number of components is equal to the number of nodes. There are also connection strengths w_{ij} from node j to node i, for all pairs where i≠j. (Not all pairs of nodes have to be connected; for those that are not connected, $w_{ij} = 0$.

Also, the w_{ij} can be positive or negative; in fact, in examples they tend to be negative more often than not.) Later versions of Hopfield's network include the possibility that connection strengths might change over time as a result of associative learning. But the energy function formulation first arose in the context of unchanging connection strengths.

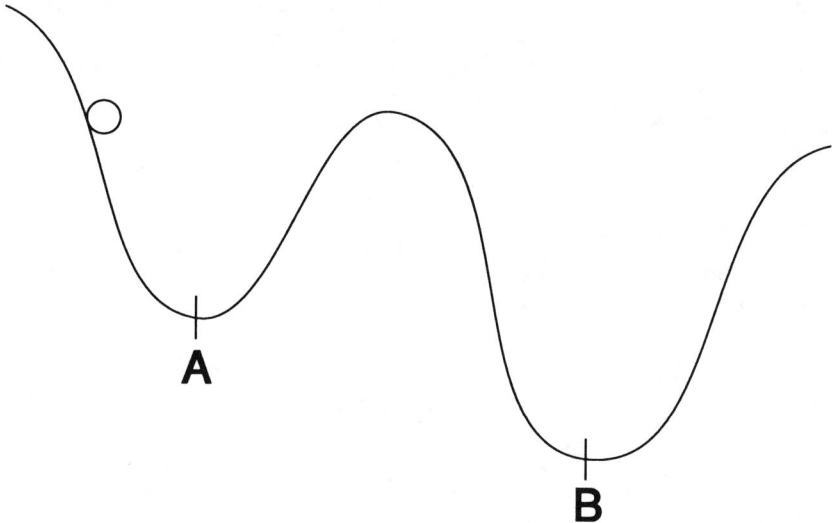

Figure 4.12. The path of the variables in a typical competitive dynamical system is analogous to the path of a ball-bearing along a curve (representing the system "energy function"). Like the ball bearing, the system eventually reaches a local minimum state of the energy (either A or B in this figure).   (Reprinted from Rumelhart & McClelland, 1986, Vol. 1, with permission of MIT Press.)

The state changes over time are governed by a linear threshold algorithm reminiscent of some of those used in Rosenblatt (1962). If $x_i(t)$ denotes the state of the ith node at time t (t an integer), this state readjusts at random times, according to the rule

$$x_i(t+1) = 1 \; if \sum_{j \neq i} w_{ij}x_j(t) > 0$$
$$x_i(t+1) = 0 \; if \sum_{j \neq i} w_{ij}x_j(t) < 0$$

(4.4).

(In later articles of Hopfield and Tank, the thresholds of 0 in Equations (4.4) were replaced by more general threshold parameters. The energy function formulation easily generalizes to that version; see Section 4.5).

The algorithm described by (4.4) means that each time x_i changes, the increment Δx_i (using the character Δ for change in any variable) is either 0 or 1 if $\sum_{j \neq i} w_{ij}x_j > 0$ and is either 0 or -1 if $\sum_{j \neq i} w_{ij}x_j < 0$. Hence Δx_i, if not 0, always has the same sign (positive or negative) as $\sum_{j \neq i} w_{ij}x_j$. Now impose the condition that the weights are symmetric[2], *i.e.*, $w_{ij} = w_{ji}$. If we then consider the Lyapunov or energy function

$$E = -\frac{1}{2}\sum_i \sum_j w_{ij}x_i x_j$$

(4.5),

each change Δx_i in a given node activity leads (by (4.4) and (4.5)) to an energy change of

$$\Delta E = -\Delta x_i \left(\sum_{j \neq i} w_{ij}x_j\right)$$

(4.6).

Since Δx_i is 0 or of the same sign as $\sum_{j \neq i} w_{ij}x_j$, the expression (4.6) means that $\Delta E \leq 0$ at all times.

Some extensions of the above energy function, in both discrete and continuous models, were made in Hopfield (1984) and Hopfield and Tank (1985, 1986). In Hopfield (1984), the discrete-time system of the 1982 article was extended to include external inputs to each node and arbitrary thresholds for node activation. Also, a generalization of the algorithm defined by (4.6) to the continuous-time case was introduced in Hopfield (1984) and developed further in Hopfield and Tank (1985, 1986). In this work, the output V_i of the

---

[2] This symmetry assumption is made for mathematical convenience and is not likely to be biologically realistic. On this point, the reader should refer back to the discussion of Kosko's BAM in Section 3.4.

i^{th} node is treated as a function (usually sigmoid) of the input, as in the articles of Wilson and Cowan (1972) and Grossberg (1973). All these systems have Lyapunov functions similar to (4.5).

The systems studied by Hopfield and Tank become laterally inhibitory (competitive) when all the w_{ij}, $i \neq j$, are set to be negative. (One of the examples studied by Hopfield & Tank, 1986, a network that can convert computational data from analog to binary, exhibits this lateral inhibitory structure.) Also, the theorem of Cohen and Grossberg (1983), showing the convergence of the system defined by (4.3) to some equilibrium state, does not depend on the values e_{ij} (which correspond to the $-w_{ij}$ in Hopfield and Tank's systems) being positive. Hence, that theorem covers the Hopfield-Tank systems as subcases. (This mathematical point is discussed more fully in Grossberg, 1987a, where it also shown that this theorem applies to the Boolean model of McCulloch and Pitts, 1943, and to the "brain-state-in-a-box" model of Anderson, Silverstein, Ritz, & Jones, 1977.)

The Implications of Approach to Equilibrium

A theorem stating that a system must approach a system equilibrium point does not specify *which* equilibrium is approached. In particular, if an energy function is always decreasing, the system will approach one of several states that are local minima of the energy function (see Figure 4.14). In some applications, the global minimum corresponds to the optimal state, and schemes have been studied for escaping from local minima that are not globally optimal. (One of these schemes is *simulated annealing*, introduced by Kirkpatrick, Gelatt, & Vecchi (1983) and further studied by Smolensky (1986) and Hinton and Sejnowski (1986) among others. Such optimality problems are discussed more fully in Chapter 7.) Hence, results of Grossberg, Amari, Hopfield, and others indicate that competitive neural systems often arrive at a choice as to what to store in short-term memory, but the nature of the choice is heavily influenced both by network parameters and by outside inputs.

There are also some exceptions to the choice-making property itself. Cohen (1988) found an example of a system obeying Equations (4.3), with $n = 2$, whose solution approaches a limit cycle (oscillates) instead of approaching an equilibrium (see Figure 4.8). In his example, the assumption that the excitatory coefficients $c_{ik} = 0$ for $i \neq k$ is removed, but the symmetry assumption $e_{ik} = e_{ki}$ still holds. Also, Ellias and Grossberg (1975) found oscillations in certain examples of the *unlumped* on-center off-surround network, where excitatory and inhibitory cell populations occupy separate nodes. (The unlumped case is probably closer to real neuroanatomy than the *lumped* case, including the other articles of Grossberg's group, where the same nodes are excitatory and inhibitory. As was discussed early in this chapter, the largest cell type in the cerebral cortex, pyramidal cells, typically excite smaller interneurons which in turn appear to inhibit other pyramidal cells.) Finally, approach to limit cycles was found in some subcases by Amari (1977a).

In spite of these exceptions, the choice-making property of competitive or competitive-cooperative networks has been a valuable guide for models of natural or artificial pattern recognition. Since this type of network theories have made extensive contact with psychological and neurophysiological data in vision, we devote the next section to visual modeling. Some network mechanisms used to model specific visual data are introduced to give the reader a "hands on" sense of how data can suggest theory. In particular, we consider models of early visual processes that do not include learning or reference to previously stored templates (though the parameters defining these processes could have been influenced by development). The combination of early processing effects with learning in models of multilevel visual processes, such as coding, is discussed in Chapter 6.

The type of short-term memory found in these networks has been described (*e.g.*, Hopfield, 1984; Cohen, 1988) by the term *content-addressable memory*, familiar from computer science. This term signifies that each node ("address") is distinguished by what input events it encodes.

4.3. COMPETITION AND COOPERATION IN VISUAL PATTERN RECOGNITION MODELS

Visual Illusions

The dynamics of the visual system can be illuminated by studying some of its characteristic illusory percepts. Several network models (Wilson & Cowan, 1973; Levine & Grossberg, 1976; Grossberg & Mingolla, 1985 a, b) incorporate the notion that such illusions are by-products of a lateral inhibitory network designed to correct for irregularities in the luminance data that reaches the retina. Models of visual illusions typically involve both competition and cooperation, sometimes along different dimensions (as in the aforementioned work of Dev, 1975) and sometimes within the same dimension.

Levine and Grossberg (1976) simulated some illusions in angle perception. Their model incorporated the findings of Hubel and Wiesel (1962, 1965) that most cells in the primate or cat visual cortex respond preferentially to lines or bars of some particular orientation; a characteristic tuning curve for orientation in a visual cortical cell is shown in Figure 4.13. The Levine-Grossberg model is based on a recurrent competitive-cooperative network in which each node represents a specific line orientation. The interaction coefficients of Equations

(4.3) decrease with distance between the orientations coded by given nodes, but the c_{ik} decrease faster with distance than the e_{ik}; that is, cooperative interactions are short-range and competitive interactions long-range. Percepts of one or two lines are denoted by inputs that are nonzero to the nodes corresponding to those lines. The perceived orientation of a line is interpreted as being the orientation corresponding to that node whose activity is largest after the inputs have been transformed by recurrent interactions.

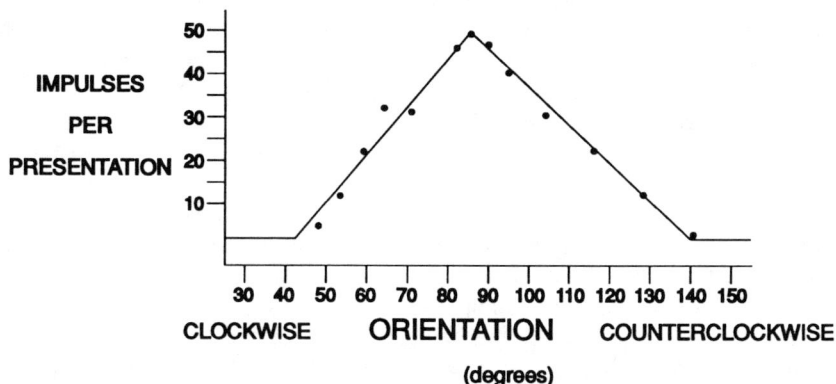

Figure 4.13. Example of the tuning curve of a complex cell (the second of three layers of visual cortical cells described by Hubel & Wiesel). Each point is the mean response, above the cell's spontaneous firing rate, for ten sweeps of a moving bar of light across an oscilloscope during a 3-second interval. (Adapted from Rose & Blakemore, 1974, with permission of Springer-Verlag.)

If all maximal node activities (B_i in Equations (4.3)) are equal, the Levine-Grossberg network can reproduce the experimental result (Blakemore, Carpenter, & Georgeson, 1969) that an acute angle is seen as up to one degree larger than it really is. In the network, an acute angle stimulus excites two orientation-sensitive network nodes. Recurrent inhibition from either of those two nodes shifts the local peak of activity away from the other to a different node corresponding to an angle a degree or two out from it. This causes a shift in each line's perceived location. If instead the nodes responsive to vertical or horizontal orientations have larger B_i than others, the same network can reproduce the result (Gibson & Radner, 1937) that a fixed near-vertical tilted line appears closer to vertical after prolonged viewing. Recall from the discussion in Section 4.2 that differences in B_i could be influenced by experience during development. In fact, there is evidence that cultural experience affects the bias toward vertical and horizontal in humans (Annis & Frost, 1973) as well as other human visual illusions (Deregowski, 1973).

In some other competitive-cooperative neural networks, each node is interpreted as a receptor for a given visual field position rather than a line orientation. For example, Wilson and Cowan (1973) and Levine (1975) both simulated results of Fender and Julesz (1967) on the perceived location of vertical lines viewed stereoptically (see Figure 4.14). In the Fender-Julesz experiments, two parallel vertical lines are shown simultaneously, each seen only by one eye; the lines are pulled apart and then slowly pushed back together. During the stage when they are being pulled apart, the two lines are seen as one for a considerable distance (2 degrees of arc) until they suddenly appear to jump apart. But while they are being pushed back together, these lines are seen as two until the distance between them is much shorter (.1 degrees of arc), at which point they appear to fuse together. Hence hysteresis, or memory, occurs: the perceived location of the two lines depends partly on their past locations as well as their present ones.

Full understanding of the binocular hysteresis effect depends on understanding of binocular depth perception, which is discussed later in this section. But the Wilson-Cowan and Levine networks simulate the Fender-Julesz result without using depth perception, by treating the two lines as inputs to nodes representing different positions in the visual field, and interpreting perceived location as corresponding to the (one or two) nodes with largest activity.

Orientation and position are not the only two variables that are coded by cell populations in the visual cortex. Table 4.1 sums up information about these and other key visual variables, such as spatial frequency; disparity of the images on the right and left retinae, which is a measure of depth; color; and ocularity (cells may have a preference for one or another eye or else respond equally to inputs from either eye).

Some neural networks used to simulate visual data combine two or more of these variables. For example, both orientation and position information are used in the networks of Grossberg and Mingolla (1985a), which simulate some illusory percepts of visual contours. An example is the white square perceived in Figure 4.15, due to Kanizsa (1976), whose corners are formed from the boundaries of four black (or two black and two gray) "pac-man" figures.

In Figure 4.15, two white line segments that are actually present and of the same orientation are perceptually joined together by an illusory longer line segment. That insight among others suggested Grossberg and Mingolla's modeling scheme. In their competitive-cooperative network, boundaries are perceived as signals "sensitive to the orientation and amount of contrast at a scenic edge, but not to its direction of contrast" (Grossberg & Mingolla, 1985a, p. 176).

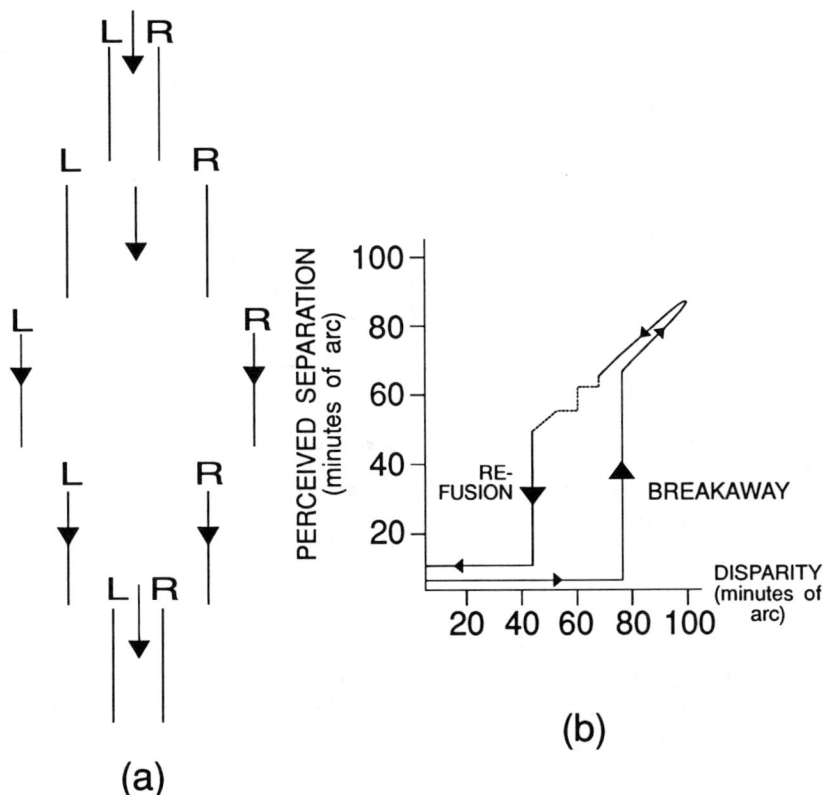

Figure 4.14. Binocular hysteresis data. (a) Schematic of successive positions of left-eye and right-eye vertical lines in the Fender-Julesz experiment. Lines denote actual positions, arrows perceived positions. (From Levine, 1975.) (b) Detailed measurement of perceived versus actual distance between lines. Dotted lines signify that perceived separation can be variable with transient fusion. Arrows denote direction of increasing time. (Reprinted from Fender & Julesz, *Journal of the Optical Society of America* **57**, 819-830, 1967, with permission of the Optical Society of America.)

Variable	Source of experimental findings
Position	Hubel and Wiesel (1962, 1965)
Orientation	Hubel and Wiesel (1962, 1965)
Ocularity (left or) right eye)	Hubel and Wiesel (1962, 1965)
Disparity (between left and right retinal images)	Barlow, Blakemore, and Pettigrew (1967)
Spatial frequency	Robson (1975)
Color	Hubel and Wiesel (1962, 1965)

Table 4.1. Stimulus variables to which single cells in the visual cortex can be differentially responsive.

Figure 4.16a illustrates insensitivity to contrast direction. Each node of the network responds to lines of a particular orientation at a particular position in the plane. There are two forms of competition, between receptors for like orientations at nearby positions (Figure 4.16b), and between receptors for widely different (especially mutually perpendicular) orientations at the same location (Figure 4.16c). The scheme of Levine and Grossberg (1976) is reversed: short-range competition is supplemented by long-range cooperation (Figure 4.16d). Such long-range cooperation enables continuous contours to form by linking together separated lines of the same orientation. One of the benefits to the organism of this linkage of contours is compensation for discontinuities (caused by blind spots) in the image on the retina.

Boundary Detection Versus Feature Detection

The importance of edge detection for understanding the form of visual percepts has been emphasized by many vision theorists. For example, Marr and Poggio (1979) and Marr and Hildreth (1980) described boundaries between light and dark areas of a scene as points of zero curvature (or inflection points) of the curve for luminance as a function of distance, as shown in Figure 4.17. This is a simple mathematical representation of the point of sharpest transition in the luminance value.

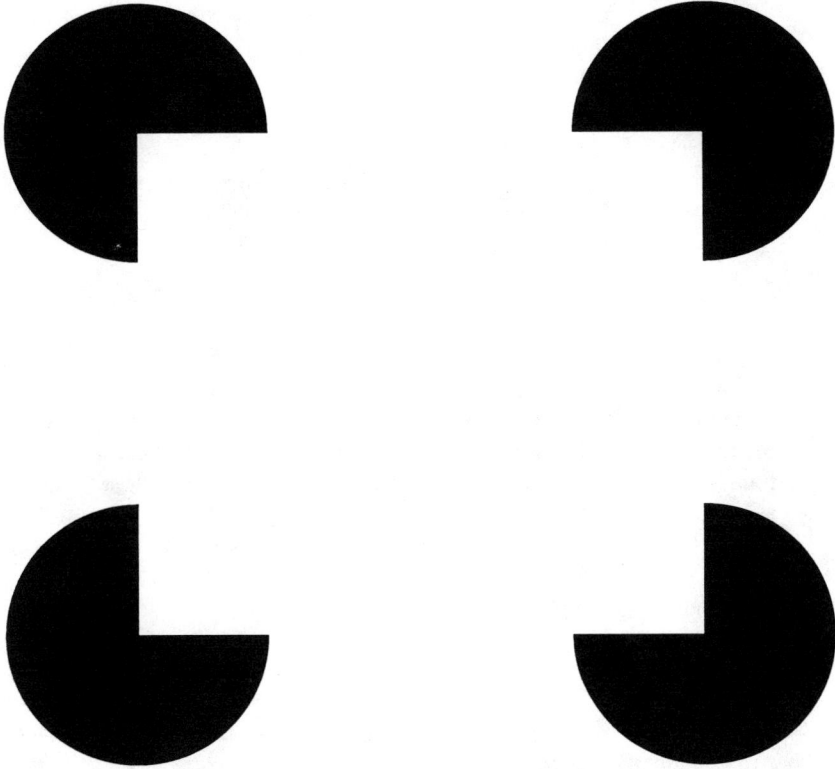

Figure 4.15. Illusory white square induced by four black "pac-man" figures. (From Kanizsa, Gaetano, Subjective contours. Copyright © 1976 by Scientific American, Inc.   All rights reserved.)

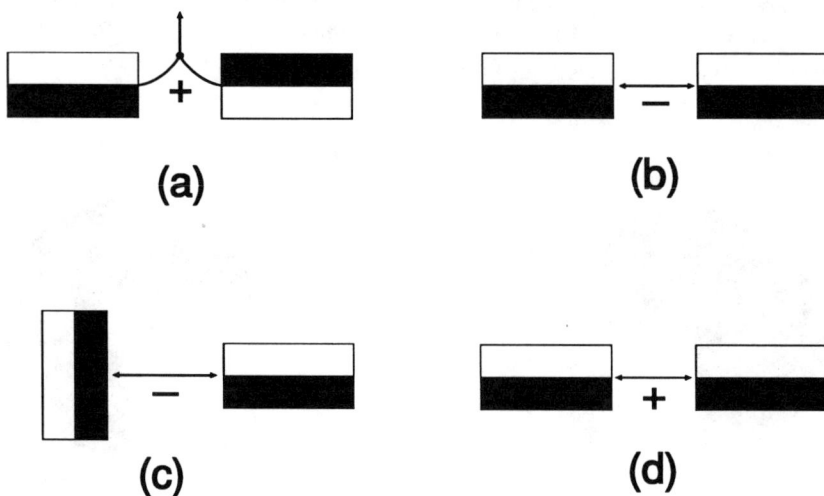

Figure 4.16. (a) *Boundary contour* signals sensitive to orientation and amount of contrast at the edge of a scene, but not to direction of contrast. (b) Like orientations compete at nearby perceptual locations. (c) Different orientations compete at each perceptual location. (d) Once activated, aligned orientations cooperate across a larger visual domain to form "real" and "illusory" contours. (From Grossberg & Mingolla, *Psychological Review* **92**, 173-211, 1985. Copyright 1985 by the American Psychological Association. Reprinted by permission)

But the mechanism for perceiving boundaries must be supplemented by another mechanism for perceiving the form of what is *inside* those boundaries, a so-called *feature contour mechanism*. The feature contour mechanism, unlike the boundary contour mechanism, should be sensitive to the direction of contrast.

One possible combination of boundary and feature contour mechanisms, using both lateral inhibition and excitation, is discussed in Grossberg (1983). In the feature contour mechanism, the excitatory and inhibitory spread coefficients c_{ik} and e_{ik} (as in (4.3)) determine structural scales of the on-center off-surround network. The network's recurrent interactions transform structural scales into functional scales.

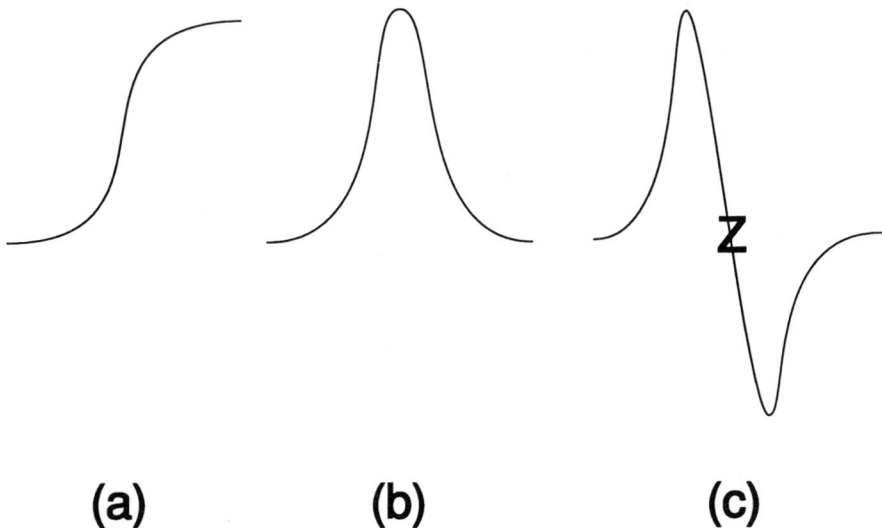

(a)   **(b)**   **(c)**

Figure 4.17. The notion of a zero-crossing. (a) A transition (edge) between dark and light regions is shown by a sharp rise in the graph of luminance as a function of distance. (b) The first derivative of this function has a peak. (c) The second derivative of this function has a *zero-crossing* (transition from positive to negative) at the point Z. (Adapted from Marr, 1982, with permission of W. H. Freeman and Company.)

Figure 4.18 describes one scheme for functional scaling. A linear nonrecurrent mechanism that can only generate boundaries (Figure 4.18b) is contrasted with a nonlinear recurrent mechanism that can generate perceptions of both boundaries and interiors (Figure 4.18c). Initially, all nodes excited by the rectangular input of Figure 4.18a receive equal inputs. Since the inhibitory interaction coefficients e_{ik} are distance-dependent, once recurrent inhibition has time to be established, nodes excited by the part of the rectangle near its boundary receive less inhibition than those nodes nearer the rectangle's center. As time goes on, those selectively enhanced boundary nodes inhibit other nodes whose preferred positions are immediately contiguous to those boundaries but closer to the center. This in turn disinhibits some nodes still nearer to the center, leading to the wave-like pattern shown in Figure 4.18c. The distance between peaks of the wave ("functional scale") is dependent in a complex nonlinear way on the excitatory and inhibitory interaction coefficients c_{ik} and e_{ik}; see Grossberg (1983, p. 646) for details.

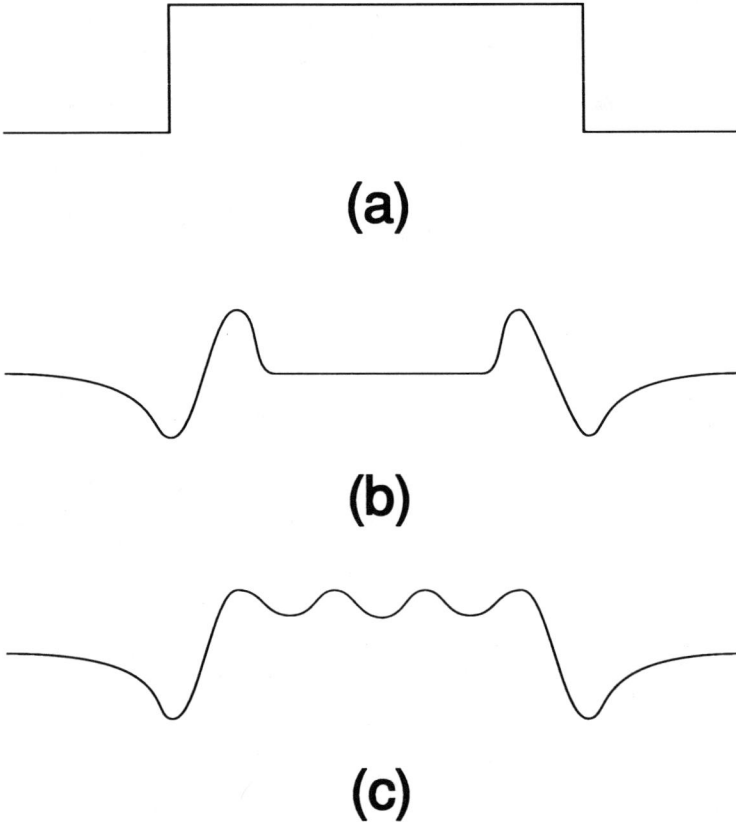

(a)

(b)

(c)

Figure 4.18. (a) Input pattern whereby a region is activated uniformly and the rest of the visual field not at all. (b) Response of a feedforward competitive network to pattern (a); edges of the activated region are enhanced and its interior suppressed. (c) Response of a feedback competitive network to pattern (a); the interior is activated in a spatially periodic fashion. (Reprinted from Grossberg, 1983, with permission of Cambridge University Press.)

Figure 4.18 and the accompanying mathematics also provide one possible explanation for the experimental result (*e.g.*, Robson, 1975) that many visual cortical neurons fire preferentially to some specific spatial frequency. From this result, many theorists have concluded that spatial frequency is one of the primitives of the visual system, or, more speculatively, that the visual system performs Fourier analysis of patterns into frequency components (*e.g.*, Wilson & Bergen, 1979; Graham, 1981; Pribram, 1991). In Grossberg's scheme, by contrast, Fourier analysis (which is a linear transformation) does not occur, and spatial frequency detection is a by-product of more fundamental nonlinear interactions.

Binocular and Stereoscopic Vision

Competitive-cooperative neural networks have been particularly fruitful in modeling the construction of a three-dimensional image from the disparate images received by the left and right retinae. Many binocular vision theorists (Sperling, 1970; Julesz, 1971; Dev, 1975) have explained the formation of depth percepts using networks whose nodes detect specific disparities between the two retinal images. The relationship between disparity of the two images and depth of the actual object viewed is illustrated in Figure 4.19. The existence of cells in the visual cortex responding preferentially to given disparities was demonstrated experimentally by Barlow, Blakemore, & Pettigrew, 1967. Typically, these models are based on cooperation between detectors of the same disparity at different positions, and competition between detectors of different disparities at the same position. (Amari & Arbib, 1977 discuss the mathematical dynamics of systems that combine competition and cooperation in this fashion.)

Modeling of binocular vision has been advanced by the study of *random-dot stereograms*, introduced by Julesz (1960, 1971). These are pairs of patterns presented separately to the two eyes, each of which alone consists of incoherent random dots but which together can lead to sensations of depth (Figure 4.20). As described by Dev (1975, p. 524): "The two patterns are identical to each other in some regions and differ in others. In the regions where they differ, the difference simply consists of a lateral shift of one pattern with respect to the other." Hence, "the region of the pattern that requires lateral shift is perceived as at a depth other than the depth at which the observer is fixated." For example, if the laterally shifted region is square-shaped, the observer sees a square with some of the random dots lying above the rest of the figure. Dev (1975) developed a computational procedure, involving cooperation and competition between disparity detectors as described above, to analyze the perception of depth surfaces from these stereograms (see Figure 4.21). This computational procedure has since been refined by Marr and Poggio (1977a) among others.

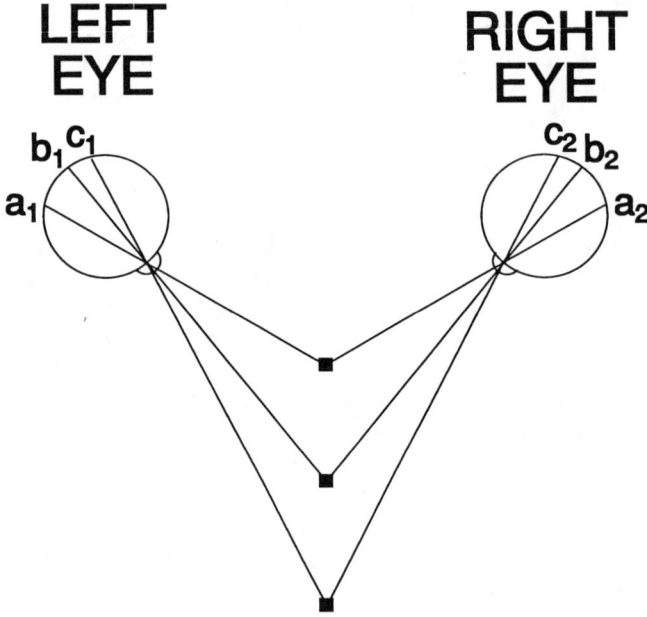

Figure 4.19. Depth of a binocularly viewed object is encoded by disparity of positions along the circles denoting the two retinae (a_1 to a_2 for the nearest object, b_1 to b_2 for the next nearest, c_1 to c_2 for the farthest).

The basic computational problem involved in stereo vision was described in Marr and Poggio (1979) as *the elimination of false targets*. That is, given a point in the left-eye image, the eyes and brain first calculate its disparity with respect to many different points on the right-eye image. Hence, a variety of different depth measurements is possible, and a choice must be made (using a competitive mechanism) as to which is the correct corresponding point in the right-eye image.

Figure 4.20. Example of a random-dot stereogram. If one square is presented to each eye through a suitable stereoscopic viewer, a 40x40 pixel square is seen in depth at the center. (Reprinted from Fender & Julesz, 1967, with permission of the Optical Society of America.)

Marr and Poggio (1979) also summoned various evidence to show that retinal image disparity measures as in the earlier models are insufficient to compute perceived depth. Hence, achieving a coherent three-dimensional percept (whether of a random-dot stereogram or a natural, binocularly viewed scene) involves integrating disparity information with orientation and spatial frequency information. In Marr and Poggio's model, a three-dimensional scene is filtered through channels ("masks") that select particular orientations. Boundaries can be located by taking the image through given orientation masks and locating the edges at zeros of the second derivative of perceived luminance (*cf.* Figure 4.17). Similar filtering is done through spatial frequency channels.

Given all the disparity, orientation, and spatial frequency information, Marr and Poggio showed how to construct a coherent three-dimensional approximation of a given 3-D scene; they called this approximation the *2 1/2-D sketch* of the scene. An example of a 2 1/2-D sketch is shown in Figure 4.22; the concept was developed further in Marr's now classic book (1982) on vision.

More recent work on random-dot stereograms has been done by Hinton and Becker (1990). These researchers combined perception with learning to discover planar or curved surfaces in stereograms. Their learning procedure was designed to maximize the coherence of information from spatially adjacent patches of the images.

Grossberg (1983) compared his approach to visual (including stereopsis) modeling with the approach of Marr and other members of his school. His article includes commentaries by two members of that school, Grimson (1983) and Stevens (1983). Some of this dialogue is summarized here because of its general implications for neural modeling.

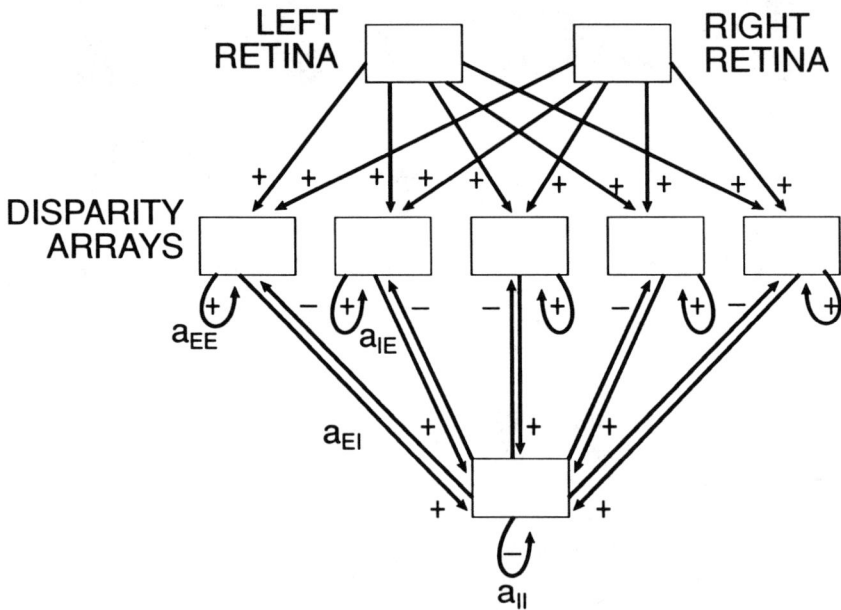

Figure 4.21. Network for simulation of depth perception in random-dot stereograms. Each "disparity array" is a group of nodes sensitive to illumination of a point on the left retina combined with illumination of the point on the right retina that is laterally displaced by the given amount (*cf.* Figure 4.20). a_{EE}, a_{EI}, a_{IE}, a_{II} are functions describing interactions among node groups. (Adapted from Dev, 1975, copyright © 1975 IEEE, with permission of the publishers.)

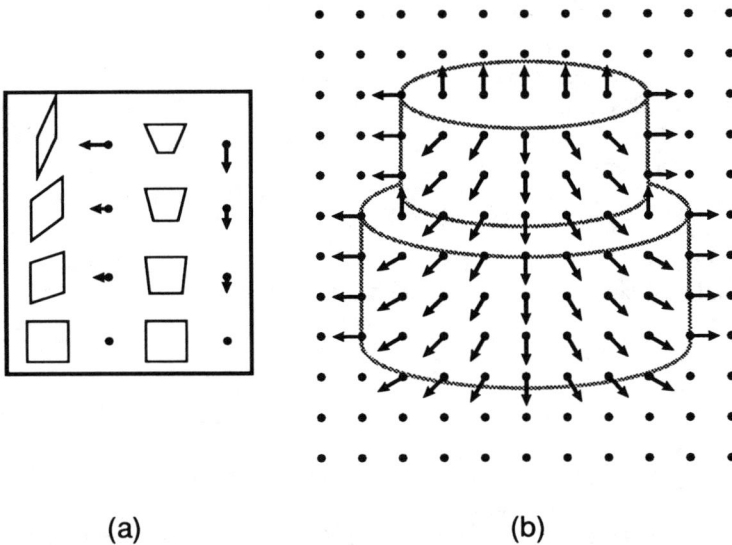

(a)   (b)

Figure 4.22. (a) Representation of surface orientation. Orientation of arrows is determined by the projection of the surface perpendicular to the image plane, and length of arrows represents the dip out of that plane. (b) 2 1/2-D sketch including surface orientations and their discontinuities. (Reprinted from Marr and Poggio, 1977b, with author's permission.)

Comparison of Grossberg's and Marr's Approaches

In both sets of models, disparity information is influenced by orientation and spatial scale information. But Grossberg's model involves nonlinear feedback mechanisms, whereas Marr and Poggio's model is linear and feedforward. Grossberg argued that nonlinear and feedback mechanisms are necessary for accurate representation of many kinds of visual information, such as reflectances. In response, Grimson (1983, p. 666) posed the following question: "Can early visual processing be considered as a system of roughly independent modules which interact loosely to create a global perception, or is the processing so tightly interconnected that the simplest possible description of the process is in terms of its interactions?" Grimson went on to answer his own question on the side of the first, "modular" approach. Citing the large number of psychophysical predictions made by Marr and Poggio (1979), he said that tightness of interaction as posited by Grossberg is not needed to account for psychophysical data.

Another approach to binocular vision has been developed by Cohen and Grossberg (1984) and Grossberg (1987b). In the networks of Cohen and Grossberg, there is extensive feedback between monocular and binocular representation areas, each with its own separate on-center off-surround network (see Figure 4.23 for an example). This approach has both similarities to and differences from Marr's, as described next.

Grimson's comment appears to be influenced by the approach of traditional artificial intelligence, which tends toward separate heuristic programs for separate tasks. This heuristic-programming flavor also permeates some comments by Stevens (1983, p. 675): "As Marr and Poggio (1977) eloquently argue, complex information processing requires satisfactory descriptions at several levels, of which a mechanism description is but one. They distinguish the *computational theory* (What is the goal of the computation, why is it appropriate, and what is the logic of the strategy by which it can be carried out?), the level of *representation and algorithm* (What is the representation for the input and output, and what is the algorithm for the transformation?) and the level of *implementation* (How can the representation and algorithm be realized physically?)."

Stevens went on to say: "Grossberg's descriptions of visual computations ... are primarily at the level of mechanism, of patterns of neural activity within networks. There is no notion, for instance, of symbolic information processing. For those of us interested in understanding vision, the real problems seem to lie here." In Grossberg's approach, the computational theory and algorithm levels are not apparent but are assumed to be represented by other nodes and connections, outside and interacting with the network for early visual processing. Some of these nodes and connections are shown in other articles, while some have yet to be developed.

Related issues arise in this book's discussion of knowledge representation (Section 7.2). We shall see that while connectionism has had a major impact on cognitive science, there is still a strong school of cognitive scientists arguing against connectionism. This school (for example, Pylyshyn, 1984; Fodor & Pylyshyn, 1988) contends that connectionist models cannot be used in the understanding of some levels of a cognitive task — typically, those levels involving purposes and goals. But their argument appears to be based on an overly narrow sense of what constitutes a connectionist model. While *current* connectionist networks fall short of modeling purposes and goals, it is likely that one or more of the classes of networks now in existence will *evolve* toward a network that can model these phenomena.

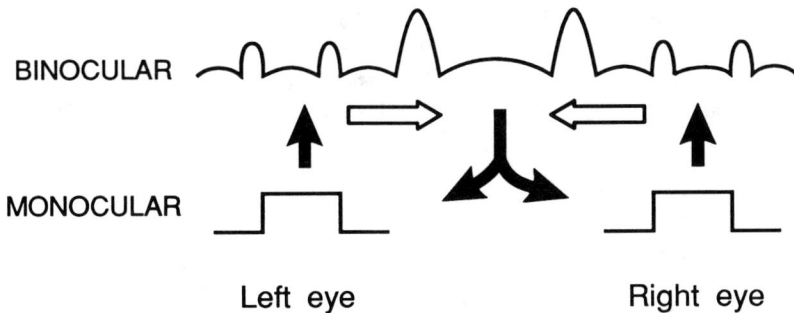

BINOCULAR

MONOCULAR

Left eye   Right eye

Figure 4.23. Monocular processing of patterns through feedforward competitive networks is followed by binocular matching of the two transformed monocular patterns. Pooled binocular edges are then fed back to both monocular representations. (Reprinted from Cohen & Grossberg, 1984, with permission of Lawrence Erlbaum Associates.)

4.4. USES OF LATERAL INHIBITION IN HIGHER-LEVEL PROCESSING

Kilmer *et al*. (1969) developed an early computational model of the reticular formation, consisting of connected modules, each receiving independent sensory information. Based on this information, each module "votes" for inclining the organism toward one of several gross behavioral modes (eating, sex, exploration, etc.). The model of Kilmer *et al*. includes competition, between nodes that represent different drives, and cooperation, between modules that incline the organism toward satisfying the same drive. Montalvo (1975) noted that the structural features of such a two-dimensional competitive-cooperative network of drive representations are similar to those of Dev's (1975) competitive-cooperative network of binocular disparity detectors.

The idea that competition between drive representations is biased by the current sensory environment also appears in Grossberg (1975). Grossberg proposed a *sensory-drive heterarchy* (see Figure 4.24) in which each drive representation is activated by a combination of internal drive level and external sensory inputs compatible with the given drive. (In one variant, drive and sensory influences combine multiplicatively rather than additively, so that neither can activate the representation without the other.) In this way, even if one drive is strongest, another drive can be satisfied if cues compatible with the first drive are unavailable; for example, one can eat meals in spite of prolonged absence of a sexual partner.

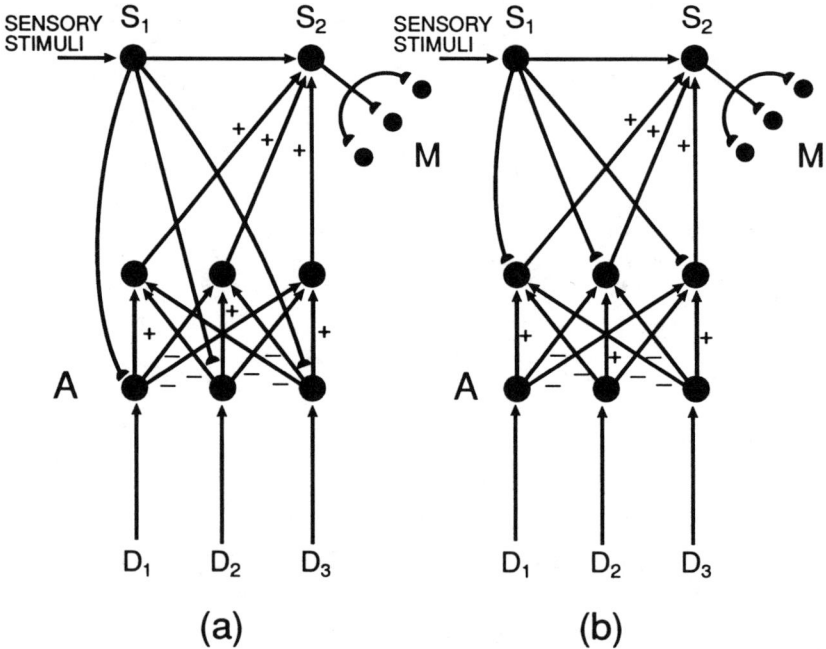

Figure 4.24. Competition between drives, denoted D_i. In a conditioning network (see Chapter 5 for details), sensory stimulus representations, collectively denoted S_1, are conditioned to drive arousal sources A. The strongest drive sends feedback to S_2, allowing S_1 nodes to be conditioned to motor responses at M. (a) *Sensory-drive heterarchy*: the "winning" drive is determined by a combination of internal drive level and compatible cues. (b) *Drive heterarchy*: the "winning" drive is determined only by internal drive level. (Adapted from Grossberg, 1975, with permission of Academic Press.)

Competitive neural processes, as noted in Chapter 1, are also likely to be important in category formation, which is discussed in Chapter 6. In order to make choices when classifying a sensory pattern, many modelers (for example, Bienenstock *et al.*, 1982; Rumelhart and Zipser, 1985; Carpenter and Grossberg, 1987a, b) use an on-center off-surround field among cell populations within a level of nodes that detects categories (presumably at intramodal or intermodal association areas of the cerebral cortex).

Finally, there can be competition between representations of time sequences of sensory stimuli or motor actions. That idea was propounded by Grossberg (1978b) in a model of goal-directed behavior. In this model, competition between sequence representations is biased in favor of longer over shorter time sequences. This bias causes the network to respond not just to the most recent events but to a longer sequence of events and to the consequences of its own recent past actions. The representation of stimulus or motor sequences is developed further in the masking field model of Cohen and Grossberg (1987) and related to functions of the frontal cortex in Levine (1986). These issues are discussed more fully in Chapter 7.

Hence, the on-center off-surround or competitive-cooperative network is one of the most versatile and most widely used neural architectures. Along with associative synaptic modification, it is a prime component of networks that replicate complex cognitive processes, to be discussed in the next three chapters.

4.5. EQUATIONS FOR VARIOUS COMPETITIVE AND LATERAL INHIBITION MODELS

Equations of Sperling and Sondhi

Sperling and Sondhi (1968) developed a lateral inhibitory model of effects in the mammalian retina, in order to explain certain data on luminance and flicker detection. Their model includes both feedback and feedforward stages. In the feedback stage, as shown in Figure 4.5, the jth node is excited by the $j-1$ st node, for $j < n$, and inhibited by feedback from the nth node. Hence, the activity x_j of the jth node is described by the differential equation

$$\frac{dx_j}{dt} = -x_j(1+x_n) + x_{j-1} \qquad (4.7).$$

The nonlinear term $-x_j x_n$ in Equation (4.7) corresponds to multiplicative (shunting) inhibition exerted by the nth stage of cells on the jth stage. This type of inhibition has a strength proportional to the present activity of the area being inhibited. Shunting inhibition, and the related process of shunting excitation, whose strength is proportional to the difference of a cell's activity from its maximum possible level, can be related to the effects of a presynaptic neuron on various postsynaptic ionic conductances.

Equations of Wilson and Cowan

The variables used in Wilson and Cowan (1972) are E(t) and I(t), the proportion of excitatory and inhibitory cells active at time t. The equations incorporate refractory periods of individual cells, in which a cell is prevented from becoming active a short time after firing (see Section 2.2). This leads to time-integral terms that are then removed by averaging the node activities over a suitable time interval. The refractory period is figured in as a linear factor which decreases as the number of active cells increases. If $I_E(t)$ and $I_I(t)$ are the inputs to excitatory and inhibitory nodes, respectively, the resulting differential equations are of the form

$$\Gamma_E \frac{dx_E}{dt} = -x_E + (k_E - r_E x_E) f_E(c_1 x_E - c_2 x_I + I_E)$$
$$\Gamma_I \frac{dx_I}{dt} = -x_I + (k_I - r_I x_I) f_I(c_3 x_E - c_4 x_I + I_I)$$

(4.8)

where f_E and f_i are sigmoid functions that transform the linearly combined excitatory and inhibitory signals, and Γ_E and Γ_I are positive constants (reciprocals of decay rates).

The shunting interactions in Equations (4.8) place bounds on the network's activity. For if x_E and x_I are positive, and x_E reaches the value k_E/r_E, then (4.8) shows that dx_E/dt will be negative; hence, x_E can never exceed the value k_E/r_E. Similarly, x_I can never exceed the value k_I/r_I.

The model defined by (4.8) was extended in Wilson and Cowan (1973) to include distance-dependent interactions. In the later article, the variables $x_E(t)$ and $x_I(t)$ of Equations (4.8) are replaced by $x_E(s,t)$ and $x_I(s,t)$, the average strength of excitation and inhibition at location s at time t. Otherwise, the equations are essentially the same as (4.8) with the terms for x_E and x_I inside the sigmoid functions being replaced by the spatial convolutions of x_E and x_I with distance-dependent connectivity functions. (Convolution of two functions is the operation "*" defined by

$$(f*g)(x) = \int f(x')g\,(x-x')dx',$$

which provides a moving average of one function weighted by another.) This leads to integrodifferential equations which are not shown here. These equations include four different connectivity functions — excitatory-to-excitatory, excitatory-to-inhibitory, inhibitory-to-excitatory, and inhibitory-to-inhibitory.

Equations of Grossberg and Co-workers: Analytical Results

Grossberg (1973) considered a pure on-center off-surround network, with shunting interactions, in which all nodes have the same maximum activity B and minimum activity 0. The equations for the node activities x_i in this network are

$$\frac{dx_i}{dt} = -Ax_i + (B-x_i)f(x_i) - x_i\sum_{k \neq i} f(x_k) + I_i \qquad (4.9).$$

As in (4.8), shunting interactions cause a bound in each node's activity, since (4.9) shows that if all $x_k > 0$ and some $x_i = B$, then that dx_i/dt is negative; hence, no x_i can exceed B.

An important subcase of Equations (4.9) is the case where the inputs I_i are all equal to 0. This was interpreted to mean that the inputs are encoded by the starting values $x_i(0)$ for the node activities. The pattern that is ultimately stored in short-term memory, after transformation by the recurrent interactions, was interpreted as $x_i(\infty)$. This limiting, or equilibrium, pattern was shown to exist in a wide variety of cases; later (in Grossberg, 1978) limits were shown to exist for a more general system which included as subcases the systems of both Grossberg (1973) and Grossberg and Levine (1975).

What the limiting pattern is, that is, which parts of the input pattern are stored in short-term memory, depends on the choice of the function f. That function, which has to be monotone increasing, is a signal function representing input-output transformations at the neuronal level, reminiscent of similar constructions in Wilson and Cowan (1972, 1973).

The classes of functions f that were studied closely are the ones shown in Figure 4.10. The dynamics of the network for these functions are as shown in Figure 4.11:

a) f LINEAR. In this case, the limiting pattern has a *fair distribution*: the values $x_i(\infty)$ are proportional to the values $x_i(0)$, thus representing faithful storage of the original pattern.

(b) f GROWS SLOWER THAN LINEARLY. The limiting pattern has a *uniform distribution*: all $x_i(\infty)$ are equal, regardless of the distribution of the $x_i(0)$.

(c) f GROWS FASTER THAN LINEARLY. The limiting pattern has a *0-1 distribution*: for those x_i where $x_i(0) = \max \{x_i(0): i=1, ..., n\}$, $x_i(\infty)=1$. For all other nodes, $x_i(\infty)=0$.

(d) f SIGMOID. Since a sigmoid function is linear, faster than linear, and slower than linear over different ranges of its argument, the limiting distribution combines fair, uniformizing, and 0-1 tendencies. Inequalities were found which prevent uniformization from occurring. Hence, fair and 0-1 effects combine into an effect described as *contrast enhancement with noise suppression*. The limiting $x_i(\infty)$ are proportional to $x_i(0)$ if $x_i(0)$ is above a threshold value (*quenching threshold*) and equal to 0 if $x_i(0)$ is below that value.

The equations in Grossberg and Levine (1975) are the same as (4.9) except that B is replaced by B_i which might be different for each i. The set of nodes with the same B_i, called a *subfield*, can be interpreted as the set of neuron populations responsive to a particularly sensory feature (such as the color red or the vertical orientation). The results of Grossberg (1973) generalized to distributions that were fair, uniform, 0-1, or contrast enhancing *within each subfield*, as shown in Figure 4.11.

The equations of Grossberg and Levine (1975) can be rewritten

$$\frac{dx_i}{dt} = -Ax_i + B_i f(x_i) - x_i \sum_k f(x_k) + I_i \qquad ,$$

the sum now being over all k *including* i (see Exercise 2 of this chapter). Those equations are a subcase of the more general system

$$\frac{dx_i}{dt} = A_i(\vec{x})[B_i(x_i) - C(\vec{x})] \qquad (4.10)$$

where $\vec{x}$ denotes the vector $(x_1, x_2, ..., x_n)$. Grossberg (1978a) proved that every solution of the system of equations (4.10) approaches an equilibrium, the only restrictions on the functions in the equation being that A_i are nonnegative,

B_i are bounded, and C is nondecreasing with respect to each x_i. Since C($\vec{x}$) has a negative influence on the growth of x_i, the latter condition ensures that each node will tend to inhibit other nodes. The proof of approach to equilibrium did not use Lyapunov functions; rather, competitive interactions were shown to restrict the possible number of changes in which variable is growing fastest at a given time. This technique had been used before in Grossberg and Levine (1975) and was carried further in a general mathematical study of competitive dynamical systems by Hirsch (1982, 1984).

But Equation (4.10) generalizes only the form of Grossberg's equations in which inhibition is distance-independent. It does not encompass the distance-dependent equations

$$\frac{dx_i}{dt} = -Ax_i + (B_i - x_i)\sum_{k=1}^{n} f(x_k)c_{ik}$$
$$- (x_i - C_i)\sum_{k=1}^{n} f(x_k)e_{ik} + I_i \tag{4.3}$$

which were mentioned in Section 4.2. As Cohen (1988) proved, solutions of Equations (4.3) do *not* always converge to an equilibrium, even when n=2. But Cohen and Grossberg (1983) showed that these equations do converge to an equilibrium in the case where there is no internode cooperation, that is, excitatory interaction coefficients c_{ik} are 1 when i=k and 0 otherwise, and inhibitory interaction coefficients obey the symmetry condition $e_{ik} = e_{ki}$.

The Cohen-Grossberg theorem does, however, apply to a system which is more general than (4.10). That system is

$$\frac{dx_i}{dt} = a_i(x_i)\left[b_i(x_i) - \sum_{k=1}^{n} c_{ik}d_k(x_k)\right] \tag{4.11}$$

where the interaction coefficients c_{ik} are symmetric ($c_{ik} = c_{ki}$), the functions a_i are nonnegative, b_i are arbitrary, and d_k are differentiable with nonnegative derivative (which indicates the competitive nature of the system). Equation (4.11) includes as subcases not only (4.3) but also the continuous form of the equations due to Hopfield (1984) and Hopfield and Tank (1985, 1986), as we shall see in the next subsection. Cohen and Grossberg showed that the Lyapunov function

$$V(\vec{x}) = -\sum_{i=1}^{n} \int_0^{x_i} b_i(y)d_i'(y)dy + \frac{1}{2}\sum_{j,k=1}^{n} c_{jk}d_j(x_j)d_k(x_k) \tag{4.12}$$

is nonincreasing along trajectories of the system (4.11). We will not give their demonstration here, but in the next subsection we will give an analogous demonstration for the Hopfield-Tank network. The more general demonstration will be left as Exercise 3 of this chapter.

Equations of Hopfield and Tank

Recall from Section 4.2 above that Hopfield (1982) developed a linear threshold algorithm in which nodes had states that can take on the value 1 or 0. The ith node readjusts its state, at random moments in time, according to the rule

$$x_i(t+1) = 1 \ \textit{if} \ \sum_{j \neq i} w_{ij} x_j(t) > 0$$
$$x_i(t+1) = 0 \ \textit{if} \ \sum_{j \neq i} w_{ij} x_j(t) < 0$$

(4.4).

Then Hopfield considered the energy (Lyapunov) function

$$E = -\frac{1}{2} \sum_i \sum_j w_{ij} \ x_i x_j$$

(4.5).

He then showed that if energy changes by an amount ΔE whenever x_i changes by Δx_i, then

$$\Delta E = - \Delta x_i (\sum_{j \neq i} w_{ij} x_j)$$

(4.6).

Since Δx_i is 0 or of the same sign as $\sum_{i \neq j} w_{ij} x_j$, (4.6) implies that $\Delta E \leq 0$ at all times.

Some extensions of the above energy function, in both discrete and continuous models, were made in Hopfield (1984) and Hopfield and Tank (1985, 1986). In Hopfield (1984), the discrete system of the 1982 article was extended to include external inputs I_i to each node and thresholds θ_i that were not necessarily equal to 0. Hence, (4.4) was replaced by the criterion

$$x_i(t+1) = 1 \ if \ \sum_{j \neq i} w_{ij}x_j(t)+I_i>\theta_i$$
$$x_i(t+1) = 0 \ if \ \sum_{j \neq i} w_{ij}x_j(t)+I_i<\theta_i \qquad (4.13).$$

Under the condition (4.13), there will always be a decrease in the Lyapunov function

$$E = -\frac{1}{2}\sum_i\sum_j w_{ij}x_ix_j - \sum_i I_ix_i + \sum_i \theta_ix_i \qquad (4.14).$$

A generalization of the algorithm defined by (4.13) to the continuous-time case was introduced in Hopfield (1984) and developed further in Hopfield and Tank (1985, 1986). In this work, the output x_i of the ith node was treated as a function (usually sigmoid) of the input u_i, as in the articles reviewed earlier in this section by Wilson and Cowan (1972) and Grossberg (1973). The equation for u_i is then

$$C_i\left(\frac{du_i}{dt}\right) = \sum_j w_{ij}x_j - \frac{u_i}{R_i} + I_i \qquad (4.15)$$

where $x_i = g_i(u_i)$ for some increasing, differentiable functions g_i. Since g_i is increasing, it has an inverse function g_i^{-1}; hence, one can write $u_i=g_i^{-1}(x_i)$. C_i and R_i are analogs of capacitance and resistance across the membrane of a single neuron (for more details on membrane electrical flows, *cf.* Katz, 1966).
    The system (4.15) also has a Lyapunov function similar to (4.14). This function is

$$E=-\frac{1}{2}\sum_i\sum_j w_{ij}x_ix_j+\sum_i\frac{1}{R_i}\int_0^{x_i}g_i^{-1}(V)dV+\sum_i I_ix_i \qquad (4.16).$$

If the matrix of weights is symmetric ($w_{ij}=w_{ji}$), then differentiating (4.16) with respect to t yields

$$\frac{dE}{dt} = \sum_i\frac{\partial E}{\partial x_i}\frac{dx_i}{dt} = -\sum_i\frac{dx_i}{dt}\left[\sum_j w_{ij}x_j - \frac{u_i}{R_i} + I_i\right] \qquad (4.17).$$

But the expression in brackets on the right-hand side of (4.17) is just the right-hand side of (4.15). Hence

$$\frac{dE}{dt} = - \sum_i C_i \frac{dx_i}{dt}\frac{du_i}{dt} = - \sum_i C_i \frac{dx_i}{dt}\left[\left(\frac{d}{dt}\right)g_i^{-1}(x_i)\right]$$

$$= - \sum_i \frac{C_i\left(\frac{dx_i}{dt}\right)^2}{g_i'\left(g_i^{-1}(x_i)\right)}$$

which means that $dE/dt \leq 0$, because C_i is a positive constant and g_i, being an increasing function, has a positive derivative.

The system (4.15) arises from the Cohen-Grossberg system (4.11) with the following substitutions: $a_i(u_i)$ in (4.11) is the constant function equal to $1/C_i$; the coefficients c_{ij} in (4.11) are the *negatives* of the coefficients w_{ij} in (4.15); the functions d_j and g_j of the two respective systems are identified; and the $b_i(u_i)$ are set equal to $-(u_i/R_i) + I_i$. With these substitutions, and the Hopfield-Tank identity $x_i = g_i(u_i)$, it can be seen (*cf.* Exercise 3 of this chapter) that the Lyapunov function (4.12) reduces to (4.16).

Equations of Amari and Arbib

Amari (1977a) developed an equation for a single-layer neural field of lateral inhibition type. His variable is an activity u(s,t) that depends both on (unidimensional) visual field location s and time t. This equation is

$$\tau\frac{\partial u(s,t)}{\partial t} = - u + \int_y w(s-y)f[u(y)]dy + h + I(s,t) \qquad (4.18).$$

In Equation (4.18), w(x) is a distance-dependent weighting function, one that typically combines short-range excitation and long-range inhibition in an additive fashion, as shown in Figure 4.25. The function f is a step function (1 for u above a threshold, 0 for u below) which is, of course, an approximation of a sigmoid. The constant h denotes baseline activity level, and I(s,t) denotes outside inputs.

In Amari and Arbib (1977), Equations (4.18) were elaborated into equations for a two-dimensional competitive-cooperative field. The two dimensions are position and binocular disparity. There are separate excitatory and inhibitory weighting functions that combine multiplicatively and separate excitatory and inhibitory nodes. (Only excitatory and not inhibitory activity is disparity-dependent.)

EXCITATORY
SPREAD

c(x)

DISTANCE

INHIBITORY
SPREAD

e(x)

DISTANCE

INTERACTION
WEIGHT

w(x)
= c(x) - e(x)

DISTANCE

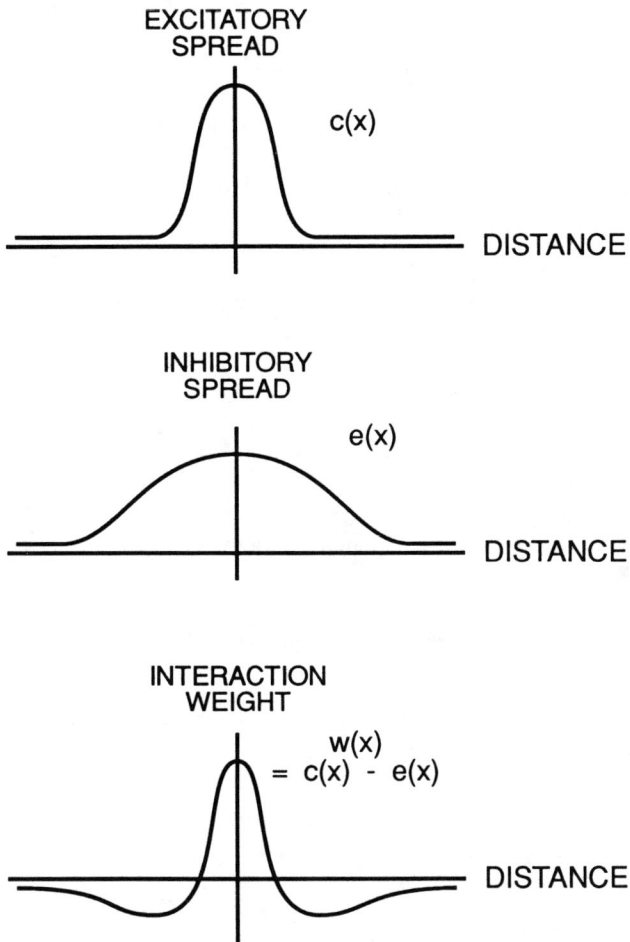

Figure 4.25. If excitatory and inhibitory spreads as in Figure 4.4 combine subtractively, a distance-weighting function arises like the one in the bottom graph. This function is called a *difference-of-Gaussians* (DOG) because the functions c(x) and e(x) can arise from a Gaussian (normal) probability distribution. The DOG is used in model equations by Amari (1977a) and many others.

EXERCISES FOR CHAPTER 4

1. Consider Grossberg's differential equation for shunting without lateral inhibition:

$$\frac{dx_i}{dt} = - Ax_i + (B-x_i)I_i \qquad (4.1).$$

Let the inputs I_i form a constant spatial pattern, $I_i = \theta_i I$. The steady state solution of a system of differential equations is obtained by setting the derivatives equal to 0. Hence, in (4.1), at the steady state values $x_i(\infty)$,

$$0 = - Ax_i(\infty) + (B - x_i(\infty))\ I_i = - Ax_i(\infty) + (B - x_i(\infty))\theta_i I,$$

so that, by algebra, $x_i(\infty) = B\theta_i I/(A + \theta_i I)$. This leads to a distortion of the relative pattern weights θ_i. Such distortion has been called the *noise-saturation problem* (Grossberg, 1973; Dalenoort, 1983), because insignificant inputs ("noise") are amplified, while distinctions between intense inputs are blurred ("saturated").

(a) Show that in the shunting equations *with* lateral inhibition,

$$\frac{dx_i}{dt} = - Ax_i + (B-x_i)I_i - x_i\sum_{k\neq i} I_k \qquad (4.2)$$

with $I_i = \theta_i I$, the noise-saturation problem disappears. That is, the steady state values $x_i(\infty)$ are proportional to θ_i, the relative pattern weights.

(b) Find the steady state values of x_i if (4.2) is replaced by the same equation with minimum activities equal to C instead of 0, namely

$$\frac{dx_i}{dt} = -Ax_i + (B-x_i)I_i - (x_i-C)\sum_{k\neq i} I_k \qquad (4.19).$$

Show that for a network defined by (4.19), the $x_i(\infty)$ are proportional to the θ_i - K for K a constant.

(c) Find the steady state values of x_i if (distance-dependent) excitatory and inhibitory interaction coefficients are included, *i.e.*,

$$\frac{dx_i}{dt} = -Ax_i + (B - x_i)\sum_{k=1}^{n} I_k c_{ki} - (x_i - C)\sum_{k=1}^{n} I_k e_{ki} \qquad (4.20).$$

Note that for the network defined by (4.20), steady state values are no longer proportional to θ_i, and study how they vary with total intensity I for different choices of c_{ki} and e_{ki}.

2. The following problems deal with simulation of the shunting recurrent on-center off-surround equations with attentional biases

$$\frac{dx_i}{dt} = -Ax_i + B_i f(x_i) - x_i \sum_{k=1}^{n} f(x_k) \qquad (4.21)$$

as studied by Grossberg and Levine (1975). (Note: Equations (4.21) are a subcase of (4.3) from the text.) Let n=3.

(a) Let A=1, B_1=5, B_2=4, B_3=3, and let the signal function f(x) be x^2. Note this means f is "faster than linear"; see Section 4.2 and Figure 4.13. Choose 5 different sets of initial conditions in which the x_i are in an order opposite to that of the B_i, that is, $x_1(0) < x_2(0) < x_3(0)$. Verify that the values of x_i reach nearly steady state values after several hundred iterations. Vary the ratios between the initial values of x_i and study how that affects which node "wins" the competition.

(b) Do the same as part (a) with a sigmoid signal function, and with a slower-than-linear signal function such as f(w) = aw/(b+w) for a suitable a and b. The sigmoid should be chosen so that f(0) = 0 and so that its inflection point occurs at a positive x value; initial node activities should then be chosen within the range where f is faster-than-linear or nearly linear.

**3. Show, by using the chain rule for differentiation, that the function V(x) defined by Equation (4.12) is in fact nonincreasing along solutions of the Cohen-Grossberg differential equations (4.11).

o4. One of the most studied visual illusions is the *Muller-Lyer* illusion. In this illusion the perceived length of a line segment can be influenced by the directions of arrowheads to its side; for example, the line segment in (a) looks longer than the line segment in (b), even though objectively they are of the same length (see Figure 4.26). Construct a neural network that explains the Muller-Lyer illusion. Use a recurrent competitive-cooperative network of position and orientation detectors, as discussed in Section 4.3 above.

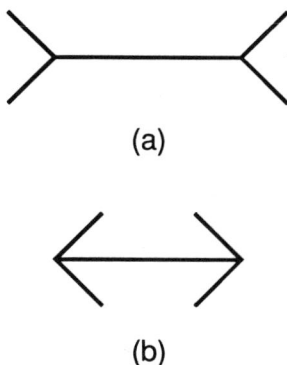

(a)

(b)

Figure 4.26. Muller-Lyer illusion (see text for details).

5. This problem relates to using a part of the boundary contour system of Grossberg and Mingolla (1985b, 1987) to generate *end cuts*, or induced contours perpendicular to perceived orientations. Specifically, it relates to simulating the perceived horizontal end of a thickened vertical line.

The network used is first organized into contrast-sensitive oriented receptive fields, called *masks*. It includes a lattice of locations, and at each location there are 12 nodes with equally spaced preferred orientations. Since any given direction and the direction 180 degrees opposite to it count as the same orientation, these preferred orientations are $180/12 = 15$ degrees apart, as shown in Figure 4.27:

k=0 —   0 degrees (horizontal)   k=1 —   15 degrees (tilted to right)
k=2 — 30 degrees   k=3 —   45 degrees
k=4 — 60 degrees   k=5 —   75 degrees
k=6 — 90 degrees (vertical)   k=7 — 105 degrees
k=8 — 120 degrees   k=9 — 135 degrees
k=10 — 150 degrees   k=11 — 165 degrees

The details of the oriented masks, given in Grossberg and Mingolla (1985b), are neglected here. The line in Figure 4.27 is simply treated as a set of inputs J_{ijk} to a competitive stage of network nodes with activities x_{ijk}. For each triple of positive integers i, j, k, J_{ijk} is 1 if there is an input with orientation k at position (i,j), and 0 otherwise.

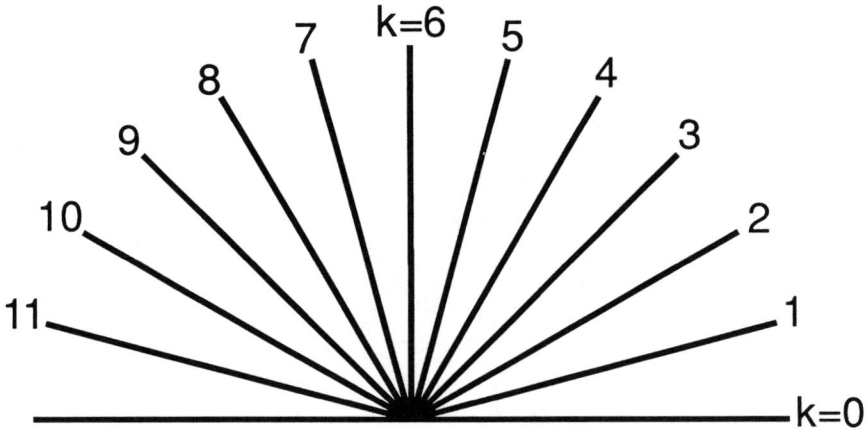

Figure 4.27. Preferred orientations of different nodes in the network of Grossberg and Mingolla (1985b). Actual angles are given in the text.

The steady state values of these x_{ijk} are

$$x_{ijk} = \frac{1 + B J_{ijk}}{1 + \sum_{p,q} B J_{pqk} A_{pqij}} \tag{4.22}$$

where I is a tonic input, B is a positive constant, and A_{pqij} is the interaction strength between positions (p,q) and (i,j). A_{pqij} is set to equal a positive constant F if (p,q) = (i,j) or (p,q) is one of the four neighbors of (i,j), that is (i+1,j), (i-1,j), (i,j+1), and (i,j-1), and 0 otherwise. Then the x_{ijk} undergo opponent processes at each position between the nodes corresponding to perpendicular orientations. Hence, if k and K are the indices corresponding to mutually perpendicular orientations, the output of the kth node becomes

$$z_{ijk} = C[x_{ijk} - x_{ijK}]^+ \qquad (4.23)$$

for C another positive constant. Finally, the z_{ijk} are put through a second stage of competitive interactions, whose final outputs are

$$y_{ijk} = \frac{E z_{ijk}}{D + \displaystyle\sum_{m=1}^{12} z_{ijm}} \qquad (4.24)$$

with D and E two more positive constants.

Find a set of parameter values that reproduces the perceived end cut. (Note: for many parameter settings, the network will also act as though it sees a horizontal orientation to the sides of the line. Grossberg and Mingolla believe that this side horizontal percept is eliminated by other interactions not shown in the network simulated here.)

*6. Read the article of Dev (1975), from which Figure 4.21 is taken. Build a network according to the lines she suggested with competition and cooperation in the position dimension, and competition in the disparity dimension. Attempt to reproduce the results of Figure 7 of that article, where a random-dot stereogram is partitioned into segments in which different disparities are detected. (Note: inhibition between detectors of different disparities is mediated by "inhibitory arrays" at another level of Figure 4.21.)

○7. After reading appropriate references discussed in Grossberg (1983), give an argument for one or another side of the controversy over whether the visual system performs Fourier analysis, or whether spatial frequency perception is a by-product of more primitive interactions.

8. This exercise is designed to simulate the process of boundary completion, as in the illusory square of Figure 4.15. Specifically, the object is to qualitatively reproduce the graphs in Figure 4.28, which is based on Grossberg and Mingolla (1985a).

The "Y Field" and "Z Field" in Figure 4.28 correspond to two different layers of nodes that respond to a given orientation (say, horizontal) at different visual field positions. They are part of a competitive-cooperative feedback loop, most of which does not need to be reproduced to get the desired effects.

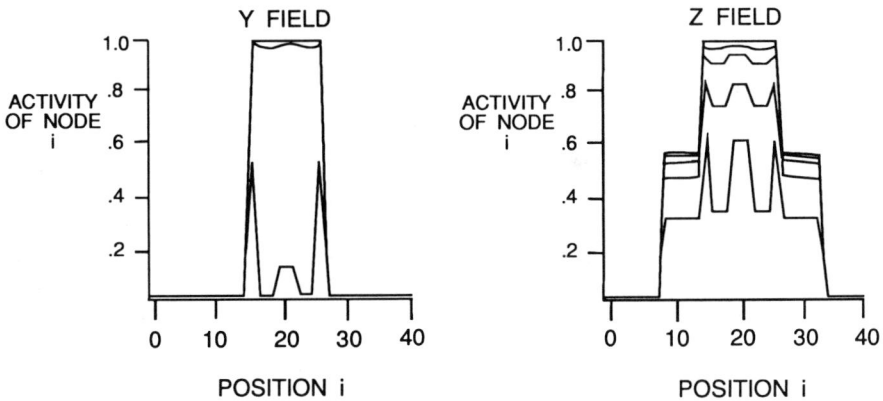

Figure 4.28. Activities of two layers of nodes in response to lines of a given orientation; see text for details. (From Grossberg & Mingolla, *Psychological Review* **92**, 173-211, 1985. Copyright 1985 by the American Psychological Association. Adapted by permission.)

Index the nodes of the lower level (Y field) from 1 to 40. Let nodes 15 and 25 receive sustained inputs I_i and not the others. The equations for feedback between higher and lower levels are:

$$\frac{dy_i}{dt} = -By_i + (C - y_i)E[z_i - \delta]^+ + I_i$$

$$\frac{dz_i}{dt} = -Az_i + \frac{.5(S_i^{left})^2}{\gamma^2 + (S_i^{left})^2} + \frac{.5(S_i^{right})^2}{\gamma^2 + (S_i^{right})^2}$$

where the signals from left and right neighbors are

$$S_i^{left} = y_i + D\sum_{j=u}^{i-1} y_j, \ u = \max(i-6, 1)$$

$$S_i^{right} = y_i + D\sum_{j=i+1}^{v} y_j, \ v = \min(i+6, 40)$$

Find settings of the parameters A, B, C, D, E, Γ, and δ that will yield the "filling-in" process described by the two graphs. The separate curves on each graph denote the activities of the nodes in the appropriate field at successive times, with the higher curves occurring at later times. (The reason that there are only three curves on the Y graph but 5 on the Z graph is that the Y field converges much faster. Hint: the simulations that have worked best have used a very high value for the feedback parameter E, a value 50 to 200 times as high as those of the other parameters.)

5

Conditioning, Attention, and Reinforcement

The true art of memory is the art of attention.

Samuel Johnson (*The Idler*)

"Attention," the mynah chanted in ironical confirmation.
"Do you have many of these talking birds?"
"There must be at least a thousand of them flying about the island.
It was the Old Raja's idea. ... And now," he added in another
tone, "you'd better start listening to our friend in the tree."

Aldous Huxley (*Island*)

Every animal or intelligent machine requires an attentional mechanism in order to make sense of events in its environment and predict their consequences on the basis of past contingencies. If a neural network has no criterion for selecting which stimuli to attend to and which to ignore, it becomes so overwhelmed with stimuli as to make functioning impossible. Solutions to the problem of selective attention in biological organisms are likely also to have profound implications for the control of adaptive machines, through the incorporation into such machines of goal direction or planning (see, *e.g.*, Barto, Sutton, & Anderson, 1983; Athale, Friedlander, & Kushner, 1986; Cruz, 1991).

The psychological concept of attention is complex and protean; some of its aspects are summarized in the collection of articles edited by Parasuraman and Davies (1984). As described by Kahneman and Treisman (1984) in that same collection, most theories of attention have emphasized either the need to choose appropriate percepts to attend to, in order to avoid confusion and overload, or the need to choose appropriate responses, in order to avoid paralysis and incoherence. Even among those psychologists who emphasize the perceptual aspects, the study of attention is complicated by contextual issues such as priming; this is discussed briefly in the word recognition subsection of Section 7.2.

In this chapter, we will emphasize perceptual choice more than response choice. As the second thought experiment in Chapter 1 suggests, a mechanism for perceptual selective attention should include some form of decision or competition between stimuli; this suggests lateral inhibition as discussed in Chapter 4. Attention can also include competition between parts or aspects of a single stimulus; examples of this arise in various networks discussed in Chapters 6 and 7, networks that model aspects of categorization and knowledge representation.

Recall the other part of our thought experiment: in a competitive network, how does one stimulus achieve an "advantage" over others? One possible way is for the neural representation of that stimulus to have somehow been associated with another stimulus or internal pattern that had prior significance for the system. This suggests that an attentional mechanism should also include associative synaptic modification as discussed in Chapter 3.

This chapter cannot do justice to the vastness of the attentional literature. It concentrates, therefore, on one form of attention that has been modeled quantitatively, the form that occurs during conditioning. This leads to discussion of various neural network models for *Pavlovian* or *classical* conditioning, the type of learning that occurs, for example, when a dog learns to salivate in response to a bell that has been repeatedly paired with food (Pavlov, 1927). Brief mention is made of the other general type of conditioning, *operant* or *instrumental* conditioning, such as occurs when an animal learns a movement such as lever-pressing when that movement is rewarded with food (Skinner, 1938). This book lacks space to cover in depth the controversy over whether classical and operant conditioning are based on the same or different underlying mechanisms (*e.g.*, King, 1979; Mackintosh, 1983), but as a first approximation for modeling purposes, I take the position that there is a unified mechanism for both types of conditioning.

Classical conditioning is one of the building blocks of more complex forms of associative learning. For that reason, it is a major current focus of investigation both by neural network modelers (*e.g.*, Sutton & Barto, 1981, 1991; Grossberg, 1982a, b; Klopf, 1988; Grossberg, Levine, & Schmajuk, 1991) and by single-cell neurophysiologists (*e.g.*, Byrne, 1987; Anderson *et al.*, 1989). In fact, there have been some promising efforts at incorporating both psychological and physiological data into computational models of Pavlovian conditioning (*e.g.*, Hawkins & Kandel, 1984; Gelperin, Hopfield, & Tank, 1985; Schmajuk & Moore, 1985; Moore, Desmond, Berthier, Blazis, Sutton, & Barto, 1986; Zipser, 1986; Gluck & Thompson, 1987; Aparicio & Strong, 1991; Schmajuk & DiCarlo, 1991).

5.1. NETWORK MODELS OF CLASSICAL CONDITIONING

Early Work: Brindley and Uttley

The first neural networks for Pavlovian conditioning were developed in the 1960's within the framework of all-or-none neuronal models. Brindley (1967, 1969) modeled some conditioning data using the all-or-none, symbolic logic framework of McCulloch and Pitts (1943), with the addition of modifiable synapses as proposed by Hebb (1949).

Brindley (1967) discussed ten types of modifiable synapses and the logic of their operation. Of the ten, the most important was the "Hebb synapse," whose facilitation depends on correlated pre- and postsynaptic activities. Also, it was possible for two neurons in the network to be connected by two different synapses, one modifiable and one unmodifiable.   This was an early approximation to a connection between continuous neural elements where synaptic strength can take on a range of values.

Figure 5.1 shows Brindley's networks both for classical conditioning and for operant, or instrumental, conditioning. In spite of oversimplifications resulting from the all-or-none framework, the networks used here had structural details in common with later models.   For example, in Figure 5.1a, strengthening of the "semicircular" synapse makes the conditioned stimulus (CS) able to activate a cell that at first needed the unconditioned stimulus (US) to be activated.   This takes place through the action of a network of interneurons, one of which is *polyvalent*, that is, responding to a combination of conditioned and unconditioned stimuli. The ideas of polyvalence and of the CS "gaining control" of a US-activated arousal area were prominent in the continuous model of Grossberg (1971), which formed the basis for several later articles. Brindley (1969) also extended some of these ideas to the learning of sequences of three words.

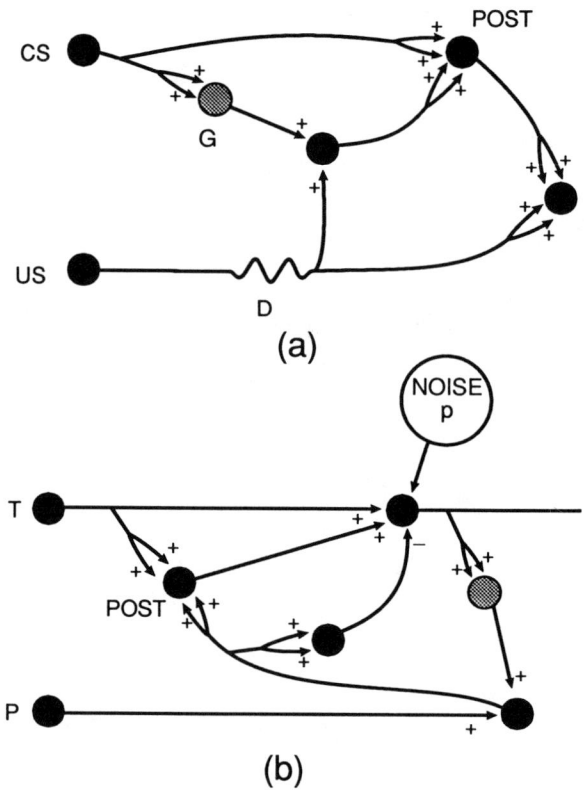

Figure 5.1. (a) Classical conditioning network. CS is followed by US. Node G fires repeatedly in response to a single impulse, and D is a delay.   "POST" means the synapse is modified by postsynaptic activity.   (b) Operant conditioning network, with trigger stimulus T.   Before learning, T evokes a response with (low) probability p.   After the response has been repeatedly rewarded with reinforcer P, that probability becomes close to 1. (Adapted by permission of the publisher from Levine, *Mathematical Biosciences* **66**, 1-86.   Copyright 1983 by Elsevier Science Publishing Co., Inc.)

Another early set of theories relevant to classical conditioning was developed by Uttley (1970, 1975), using mathematical information theory. These articles were based on a pattern discrimination model of Uttley (1966), which was in turn based on a linear threshold network (see Chapter 2) with binary inputs x_i, and an output

$$(\sum_{i=1}^{m} w_i x_i) + \Gamma$$

where the w_i are synaptic connection strengths and Γ is the negative of a response threshold (see Figure 5.2). The output cell responds if and only if the output is positive; its binary signal is called y. The connection strengths w_i are in turn calculated from relative probabilities of the co-occurrence of events. If the binary value x_i is 1, then w_i is set equal to

$$P(x_i,y) = \log_2 \frac{p(x_i \cap y)}{p(x_i)p(y)}$$

with p denoting the probability of a given event and $\cap$ denoting the co-occurrence of two events. If x_i is 0, then w_i is set equal to $P(x_i',y)$, x_i' denoting the opposite or complementary event to x_i. Hence, synaptic weights increase with co-occurrence of pre- and postsynaptic events, in accordance with Hebb's postulate (see Chapters 2 and 3).

Uttley, however, found this Hebbian learning could cause connection weights to increase without bound. He solved this problem by reversing the sign of the synaptic change, making conductivities proportional to the *negative* of the cross-correlation or information function P(x,y). Networks based on such negative feedback laws he called *informons*. Changing the sign of the effect of cross-correlation on synaptic efficacy is not the only possible method of stabilizing a network with modifiable synapses. Other methods for bounding total activity include exponential decay with time of individual node activities and connection strengths (Grossberg, 1968 a, b); exponential decay combined with normalization of pattern weights (Grossberg, 1976b); and conservation of total synaptic efficacy (Malsburg, 1973; Wilson, 1975). Uttley's method of negative-information synapses, however, has some interesting psychological consequences which we shall see below and which foreshadow a class of more recent models.

Uttley (1975) studied a class of informon networks in classical conditioning. In the case of a single CS-US connection, the CS, called A, excites a pathway of variable weight w_A, and a US, called U, excites a pathway of fixed weight w_U. The output of the informon which includes these pathways is the conditioned response (CR). Due to the negative information synapses, it was shown (the demonstration is not given here) that if A is always reinforced by U, the steady-state equation for the weights reduces to $w_A + w_U = 0$. Since $w_A > 0$, because A develops a positive associational strength, this means that $w_U < 0$, that is, the pathway from U to the CR is inhibitory.

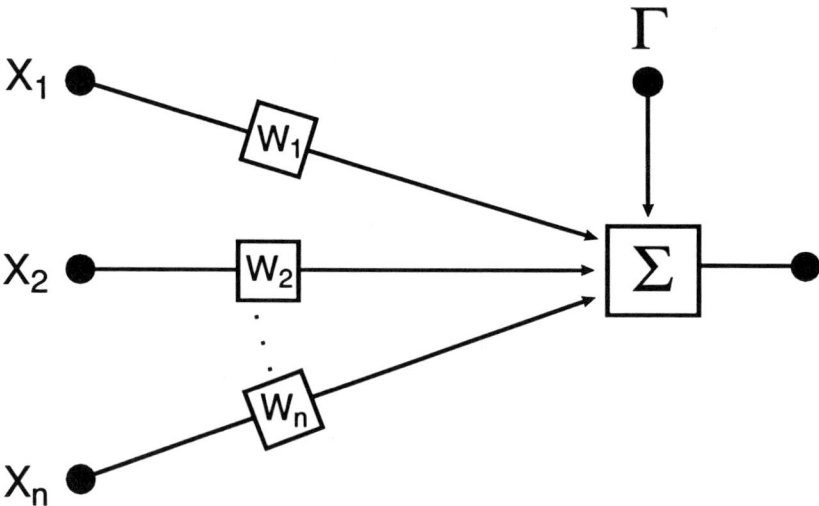

Figure 5.2: Schematic linear informon network. (Adapted by permission of the publisher from Levine, *Mathematical Biosciences* **66**, 1-86.   Copyright 1983 by Elsevier Science Publishing Co., Inc.)

Hence, in Uttley's model, there is a given response, and different stimuli compete for ability to be associated with that response. Enhanced associational strength of one stimulus tends to weaken the associational strength of another. This is also true in the case of two CS's; hence his theory can account for such classical conditioning data as the *blocking* and *overshadowing* paradigms (Kamin, 1969), which are illustrated in Figure 5.3. In blocking (recall the discussion in Section 3.3 above), the animal is first given many presentations of one CS (A), each followed by a US at a given time interval. The CS A is then presented many times in combination with another stimulus X, each pair followed by the US at the same time interval as before. On recall trials, the animal has developed a CR to A alone or to the AX combination, but not to X alone. In overshadowing, the US is associated with the AX combination but cue A is more *salient* than cue X — that is, either more intense or more important to the organism's survival. Again, no CR has been learned to X alone. The competition for associability between different, previously neutral stimuli provides a compact explanation for both of these effects. (In Uttley's model, salience is an arbitrarily set parameter.)

The inhibitory US-to-CR connection, however, seems paradoxical because the US reliably evokes the unconditioned response (UR). (The question whether the CR and UR are the same response, or at least similar, is discussed later in this section.) A series of later articles (Uttley, 1976a, b, c) illuminated this inhibitory connection further. These articles introduced an additional input called Z to the informon with a binary signal F(Z) of fixed conductivity w_i. The equation for the output signal became

$$F(Y) = (\sum_i F(X_i)w_i) + w_Z$$

and the equation for change in conductivity with stimulus pairing became

$$\Delta w_i = -kF(x_i)(\sum_j F(x_j)w_j + F(Z)w_Z) \tag{5.1}$$

where k is a positive constant. If there is to be a positive term in (5.1), w_Z has to be constant and negative.

What is the significance of the fixed inhibitory Z pathway? Uttley (1976a) called it a *classifying* pathway. He explained (1976a, p. 28): "The function of F(Z) must be to signal whether the total stimulus to all the variable pathways is a member or non-member of some class." In other words, correlated stimuli tend to become negatively associated unless the particular stimulus possesses a preassigned significance that overrides the negative feedback at the synapses.

Rescorla and Wagner's Psychological Model

Uttley's informon model is mathematically similar to the learning model of Rescorla and Wagner (1972), which is described in psychological rather than neural terms. It was noted in Chapter 3 that Rescorla and Wagner based their theory on the general principle that learning only occurs when events violate expectations.

Rescorla and Wagner expressed their psychological principle as a system of difference equations. Their variables are CS-US associative strengths as defined in Hull (1943; *cf.* Section 2.1). Let the CS be labeled A, and the associative strength between A and the US be labeled w_A. Then the equation for the change in w_A over time is

$$\Delta w_A = \alpha_A \beta (w^{max} - w_{AS}) \tag{5.2}$$

where α_A is the intensity of the CS; ß is a learning rate associated with the given US; S refers to one or more stimuli present along with A; and w^{max} is the asymptotic value of associative strength, which is a function of the current reinforcement strength of the US. (w^{max} is 0 if reinforcement by the US does not occur). The compound associative strength w_{AS} is assumed to equal $w_A + w_S$.

Rescorla and Wagner used their equations to explain blocking (see Figure 5.3). Their model predicts that if A is strongly connected with the US, such as shock, the associative strength between A and the US reaches its asymptote. In terms of Equation (5.2), while A alone is being conditioned to the US, $w_{AS} = w_A$ becomes very close to w^{max}. The analog of (5.2) for S is

$$\Delta w_s = \alpha_s \beta (w^{max} - w_{AS}) \tag{5.3}.$$

Since at the time of presentation of the compound stimulus AS, $w_{AS} = w_A$ is already nearly equal to the value of w^{max} for the given US, (5.3) says that Δw_S will be nearly 0; hence, S will not become significantly conditioned to the US.

The Rescorla-Wagner model is still widely used by psychologists because it explains many of the basic conditioning paradigms with a few simple equations. Yet as noted before, it is not a genuine neural network (connectionist) model. Also, it is not a real-time model; changes in variables are all-or-none for each *trial* (presentation of one or more CS's followed by a US) and ignore temporal relationships within a trial. As we shall see below, several other modelers incorporated insights of Rescorla and Wagner into real-time neural network models (some also inspired by invertebrate neurophysiological data). These modelers include Sutton and Barto (1981, 1991), Klopf (1982, 1988), and Hawkins and Kandel (1984).

Grossberg: Drive Representations and Synchronization

Rescorla and Wagner noted that blocking, and overshadowing as well, could also be explained by an alternative model. The alternative model was based on attentional competition: a more salient cue, or one that has already strengthened an association to the US, can receive more attention than a less salient cue and thereby inhibit learning of associations to the less salient cue. This section and the next include developments of this attentional interpretation of blocking in some other network models. These models include qualitative network analyses by Grossberg (1975) and quantitative computer simulations by Grossberg and Levine (1987). These articles built on a theory of conditioning first developed in Grossberg (1971), which in some ways sharply contrasts with the Rescorla-Wagner theory.

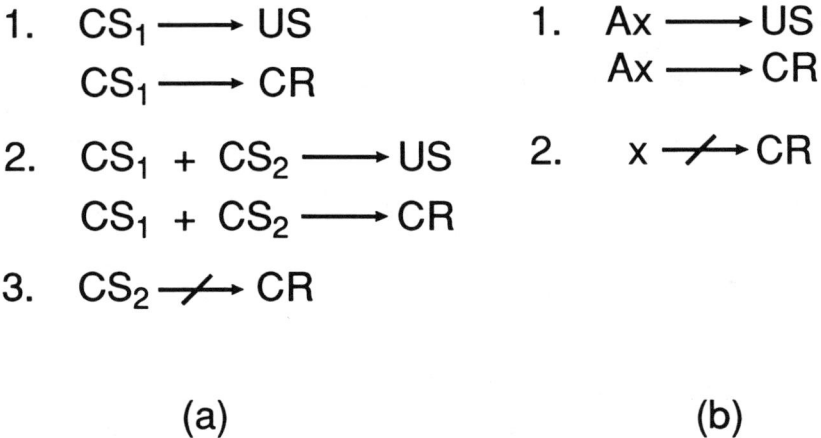

1. $CS_1 \longrightarrow US$
    $CS_1 \longrightarrow CR$

2. $CS_1 + CS_2 \longrightarrow US$
    $CS_1 + CS_2 \longrightarrow CR$

3. $CS_2 \nrightarrow CR$

1. $Ax \longrightarrow US$
    $Ax \longrightarrow CR$

2. $x \nrightarrow CR$

(a)   (b)

Figure 5.3. Schematic of experimental stages in blocking (a) and overshadowing (b); see text for explanations.

Grossberg (1971) set out to provide a unified mechanism for both classical (Pavlovian) and operant (instrumental, Skinnerian) conditioning in the framework of his earlier articles on spatial pattern learning (Grossberg, 1968a, b, 1969a, c; see Chapter 3). His fundamental postulate was: after a time period where a CS repeatedly precedes a US, the CS must be able to generate a motor response previously associated with the US. But if the US is treated as a spatial pattern, a difficulty arises if the CS-US time lag is variable, as illustrated in Figure 5.4. Unless the network is carefully designed, the CS could become associated not with the US but with a noisy mixture of the US and other patterns experienced during the same time period (*cf.* Exercise 2(b) of Chapter 3).

Before discussing Grossberg's approach to the problem of variable time lags, also known as the *synchronization problem* (Grossberg & Levine, 1987), it is important to remark that his analysis assumed, to a first approximation, that the conditioned response (CR) is the same as, or similar to, the unconditioned response (UR). This outlook is known in psychology as *stimulus substitution theory*. Mackintosh (1983, pp. 67-74) discusses whether stimulus substitution theory is correct. The experimental literature on this point is quite varied: sometimes the CR and UR are similar, and sometimes the CR involves only a small part of the usual responses to the unconditioned stimulus. As an example of the latter, "Pavlov's dogs salivated to the CS signalling food, they did not routinely lick, chew, bite, or swallow it" (Mackintosh, 1983, p. 70). Moreover, the same CS elicits a variety of orientation and approach responses not elicited by food itself. Yet for modeling purposes, the idea that learning causes the CS to elicit a response previously made to the US has been a useful starting point.

Figure 5.5 shows a network designed to address the synchronization problem. This network includes three sets of nodes. The first set, collectively called *S* for sensory, consists of short-term representations for particular sensory stimuli, both CS's and US's. These *S* nodes are connected into a competitive-cooperative subnetwork like those discussed in Chapter 4, leading both to short-term memory and to selective attention. The second set, collectively called *M* for motor, becomes activated after the US or the conditioned CS is presented, leading to particular responses that could either be skeletal (as in the rat's lever pressing) or autonomic-visceral (as in the dog's salivation). The third set of nodes, collectively called *A* for arousal, are of particular interest for later theory.

The synchronization problem was solved by having a strong preexisting connection from the US representation to a particular arousal locus, and then allowing repeated CS-US pairing to strengthen the ability of the CS to activate that same arousal locus. In this model, other patterns do not weaken the development of CS-US pairing unless they are associated with the same arousal locus; for example, intervening presentation of a sexual stimulus does not interfere with a bell-food pairing.

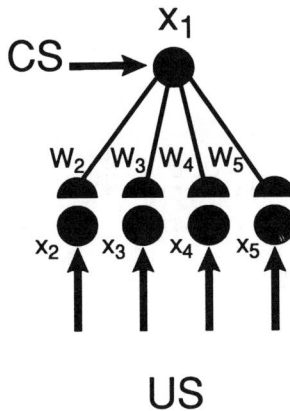

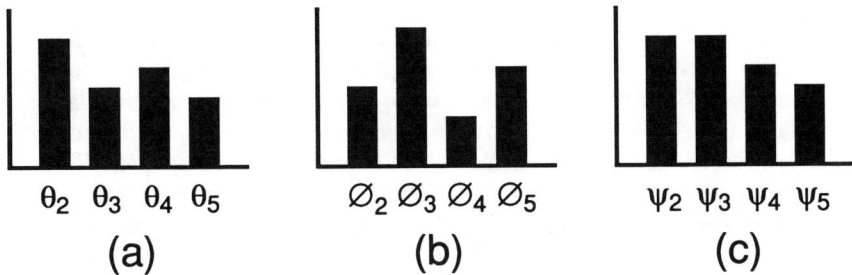

Figure 5.4. Schematic of difficulty arising, *e.g.*, in an outstar network, when CS-US time lag is variable. (a) Spatial pattern $\{\theta_i\}$ representing the US to be learned.   (b) Spatial pattern $\{\phi_i\}$ presented randomly at times between US presentations. (c) Noisy mixture $\{\psi_i\}$ of $\{\theta_i\}$ and $\{\phi_i\}$, which is learned by the CS node x_1 (at the synapses w_i; *cf*. Exercise 2(b) of Chapter 3).

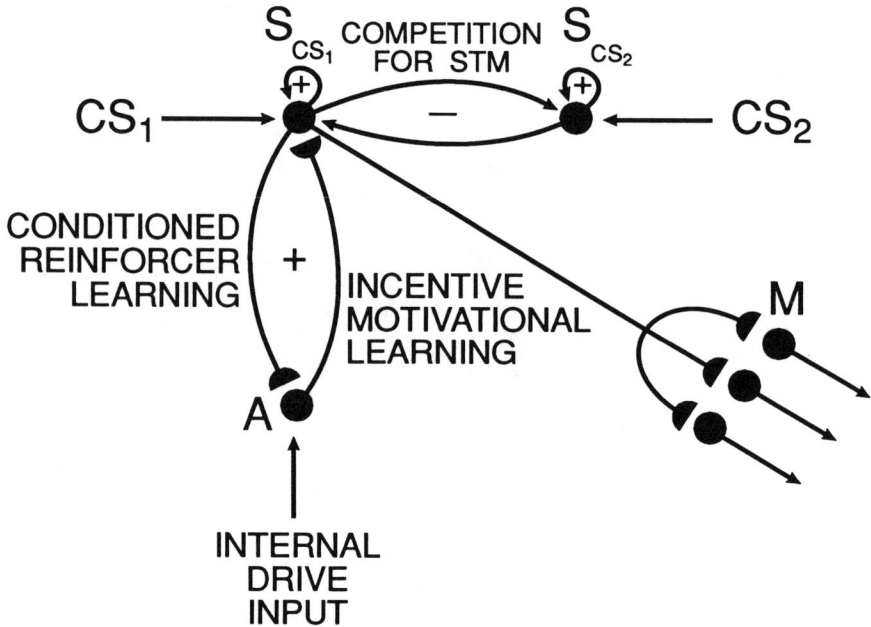

Figure 5.5. Schematic conditioning network. Conditioned stimuli (CS_i) activate sensory nodes ($S_{CS,i}$) that compete for short-term memory (STM) storage. Activated $S_{CS,i}$ send conditionable signals both to drive nodes (A) and to motor nodes (M). Conditioned reinforcer learning refers to S-to-A connections, whereby a CS repeatedly paired with a US becomes a secondary reinforcer. *Incentive motivational* learning refers to A-to-S connections (activated by internal drive combined with sensory inputs), which enhance approach to or avoidance of given stimuli. (Adapted from Grossberg & Levine, 1987, with permission of the Optical Society of America.)

The *A* loci include representations of specific drives such as hunger, thirst, and sex. Activation of a given drive representation in this network requires a combination of internal drive level *and* compatible sensory stimuli; for example, the hunger representation is activated by a combination of hunger and the presence of food. Later articles (Grossberg, 1972a, b), in order to model negative as well as positive reinforcement, expanded the concept to include "negative" arousal loci for aversive stimuli such as electric shock.

The need, in psychological theory, for drive representations that are separate from the sensory representations of particular stimuli is still a matter of controversy. Klopf (1988), for example, dismissed the need for separate drive representations by simply defining drives as sufficiently strong stimuli. But Bower (1981) and Bower, Gilligan, & Monteiro (1981) developed the idea of "emotion nodes," and Barto *et al.* (1983) showed that adaptive control can be facilitated by "adaptive critic elements"; both of these concepts are functionally similar to that of drive representations. Also, long before the development of quantitative models, Hebb (1955) argued that every sensory event has two different effects: its *cue* function, which selectively guides behavior, and its *arousal* function, which energizes behavior. This distinction corresponds in the network of Figure 5.5 to the distinction between the representation of the US or CS at S nodes and the effect of that representation, via fixed or modifiable synapses, on appropriate A nodes.

The existence of drive representations in neural networks is compatible with data on the reinforcement associated with brain stimulation. A long series of studies, starting with Olds (1955) and Olds and Milner (1954), shows that rats can learn to perform motor responses leading to stimulation of certain specific loci within the limbic system, hypothalamus, and midbrain (see Appendix 1). Rats can also learn to avoid stimulation of other loci within those brain areas.

The theory of Grossberg (1971) can also account for *secondary reinforcement* or *secondary conditioning*. This means that if a stimulus CS_1 is repeatedly followed by a US, until CS_1 evokes a conditioned response, then CS_1 can itself become a US for a new conditioned stimulus, say CS_2. If CS_2 is repeatedly followed by CS_1, then CS_2 will come to evoke the same response.

Aversive Conditioning and Extinction

The experimental literature on animal learning is vast, and this book only covers a few of its well-known paradigms. Much of this experimental work is summarized in the books of Mackintosh (1974, 1983) and Staddon (1983, 1989). This literature has, however, spawned a few terms that are in general use among experimental psychologists; these terms are introduced as they arise in the discussion of this chapter and used repeatedly thereafter. Two of the major ones are *aversive conditioning* and *extinction*.

Aversive conditioning is a term used for the suppression of a particular motor behavior by punishment. For example, the frequency of a particular motor response can be reduced by pairing that response with electric shock. *Extinction* is a term for the suppression of a motor behavior by nonoccurrence of an expected reward. For example, a response that had previously been paired with a reward such as food can be suppressed by frustration if the expected food is absent. The psychological fact that punishment and frustration have similar suppressive effects on behavior suggests that aversive conditioning and extinction might be described using similar neural mechanisms (Grossberg, 1982b, pp. 335-339).

Some controversy has existed among psychologists as to whether extinction is a passive or an active process, but the latter belief is more favored. If a dog learns to salivate in response to a bell after the bell has been repeatedly paired with food, the dog seems not to simply "forget" the salivation response if food is no longer given, but rather to counter-condition the bell to the aversive experience of frustration. That extinction is not simply passive forgetting, a return to a naive state, is suggested by the fact that reacquisition of a response by an extinguished animal is faster than initial acquisition by an untrained animal (Pavlov, 1927).

Hence, extinction is usually considered to occur as a consequence of the disconfirmation of an expectation of reward. This recalls the statement of Rescorla and Wagner (1972, p. 75) that "organisms only learn when events violate expectations." That principle leads us to seek a general mechanism for processing disconfirmed expectations in the motivational domain, whether in a positive or negative direction. For an example of the latter, a motor response associated with disconfirmation of expected *punishment*, such as a lever press which unexpectedly turns off an ongoing electric shock, can become rewarding.

Differential Hebbian Theory Versus Gated Dipole Theory

Processing disconfirmation of expectations involves comparing present with ongoing values of neural variables. Recall from Chapter 3 the discussion of two alternative methods for such a comparison. One of these methods is based on the gated dipole theory of Grossberg (1972a, b), using habituation of repeatedly presented stimuli (see Figure 3.8). The other is based on the differential Hebbian learning theory (Kosko, 1986b; Klopf, 1986, 1988), using a rule whereby synapses change in strength as a function of cross-correlated *changes* in presynaptic and postsynaptic activities (see Figure 5.6).

Neither the differential Hebbian rule nor the gated dipole rule has yet been verified in actual nervous systems. Experimental tests of these two sets of rules are likely to be as much psychological as physiological, and to be based in part on their ability to be embedded in larger networks that perform interesting cognitive tasks. Gated dipoles have been used as components of larger networks that also include associative learning rules and competitive on-center off-surround fields (*e.g.*, Grossberg & Schmajuk, 1987; Levine & Prueitt, 1989; Ricart, 1991). Differential Hebbian learning rules have not yet been tested as elements in such complex networks, but have been incorporated into models interpreted as occurring at the neuronal level.

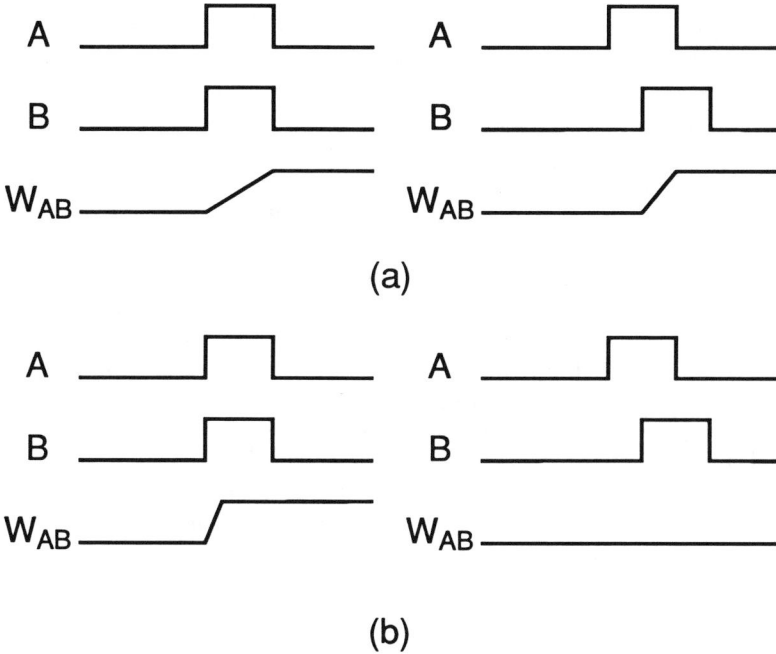

Figure 5.6. Schematic diagram of the distinction between (a) a Hebbian learning rule and (b) a differential Hebbian learning rule. In (a), the associative strength w_{AB} between A and B increases with simultaneous occurrence of inputs A and B. In (b), w_{AB} increases with simultaneous *increases* (*e.g.*, onsets) of A and B. (Note, however, that the differential Hebbian model of Klopf, 1988 obtained learning with nonsimultaneous onsets by manipulating network time lags.)

An explicit version of the differential Hebbian rule (also known as the *drive-reinforcement* rule) was developed by Klopf (1988); equations for Klopf's model are shown in Section 5.3. Klopf was led to such a rule by his earlier "hedonistic neuron" theory (Klopf, 1982) in which neurons themselves were goal-seeking devices.

Klopf's network simulated a wide variety of classical conditioning data. These data included blocking, secondary conditioning, extinction and reacquisition of an extinguished response, effects of interval between CS and US occurrences, effects of stimulus durations and amplitudes. (However, as will be seen from the equations given in Section 5.4, the simulations of CS-US interval effects depend on some weighting factors for time delays, factors that were chosen specifically to match those data. Klopf does not suggest an underlying mechanism for generating those weighting factors.) Figure 5.7 shows the basic neuronal network of Klopf's model; note the similarity to Figure 3.5 due to Sutton and Barto (1981).

Klopf's network also simulated *conditioned inhibition*. The conditioned inhibition paradigm consists of a first stage where a CS_1 is associated with a US and thereby conditioned to a CR, followed by a second stage where a combination of two stimuli CS_1 and CS_2 is associated with absence of the US. As a consequence, when CS_2 subsequently is associated with other stimuli that would otherwise evoke the CR, that CR is suppressed.

Further discussion of these alternative conditioning models will be placed in the context of data on the attentional modulation of conditioning. Such attentional modulation, including the blocking paradigm discussed above and other, more complex multistimulus experiments, is the subject of the next section.

5.2. ATTENTION AND SHORT-TERM MEMORY IN CONDITIONING MODELS

While neural network models of Pavlovian conditioning differ in their architectures, they share some common heuristic themes. Recall once more Rescorla and Wagner's principle that "organisms only learn when events violate expectations." Similarly, Grossberg (1975, p. 266), discussing blocking and similar experiments, said that "learning subjects act as minimal adaptive predictors; they enlarge the set of cues that control their behavior only when the cues that presently control their behavior do not perfectly predict subsequent events."

These heuristics, however, have led different modelers to different conclusions, which in turn have different implications for the predictions of other experimental data. For example, Pearce and Hall (1980, p. 538), developing a non-neural psychological model that is a refinement of Rescorla and Wagner's, state that "stimuli that fully predict their consequences will be denied access to the processor ... A stimulus is likely to be processed to the extent that it is not an accurate predictor of its consequences." Grossberg (1982a, p. 530) argued, however, that Pearce and Hall's statement is violated by the fact that a US is an excellent predictor of its consequences and yet is almost always processed. By way of reconciliation, he proposed that there are separate, interacting systems for processing expected events and for processing unexpected events, and that the architectures of these two systems are different. In that article and elsewhere (for example, Grossberg, 1975; Carpenter & Grossberg, 1987a), these two systems are called the *attentional* and *orienting* systems.

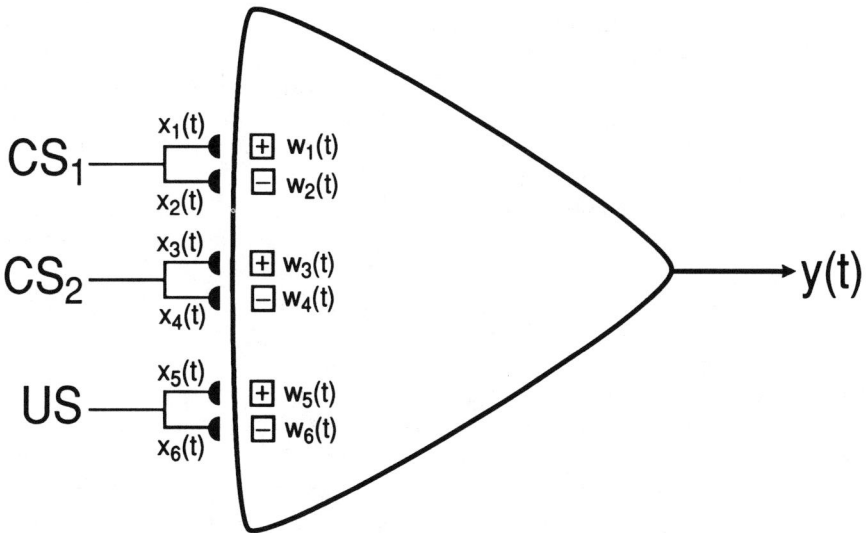

Figure 5.7. Schematic of the drive-reinforcement neuronal model. Each CS or US is represented by an excitatory and an inhibitory synapse. Synaptic weights are variable for synapses that mediate CS's and fixed for synapses that mediate US's. (Adapted from Klopf, *Psychobiology* **16**, 85-125, 1988, by permission of Psychonomic Society, Inc.)

Grossberg's Approach to Attention

A theory for the structure of the attentional system was proposed in Grossberg (1975), incorporating the type of competitive mechanism discussed in Chapter 4 for the short-term storage of patterns. This attentional mechanism is based on the network of Figure 5.5, with the additions of competition between sensory representations of different stimuli, competition between drive representations of different drives, and modifiability of feedback connections from drive to sensory representations. (Competition between drive representations had previously been utilized by Kilmer et al., 1969 to model decisions between gross behavioral modes of the organism.) One version of the resulting, more complex network is shown in Figure 5.8.

The network of Figure 5.8 is built on the sensory-drive heterarchy of Figure 4.24 (see the discussion in Section 4.4) in which each drive representation is activated by a combination of internal drive level and external sensory inputs compatible with the given drive. This enables the organism to focus attention on the particular cues that are compatible with whatever drives are relatively active at a given moment. The heterarchy is combined with an orienting system that causes motor responses to new cues in the environment, *unless* those cues are known to be associated with satisfaction of an active drive.

The attentional system in Figures 4.24 and 5.8 not only allows for the modeling of attentional effects in conditioning, but also prevents cross-conditioning of stimuli compatible with one drive to irrelevant drives. A graphic example of the consequences of such cross-conditioning is given by Grossberg (1975; see also Grossberg & Levine, 1987). In his example, suppose you are eating roast turkey for dinner with your lover. Because you are repeatedly scanning both turkey and lover, it might be expected that each sensory cue would become associated in your mind with the drive compatible with the other cue; in fact, you might learn to want to have sex with turkeys and eat your lover. Grossberg showed heuristically how such an absurd outcome is prevented by competition both between sensory loci and between drive loci.

The network explanation of attention enabled Grossberg and Levine (1987) to simulate the blocking paradigm of Kamin (1969). Grossberg and Levine (1987, p. 5016) phrased the relevant modeling issues as follows: "How does the pairing of CS_1 with US in the first phase of a blocking experiment endow the CS_1 cue with the properties of a conditioned, or secondary, reinforcer? How do the reinforcing properties of a cue ... shift the focus of attention toward its own processing? How does the limited capacity of attentional resources arise, so that a shift of attention toward one set of cues ... can prevent other cues ...

from being attended? How does withdrawal of attention from a cue prevent that cue from entering into new conditioned relationships?" These questions can be summarized by asking how an organism predicts the environment so as to maximize (optimize) positive reinforcement and minimize negative reinforcement. Some other recent modelers (Hinton, 1987; Werbos, 1988a) have termed this kind of prediction, in natural or artificial neural networks, *reinforcement learning* (*cf.* Section 7.1).

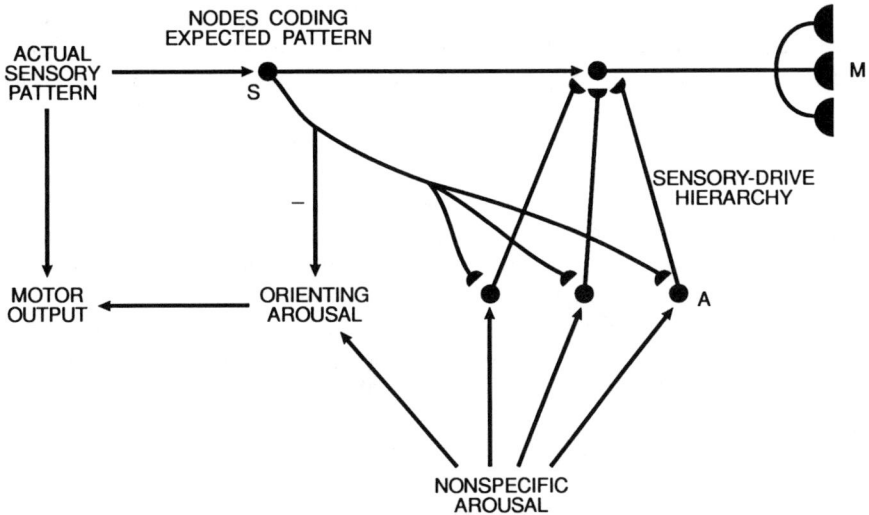

Figure 5.8. The sensory-drive heterarchy of Figure 4.24 (not shown in full) is embedded in a larger network with attentional and orienting subnetworks. An orienting ("what is it") response, is instinctive to all sensory cues, and is shut off only by expected cues; details of orienting system architecture are omitted. Expected cues are compatible with some drive, so activate the heterarchy. (Adapted from Grossberg, 1975, with permission of Academic Press.)

The first of Grossberg and Levine's "four questions" is answered by associative learning of a connection from the CS_1 representation, not to the US representation itself but to the drive representation. The second and third questions are answered by competition within the on-center off-surround subnetwork of sensory representations. Within that subnetwork, competition favors nodes corresponding to stimuli that have become conditioned reinforcers. Hence (with properly chosen parameters), the activities of other sensory nodes, such as the CS_2 node in the blocking paradigm, are suppressed, reducing the ability of those nodes to form conditioned associations. That effect causes CS_2 to be blocked, in answer to the fourth question.

Sutton and Barto's Approach to Blocking

The attentional (*i.e.*, lateral inhibitory) effects in the network of Figure 5.5 provided explanations for the blocking and overshadowing results of Kamin (1969). The model of Sutton and Barto (1981), and related models, have typically not addressed the question of which of two stimuli is more likely to attentionally overshadow the other. Barto and Sutton (1982, p. 232), discussing their own model, stated: "The model clearly does not address higher order modulatory influences such as those produced by attentional or stimulus salience factors." Mackintosh (1975), while proposing a model related to that of Rescorla and Wagner, likewise noted that such models cannot account for the fact that a more salient stimulus can block a less salient one but not vice versa.

The simulation of blocking by Sutton and Barto (1981) was based on a different mechanism, relying on a special synaptic modification rule. In addition to blocking, Sutton and Barto simulated some timing results in classical conditioning. They noted that conditioning is typically strongest when the CS precedes the US, usually by 200 to 500 milliseconds, rather than when the CS and US are presented simultaneously. An example occurs in the rabbit *nictitating membrane response* (eyeblink conditioning) data of Schneiderman and Gormezano (1964) and Smith, Coleman, and Gormezano (1969), shown in Figure 5.9. Sutton and Barto used such data to argue against a "Hebbian" learning law in which actual presynaptic and postsynaptic activities are correlated. For with Hebbian learning, all other things being equal, the optimal *interstimulus interval* (ISI)[1] should be 0 rather than 200 to 500 milliseconds.

Sutton and Barto solved the problem of simulating the ISI effect by introducing the *eligibility traces* defined by Equations (3.23) of Chapter 3. They proposed that each conditioned stimulus, in addition to having a short-term memory trace x_i, has an additional trace $\bar{x}_i$ which grows more slowly. As shown in Figure 5.10, the time course of this additional trace accounts qualitatively for the time of the optimal ISI. Sutton and Barto gave

---

[1] While the term "interstimulus interval" is commonly used in the animal learning literature, the interval is not properly "between stimuli" but rather between their onsets. In fact, the times of occurrence of the two stimuli can overlap. For this reason, some psychologists prefer the term "stimulus onset asynchrony" or SOA.

a single-cell interpretation of this trace, calling it an eligibility trace because it can be regarded as a chemical marker indicating how "eligible" the synapse is for modification. (In a larger network, however, the two sets of traces could plausibly be interpreted as being located in different brain areas, with stimulus traces being more peripheral and eligibility traces more central.)

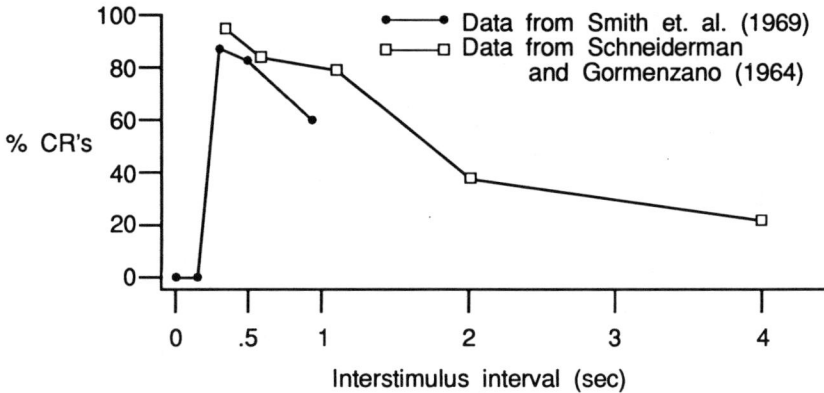

Figure 5.9. Asymptotic associative strength (measured as percentage of recall trials on which a conditioned response occurs) as a function of interstimulus interval in conditioning of the rabbit nictitating membrane response. (From Sutton & Barto, *Psychological Review* **88**, 135-170, 1981. Copyright 1981 by the American Psychological Association. Reprinted by permission.)

Sutton and Barto explained blocking by means of a synaptic modification rule whereby presynaptic activity is correlated with change in postsynaptic activity; this rule is formally analogous to the rule of Rescorla and Wagner (1972). Because the introduction of the CS_2, in this model, leads to no *increase* in US node activity, the CS_2 does not become conditioned to the US. This change in postsynaptic activity is multiplied by presynaptic eligibility, as shown in Equations (3.23). Some results of simulating this network are shown in Figures 5.11 and 5.12.

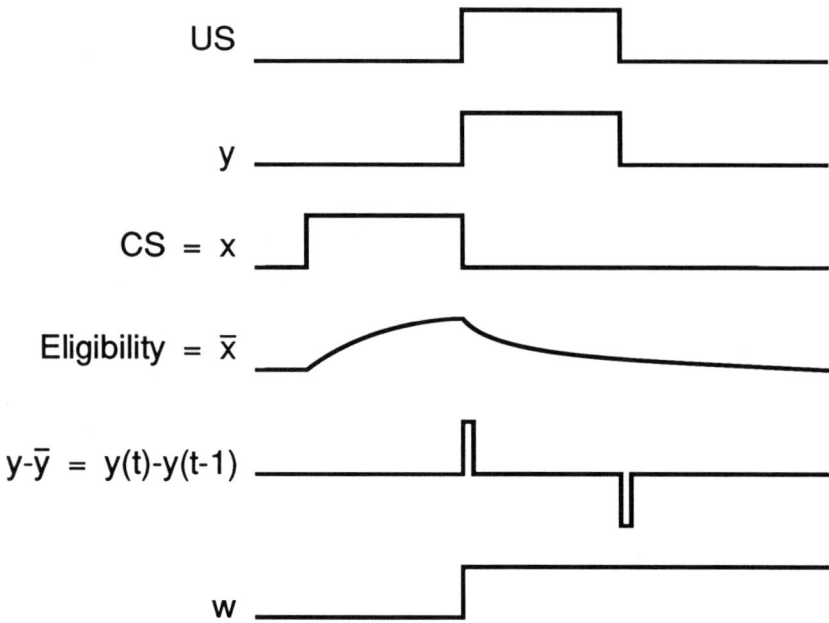

Figure 5.10. Time courses of variables in Equations (3.23) for a single trial in which a neutral CS (w=0 at the start) is followed by a US. (From Sutton & Barto, *Psychological Review* **88**, 135-170, 1981.   Copyright   1981   by   the   American   Psychological Association. Adapted by permission.)

Hawkins and Kandel (1984) proposed an explanation for blocking quite similar to those of Rescorla and Wagner and of Sutton and Barto.   The Hawkins-Kandel   model,   unlike   the   others   mentioned,   is   based   on neurophysiological data from the sea slug *Aplysia* (*cf.* the discussion of Kandel & Tauc, 1965 in Chapter 3).   These data (Hawkins, Abrams, Carew, & Kandel, 1983; Walters & Byrne, 1983) suggest that each US activates a given neuron, called a *facilitator neuron*, which influences pathways activated by CS's, and that the growth of associative strengths depends on simultaneous activation of a CS pathway and a facilitator neuron.   Within this context, Hawkins and Kandel (1984, p. 385) proposed that blocking is due to a kind of fatigue effect: "the output of the facilitator neurons decreases when they are stimulated continuously." Thus after many trials in which a CS_1 is paired with a US, the fatigue of the US's facilitator neuron prevents that US from forming associations with another stimulus CS_2.

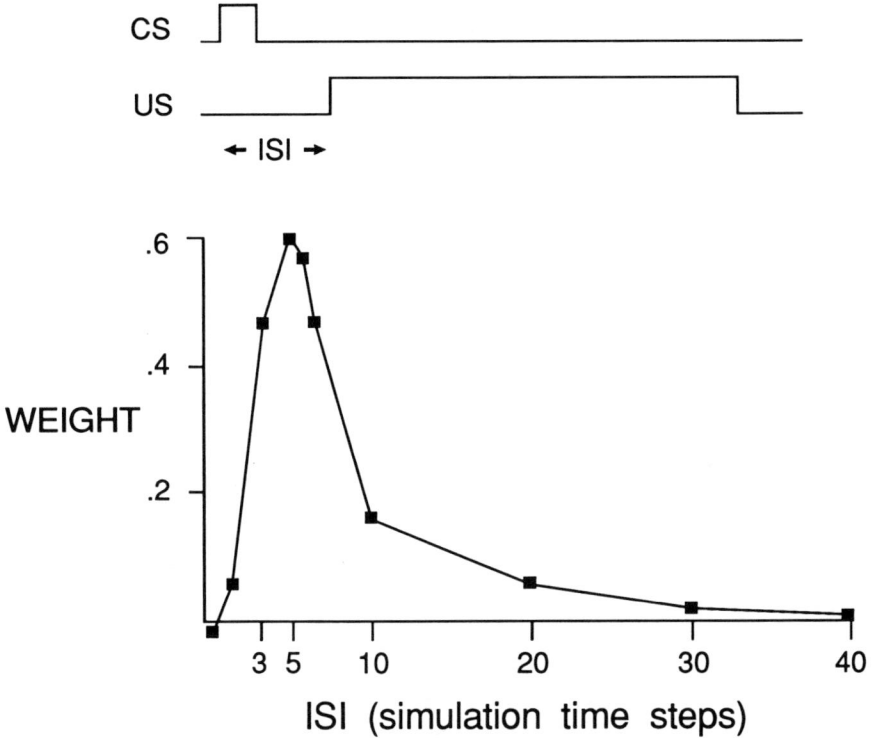

Figure 5.11. Asymptotic connection weight as a function of interstimulus interval in a simulation of (3.23). CS is on for 3 time steps, US for 30 (note how long). Trials last for 50 time steps. (From Sutton & Barto, *Psychological Review* **88**, 135-170, 1981. Copyright 1981 by the American Psychological Association. Reprinted by permission.)

STIMULUS CONFIGURATION

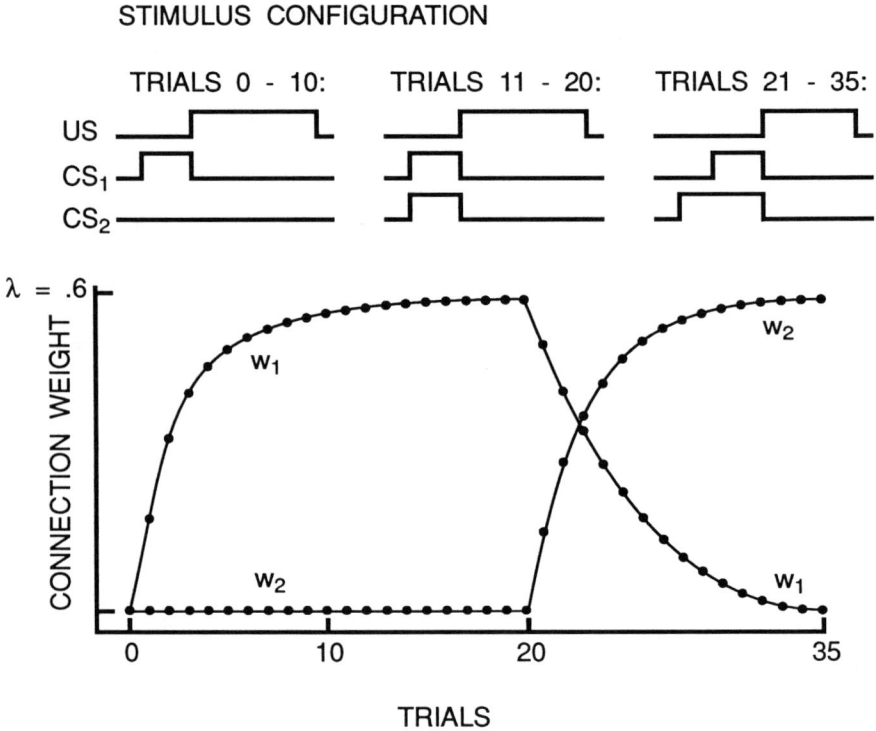

Figure 5.12. Blocking simulation. Connection weights w_1 of CS_1 and w_2 of CS_2 are shown at the end of each trial. Trials 0-10: input of CS_1 alone followed by US leads to increase in w_1. Trials 11-20: CS_1 and CS_2 presented together followed by US produces no change, *i.e.*, blocking occurs. Trials 21-35: CS_2 begins before CS_1. The output node responds to the earlier predictor and ignores the later. (From Sutton & Barto, *Psychological Review* **88**, 135-170, 1981. Copyright 1981 by the American Psychological Association. Adapted by permission.)

The model of Sutton and Barto (1981) was further elaborated in several later articles, particularly Blazis, Desmond, Moore, & Berthier (1986), Moore *et al.* (1986), and Sutton and Barto (1991). A main thrust of this later work was to fit the model to various quantitative details of the nictitating membrane response.

Some Contrasts Between the Above Two Approaches

Grossberg and Levine (1987) suggested that the Hawkins-Kandel fatigue model for blocking failed to explain the complementary phenomenon of *unblocking* (Kamin, 1969). Unblocking means that if CS_1 is first trained to a US, and then the compound stimulus $CS_1 + CS_2$ is associated with a US of a different level than before — for example, in the case of an electric shock US, the compound stimulus is associated with either a more intense or a less intense shock than is CS_1 alone — then blocking does not take place, and the animal conditions normally to CS_2. (If the shock is *much* less intense, this effect can be counteracted by a tendency to condition CS_2 to relief.)

Grossberg and Levine (1987) also discussed the invertebrate data in terms of two interrelated psychological concepts: *conditioned reinforcement* and *incentive motivation*. Conditioned reinforcement is represented, for example, by the *S*-to-*A* pathways in Figure 5.5, which cause a previously neutral stimulus to become a reinforcer. Incentive motivation is represented by the *A*-to-*S* pathways in the same figure, which yield incentives to approach or avoid particular stimuli based on their reinforcement value. That article compared the facilitator neurons found in *Aplysia* by Walters and Byrne (1983) to incentive motivational pathways.

According to Grossberg's incentive motivation theory, secondary conditioning occurs because CS presentation, after conditioning has taken place, leading to an increase in *A*-to-*S* pathway activity; hence, the CS becomes a reinforcer in its own right. However, Walters and Byrne (1983) had not demonstrated an analogous increase in activity of facilitatory pathways in *Aplysia*. Hence, the exact mechanism for secondary conditioning in that species, if it does occur, is still in question.

Grossberg and Levine (1987) simulated blocking in the context of attentional competition between stimuli as suggested by Figure 5.5. In response to the objections to "Hebbian" learning by Sutton and Barto (1981), they also simulated the ISI data of Figure 5.9 in an attentional context. With CS and US presented simultaneously, Grossberg and Levine stated, attentional competition occurs between those two stimuli, with a bias in favor of the US because it has a strong, unconditional association with the drive representation. Hence, the ISI phenomenon can be seen as a form of blocking, with the US taking the place of the CS_1 in the experiment of Kamin (1969), and the CS taking the place of Kamin's CS_2.

Figure 5.13 shows the actual network used by Grossberg and Levine (1987) to simulate the blocking and ISI effects; this network is an elaboration of the one shown in Figure 5.5. Some results from simulation of this network are shown in Figure 5.14. These results verified the efficacy of a conditioning model based on associative learning combined with attentional competition, but left open some issues in timing. (The timing issues are currently being pursued in other models; see, *e.g.*, Killeen & Fetterman, 1988; Grossberg & Schmajuk, 1989; Klopf & Morgan, 1991). In order to achieve attentional blocking of the CS by the US when the two are presented simultaneously, the threshold for an S node to increase the efficacy of an S-to-A synapse had to be set higher than the threshold for the same S node to excite activity of the corresponding A node (and thereby of the A-to-S feedback pathway). The separation of excitation of the synapse from excitation of node activities downstream is reminiscent of Sutton and Barto's separation of eligibility traces from stimulus traces.

Further Connections with Invertebrate Neurophysiology

For the sake of brevity, this book has generally concentrated on modeling at the level of large aggregates of neurons, as opposed to the level of single neurons. Yet classical conditioning is an area where the two levels of modeling have increasingly blended in the last few years. Results on the interactions of electrical potentials, transmitters, and second messengers (*cf.* the discussion of Byrne, 1987 in Section 3.4) have been incorporated into many network models of associative learning. When competing models are equally plausible from the psychological viewpoint, neurophysiological data can provide additional constraints that facilitate choosing between models.

Sutton and Barto (1981) developed some rough analogies between their eligibility variables and postsynaptic second messengers (calcium ion and cyclic AMP). This kind of analogy was developed further in the neuronal model for associative learning in Gingrich and Byrne (1987), based on the study of conditioned withdrawal reflexes in *Aplysia*. Their model was built on a previous model of *nonassociative learning* (Gingrich & Byrne, 1985). Nonassociative learning is defined as the strengthening or weakening of certain neuronal pathways without dependency on contingent stimulation of other pathways.

Gingrich and Byrne (1987) modeled cell-level phenomena analogous to classical conditioning; these phenomena had been experimentally discovered by Walters and Byrne (1983). (Recall from Chapter 3 the cautionary notes about whether such single-cell mechanisms in fact approximate mechanisms responsible for conditioned behavior of whole organisms.) In the studies of Walters and Byrne (1983), shock to the tail was used as an aversive US, while the CS was direct electrical stimulation of any one of several sensory neurons responsive to stimulation of non-tail skin areas. Stimulation of the non-tail area became aversive as a result of learning. These authors proposed a mechanism for such cell-level conditioning whereby the US nonspecifically releases a chemical modulator that strengthens synaptic pathways from sensory neurons to output areas. This type of mechanism, which they called *activity-dependent neuromodulation*, is illustrated in Figure 5.15. (Similar kinds of modulation appear in many recent models of higher-order cognitive and motivational processes, some of which are discussed in Chapter 7; see, *e.g.*, Levine & Prueitt, 1989; Hestenes, 1991; Ricart, 1991).

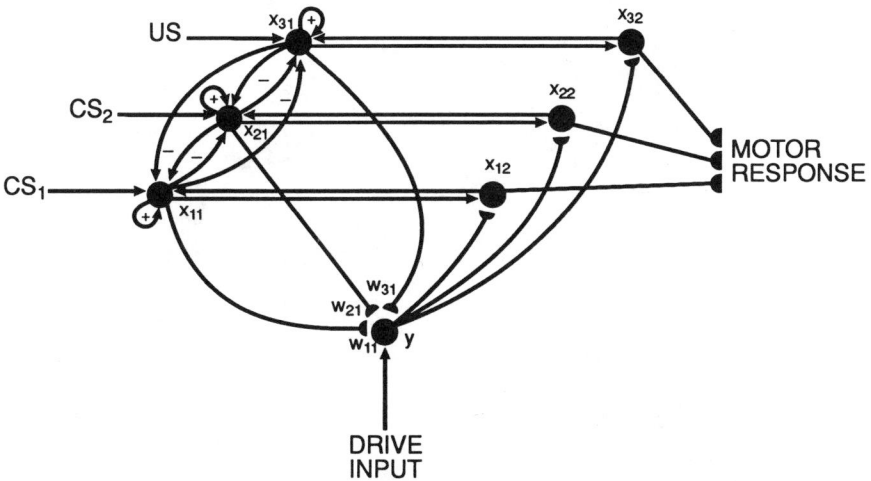

Figure 5.13. Network used to simulate blocking and ISI effects. Each CS or US sensory representation has two stages with STM activities x_{i1} and x_{i2}. Activation of x_{i1} generates unconditioned signals to x_{i2} and conditioned reinforcer signals to a drive node y. Conditioned incentive motivational signals from A activate the second sensory stage x_{i2}, which sends feedback to x_{i1}. (Adapted from Grossberg & Levine, 1987, with permission of the Optical Society of America.)

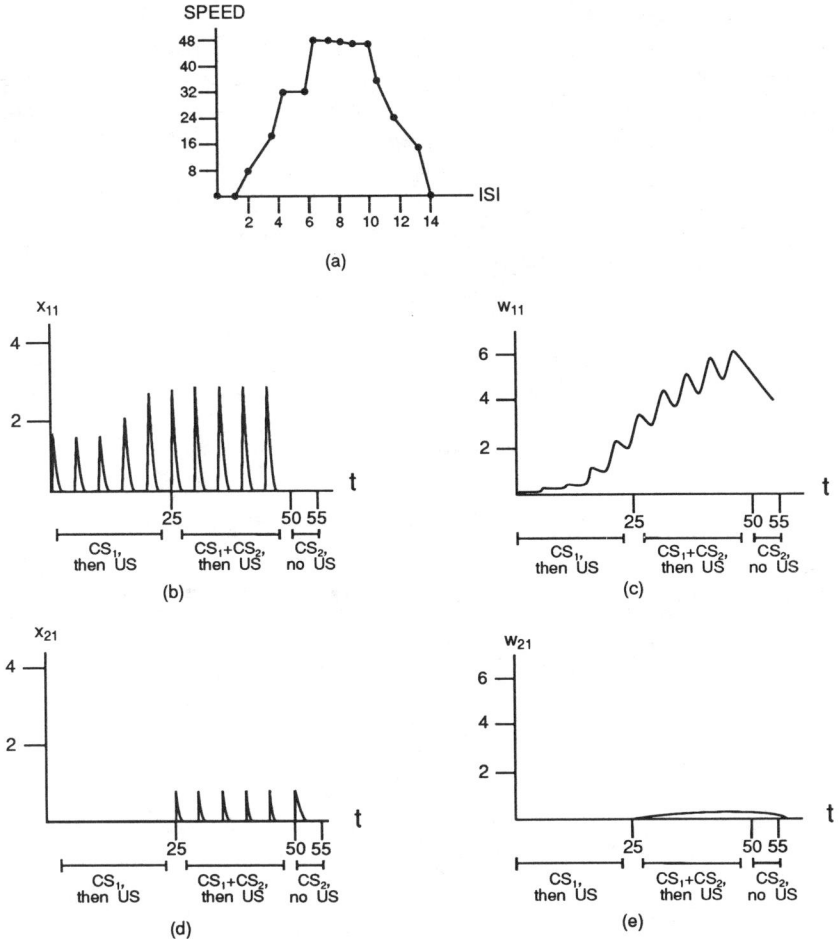

Figure 5.14. (a) Plot of CR acquisition speed as a function of ISI, in the network of Figure 5.13. A CR is said to occur when x_{12} exceeds some threshold. Speed is computed by the formula 100 x (time units per trial)/(time units to first CR). (b)-(e) are plots of variables through time in a blocking simulation, with 5 trials (each 50 time units long) of CS_1-US pairing, 5 of (CS_1+CS_2)-US pairing, 1 of CS_2 presented alone, ISI=6. (b) CS_1 STM trace x_{11}; (c) x_{11}-to-y LTM trace w_{11}; (d) CS_2 trace x_{21}; (e) x_{21}-to-y LTM trace w_{21}. (Adapted from Grossberg & Levine, 1987, with permission of the Optical Society of America.)

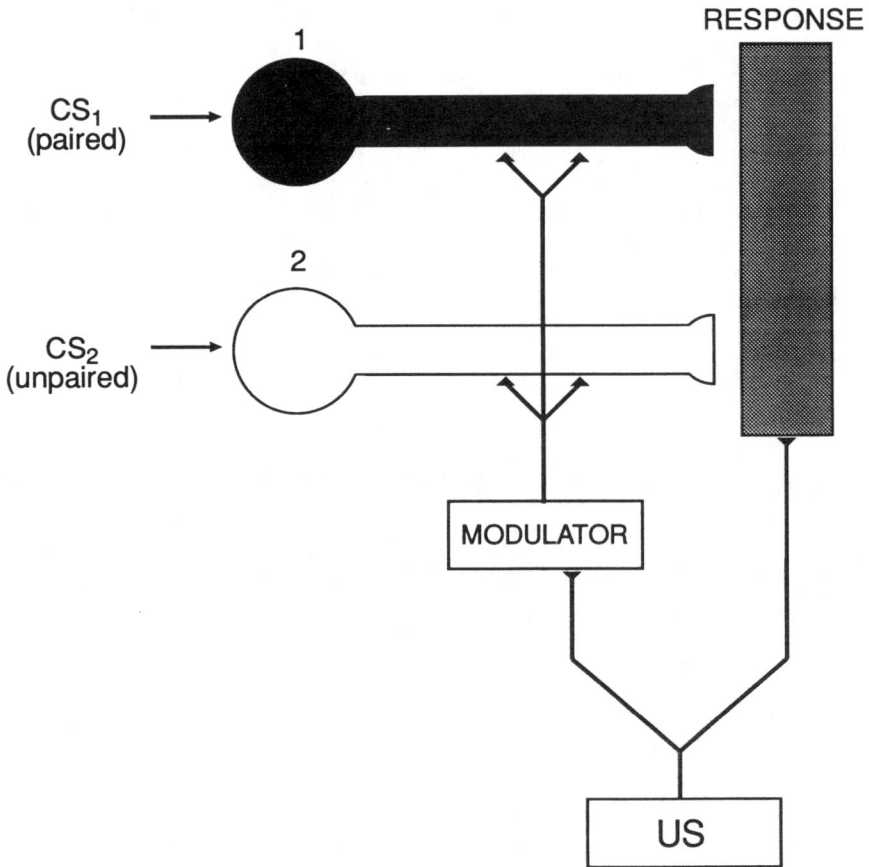

Figure 5.15. Model for activity-dependent neuromodulation in *Aplysia*. Two sensory neurons (1 and 2) make subthreshold connections to a response system. Reinforcing stimuli (US) directly activate the response system and also activate a diffuse modulatory system. Pairing of CS and US (indicated by the darkened pathway) enhances the modulatory effect over that of unpaired stimulation. (Adapted from Walters and Byrne, *Science*, **219**, 405-408, 1983, with permission of the American Association for the Advancement of Science.)

Gingrich and Byrne (1985) quantitatively simulated the roles of calcium ions and cyclic nucleotides (the commonest second messengers), along with enzymes regulating those substances, in nonassociative forms of learning (see Figure 5.16 for a possible biochemical schema). Gingrich and Byrne (1987) integrated these studies with the concept of activity-dependent neuromodulation, using equations that are discussed in Section 5.3, to model associative learning.

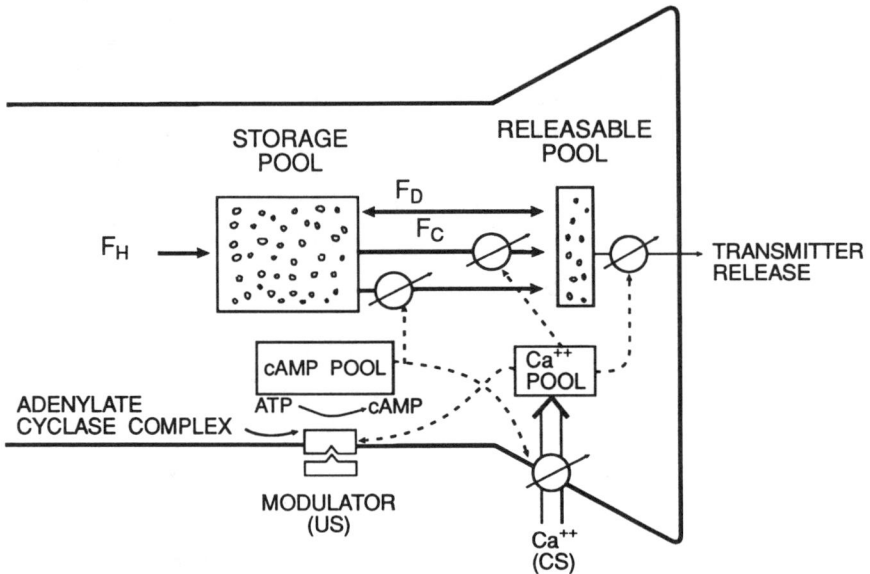

Figure 5.16. Neuronal chemistry. The CS (action potential) raises intracellular Ca^{++}; that releases transmitter, moves transmitter (F_C) from storage to a usable pool, and primes adenylate cyclase. F_D refers to diffusion between storage and usable pools. The US activates adenylate cyclase to increase cAMP, thus increasing transmitter movement (F_{cAMP}) and Ca^{++} influx, the latter via lengthened action potentials. CS-US association, *i.e.*, high Ca^{++} while the US is on, increases cAMP via adenylate cyclase priming. All these events enhance Ca^{++} influx and transmitter release occurring with the next test stimulus. Circles with arrows denote elements modulated by other variables. (Reprinted from Gingrich & Byrne, 1987, with permission of the American Physiological Society.)

Gated Dipoles and Aversive Conditioning

The quantitative study by Grossberg and Levine (1987) of attentional effects in conditioning was continued in Grossberg and Schmajuk (1987). These authors included simulations of aversive conditioning via negative reinforcement as well as *appetitive* conditioning (the opposite of aversive) via positive reinforcement. To attentional networks such as appear in Figures 5.8 and 5.13, they added the two-channel gated dipole mechanism schematized in Figure 3.8. A loop was added to the gated dipole to allow for secondary inhibitory conditioning (see Grossberg, 1975 for an explanation of why this is needed). As described in Grossberg and Schmajuk (1987, p. 197): "Secondary inhibitory conditioning consists of two phases. In Phase 1, CS_1 becomes an excitatory conditioned reinforcer (*e.g.*, source of conditioned fear) by being paired with a US (*e.g.*, a shock). In Phase 2, the offset of CS_1 can generate an off-response which can condition a subsequent CS_2 to become an inhibitory conditioned reinforcer (*e.g.*, source of conditioned relief)."

All of these considerations led Grossberg and Schmajuk (1987) to design a network, shown in Figure 5.17, called READ (for "recurrent associative dipole"). The equations for the READ circuit, combining associative learning and gated dipole equations, are shown in Section 5.3. For an appropriate range of parameter values, the network can simulate both primary and secondary excitatory and inhibitory conditioning.

Grossberg and Schmajuk (1987) went on to discuss qualitatively how extinction and conditioned inhibition can be modeled in their network. Extinction is treated as an active, not a passive, process, resulting from conditioning a CS to the "off" channel of Figure 5.17 if an expected US does not occur. This effect is not simulated explicitly in the Grossberg-Schmajuk article because it involves interaction of the READ network with another network that can measure the degree of match or mismatch of an actual with an expected stimulus. Such a match-sensitive network, based on *adaptive resonance theory* (ART), is discussed in Chapter 6, within the context of categorization and coding models. The ART network combines mechanisms of association, competition, and opponent processing with an additional design principle for adaptive interactions between network levels.

In most of the models discussed in this chapter, the simplification has been made that stimuli are represented at single nodes. Recall from Section 3.4 that different modeling considerations may arise in the learning of associations between activity patterns spanning large numbers of nodes.

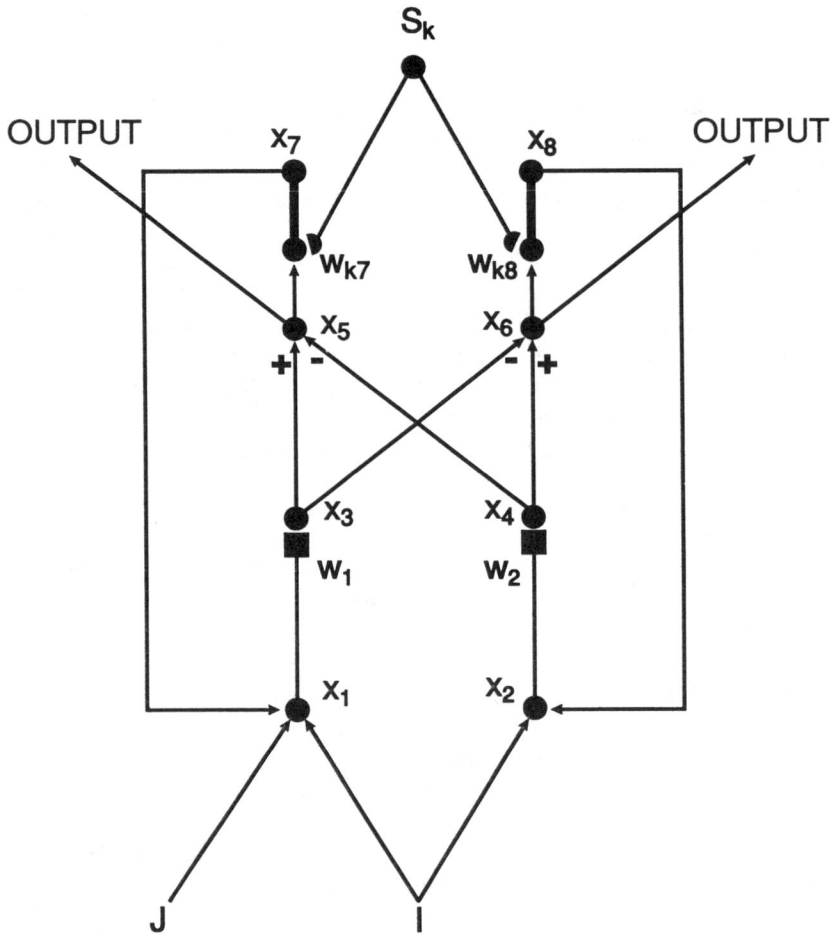

Figure 5.17. A READ (recurrent associative dipole) network, joining a recurrent gated dipole with associative learning. Learning occurs at the synapses w_{k7} and w_{k8}, from the sensory nodes S_k to the on-channel and off-channel, respectively, of the motivational gated dipole. (Adapted from Grossberg and Schmajuk, *Psychobiology* **15**, 195-240, 1987, by permission of Psychonomic Society, Inc.)

Such complex associative memory networks (*e.g.*, Kohonen, 1984; Kosko, 1987a, c, 1988) have yet to be integrated with conditioning models. Kohonen's and Kosko's networks have been the subject of mathematical theory and practical applications, but not yet of realistic brain simulations. These networks do, however, suggest possible ways that some of the subnetwork principles of earlier chapters in this book could be concatenated into larger cognitive structures. Some of these structures might have analogs in brain regions like the association areas of the cerebral cortex. These are the lobes (temporal, parietal, and frontal) which are not restricted to one sensory modality but encode some inter-modality associations.

The structure-function relations in multimodal cortical regions are largely uncharted, though Chapter 7 discusses a few tentative models bearing on these regions. However, there are also considerable connections, many of them plastic, within areas of the cortex devoted to a single modality. Intra-modality associative learning is somewhat better understood than inter-modality learning. In particular, experimental studies of coding at the level of the visual cortex provide some of the inspiration for the models of coding and categorization discussed in the next chapter. These categorization models, like the conditioning models discussed above, combine principles such as associative learning, competition, and opponent processing in complex ways.

5.3. EQUATIONS FOR SOME CONDITIONING AND ASSOCIATIVE LEARNING MODELS

Klopf's Drive-Reinforcement Model

The drive-reinforcement model of Klopf (1988), applied particularly to Pavlovian conditioning, was interpreted by its author as a single-neuron model. Hence, the $x_i(t)$ shown in Figure 5.7 can be interpreted either as presynaptic firing frequencies or as CS activity levels. The $w_i(t)$ in that figure can either be synaptic efficacies or associative weights. The single value $y(t)$ can be interpreted either as a postsynaptic firing frequency or as a level of reinforcement.

Klopf's neuronal input-output relationship is similar to the one devised by Sutton and Barto (1981) and shown in Chapter 3, with the addition of a threshold. His equation is

$$y(t) = [(\sum_{i=1}^{n} w_i(t)x_i(t)) - \theta]^+ \qquad (5.4)$$

where θ is the postsynaptic firing threshold, and the symbol "+" denotes replacing the expression by 0 if it is negative. After the value y(t) is calculated using the linear weighting function in (5.4), if it is larger than some prespecified maximum firing rate, called $y^{max}(t)$, it is replaced by $y^{max}(t)$.

The learning mechanism is based on the changes in synaptic efficacies w_i as a result of changes in both presynaptic signals x_i and postsynaptic signal y. To account for causality in conditioning, the presynaptic signal changes are averaged over a time interval, called τ, which is the longest interstimulus interval at which conditioning is effective. The contributions of presynaptic signals at preceding times are weighted by the strength of pre-to-post connections at those times and by interval-dependent learning rate constants, called c_j for time interval j. Hence

$$\Delta w_i(t) = \Delta y(t-1) \sum_{j=1}^{\tau} c_j |w_i(t-j)| \Delta x_i(t-j-1) \qquad (5.5)$$

where, as always, $\Delta y(t)$ denotes $y(t+1) - y(t)$ (hence, $\Delta y(t-1)$ denotes $y(t) - y(t-1)$), and analogous notation is used for changes in other variables. In addition, the Δx_i are not allowed to be negative, but are set to be 0 whenever they become negative.

Some Later Variations of the Sutton-Barto Model: Temporal Difference

The temporal difference model of conditioning due to Sutton and Barto (1991) includes two sets of equations, for the eligibility traces $\bar{x}_i(t)$ of the various CS's, and for $w_i(t)$, the strengths of association between those CS's and a single US. If $x_i(t)$ denotes the actual i^{th} CS input signal (1 when the stimulus is present and 0 when it is absent), then the ith eligibility trace obeys the equation

$$\bar{x}_i(t+1) = (1-\delta) \bar{x}_i(t) + x_i(t) \qquad (5.6)$$

with δ a constant between 0 and 1. If y(t) denotes the US input signal (again 0 or 1), then the associative strength $w_i(t)$ obeys the learning equation

$$w_i(t+1) = w_i(t) + \alpha z_i(t+1)\bar{x}_i(t+1) \tag{5.7}$$

where α is another constant between 0 and 1, and $z_i(t+1)$ is the temporal difference factor or change in total positive reinforcement (primary or secondary). The temporal difference factor in (5.7) is in turn given by

$$z_i(t+1) = y(t+1) + \Gamma[\sum_j w_j(t)x_j(t+1)] \\ - [\sum_j w_j(t)x_j(t)] \tag{5.8}$$

In (5.8), the two sums in brackets denote the total of all the CS signals weighted by their associative strengths. Γ, which is typically slightly less than 1 in Sutton and Barto's simulations, is a factor by which future reinforcement is discounted as compared to present reinforcement.

The READ Circuit of Grossberg, Schmajuk, and Levine

Recall from earlier discussion that the recurrent associative dipole (READ) circuit was designed by Grossberg and Schmajuk (1987). As its name implies, this circuit, shown in Figure 5.17, adds to a gated dipole some internal feedback pathways. This feedback creates the possibility of second-order conditioning, both appetitive and aversive. The network also joins the gated dipole to a mechanism for associative learning.

If an unconditioned stimulus (US) is applied to the "on" channel of Figure 5.17 (the side with odd-numbered subscripts), and nonspecific arousal to both the "on" and "off" channels, the following equations arise (*cf.* Grossberg *et al.*, 1991 for discussion):

Arousal + US + Feedback On-Activation:

$$\frac{dx_1}{dt} = -ax_1 + I + J + f(x_7) \tag{5.9}$$

where I denotes the tonic arousal input to both channels and J is the specific input to the "on" channel.

Arousal + Feedback Off-Activation:

$$\frac{dx_2}{dt} = -ax_2 + I + f(x_8) \qquad (5.10).$$

On- and Off-Transmitters (depletable):

$$\frac{dw_1}{dt} = b(1-w_1) - cg(x_1)w_1 \qquad (5.11)$$

$$\frac{dw_2}{dt} = b(1-w_2) - cg(x_2)w_2 \qquad (5.12)$$

Gated On- and Off-Activations:

$$\frac{dx_3}{dt} = -ax_3 + eg(x_1)w_1 \qquad (5.13)$$

$$\frac{dx_4}{dt} = -ax_4 + eg(x_2)w_2 \qquad (5.14)$$

Normalized Opponent On- and Off-Activations:

$$\frac{dx_5}{dt} = -ax_5 + (h-x_5)x_3 - (x_5+k)x_4 \qquad (5.15)$$

$$\frac{dx_6}{dt} = -ax_6 + (h-x_6)x_4 - (x_6+k)x_3 \qquad (5.16)$$

Total On- and Off-Activations:

$$\frac{dx_7}{dt} = -ax_7 + m[x_5]^+ + p\sum_{k=1}^{n} s_k w_{k7} \qquad (5.17)$$

$$\frac{dx_8}{dt} = -ax_8 + m[x_6]^+ + p\sum_{k=1}^{n} s_k w_{k8} \qquad (5.18)$$

where S_k denotes the amplitude of the kth conditioned stimulus (CS) and x^+, for any real number x, denotes max (x,0).

On-Conditioned Reinforcer and Off-Conditioned Reinforcer Associations:

$$dw_{k7} = S_k(-qz_{k7}+r[x_5]^+) \qquad (5.19)$$

$$dw_{k8} = S_k(-qz_{k8}+r[x_6]^+) \qquad (5.20).$$

The functions f and g in Equations (5.9) - (5.20) can either be sigmoids or *ramp* functions, that is, linear above some threshold and 0 below the threshold. All symbols not discussed so far denote positive constants.

The equations used by Grossberg and Levine (1987) to model attentional effects, such as blocking, are not shown here. They are similar to Equations (5.9) - (5.20) except that they include multiple, variable CS representations and do not include opponent pairs of on- and off-channels. Grossberg and Levine also added some terms to the CS and US short-term memory equations to allow for competition between stimuli and decay in the presence of random background noise.

The *Aplysia* Model of Gingrich and Byrne

The equations of Gingrich and Byrne (1985, 1987) were based on a model of learning and other neuronal changes in the sea slug *Aplysia*, as depicted in Figures 5.15 and 5.16. Their model is distinguished from most of the others in this book in that the variables in their equations are explicit electrical potentials and concentrations of biochemical substances. (As vertebrate brain chemistry and physiology become better understood, such quantitative cellular models are likely to become more common.)

The basic variables in Gingrich and Byrne's model are intracellular concentration of the calcium ion (C_{Ca}), intracellular concentration of cyclic AMP (C_{cAMP}), concentrations of synaptic vesicles (the synaptic "swellings" that can release chemical transmitters) in releasable form (C_R) and in nonreleasable form (C_F), and duration of action potentials at the postsynaptic cell. First, the concentration of calcium ion, as developed in the 1985 article and continued in the 1987 article, is affected positively by the action potential through a variable I_{Ca} (called an *influx*), and negatively by two variables I_U and I_D (called *effluxes*). I_U is the rate of removal of excess calcium from the cell, and I_D is the rate of calcium diffusion to other regions of the cell. These variables combine into a differential equation

$$\frac{dC_{Ca}}{dt} = \frac{1}{V_C}(I_{Ca} - I_U - I_D)$$ (5.21)

where V_C denotes the volume of an appropriate intracellular region. We now describe the calcium influx and effluxes contained in (5.21) in more detail.

The influx I_{Ca} is 0 at times when there are no action potentials. Otherwise, it is proportional to the combined strength of two processes, an activation process A and an inactivation process B. We shall use the notation k_i, with increasing numbers i, for positive constants that are not otherwise identified. Hence

$$I_{Ca} = k_1 A B$$ (5.22)

The values of A and B in (5.22) are in turn functions of the time t since the start of the action potential, specifically

$$A = 1 - \exp\left(-\frac{t}{T_1}\right)$$ (5.23)

$$B = C \exp\left(-\frac{t}{T_2}\right)$$ (5.24)

where T_1 and T_2 are *time constants* (*i.e.*, decay rates) and C is a constant between 0 and 1. During periods between action potentials, C is reset to the value $1 - (1-B') \exp(-t_1/T_3)$, where t_1 is the time since the end of the last action potential and B' is the value of B at the end of the last action potential. From some empirical data on cell biochemistry, described in Gingrich and Byrne (1985), the effluxes were assumed to be concentration-dependent functions of the forms $I_U = k_2 / [1 + (k_3 / C_{Ca}^2)]$ and $I_D = k_4 C_{Ca}$.

The synaptic vesicle concentrations increase the postsynaptic cell depolarization accompanying activation, which is called the *excitatory postsynaptic potential* (EPSP). The EPSP is the final "output" of the system. The rate of change of EPSP is described by

$$\frac{d(EPSP)}{dt} = \frac{I}{R} = k_5 I_{Ca} V_R C_R \tag{5.25}$$

where I_{Ca} and C_R are as defined above, and V_R is the volume of the releasable pool of vesicles. The concentrations C_R and C_F of releasable and nonreleasable vesicles in turn satisfy the differential equations

$$\frac{dC_F}{dt} = \frac{1}{V_F}[I_N - I_F - I_S - I_{VD}] \tag{5.26}$$

$$\frac{dC_R}{dt} = \frac{1}{V_R}[I_F + I_S + I_{VD} - I_R] \tag{5.27}$$

where V_F and V_R are both volumes, and the various fluxes (I terms) depend either on calcium concentration or vesicle concentration. The flux equations are as follows, with C+ denoting the steady-state value of C_F:

$$I_{VD} = k_6(C_F - C_R) \tag{5.28}$$

$$I_F = \left(\frac{C_F}{C+}\right)\left[\frac{k_7}{\left(1 + \frac{k_8}{C_{Ca}{}^{nl}}\right)}\right] \tag{5.29}$$

for some positive number n1;

$$I_S = \frac{C_F}{C+} PVM \tag{5.30}$$

where PVM (short for "potential for vesicular mobilization") satisfies another differential equation of the form

$$\frac{d(PVM)}{dt} = \frac{1}{T_S}\left[\frac{k_9}{\left(1 + \frac{k_{10}}{C_{Ca}{}^{n2}}\right)} - PVM\right] \tag{5.31}$$

with T_S a time constant and n2 another positive constant; and finally

$$I_N = \frac{k_{11}}{\left[1 + \dfrac{k_{12}}{(C^+ - C_F)^3}\right]} \tag{5.32}$$

Equations (5.21) - (5.32) deal with cellular interactions which do not involve associative learning. Such learning involves the postulated neuromodulator of Figure 5.15 and cyclic AMP. In the absence of the unconditioned stimulus (therefore of the neuromodulator), Gingrich and Byrne (1987) assume that cyclic AMP decays exponentially to a baseline level, hence

$$\frac{dC_{cAMP}}{dt} = -\frac{C_{cAMP}}{T_3} \tag{5.33}$$

In the presence of the US, a linear calcium-dependent term is added to (5.33), hence

$$\frac{dC_{cAMP}}{dt} = -\left(\frac{C_{cAMP}}{T_3}\right)Z + k_{13} + k_{14}C_{Ca} \tag{5.34}.$$

cAMP in turn indirectly affects the action potential variables involved in Equations (5.21) - (5.32) by lengthening the action potential (spike) duration:

$$\textit{Spike duration} = k_{15} + k_{16}C_{cAMP} \tag{5.35}.$$

C_{cAMP} in turn stimulates the transfer of vesicles from the storage to the releasable pool. Mathematically, this is represented by means of a value $F_{cAMP} = k_{16} C_{cAMP}$ which is added to the bracketed term in Equation (5.27) and subtracted from the bracketed term in (5.26).

EXERCISES FOR CHAPTER 5

○1. How might the Rescorla-Wagner model be extended to account for the phenomenon of unblocking, whereby if $CS_1 + CS_2$ predicts a different level of US stimulation than does CS_1 alone, then CS_2 conditions normally to the US? Does the Sutton-Barto model explain that phenomenon better?

2.   The following set of exercises relates to simulation of conditioning phenomena with the drive-reinforcement neuronal model, using Equations (5.4) and (5.5) with the parameters used in Klopf (1988). These parameters are:

Learning rate constants:
      $c_0=0$, $c_1=5.0$, $c_2=3.0$, $c_3=1.5$, $c_4=0.75$, $c_5=0.25$ ($\tau=5$);
CS initial synaptic weight values (*i.e.*, $w_i(t)$ at $t=0$):
      +0.1 for excitatory weights, -0.1 for inhibitory weights, unless otherwise stated;
US synaptic weight values (non-modifiable):
      +1.0 for excitatory, 0.0 for inhibitory weights.
Lower bound on weights:
      $|w_i(t)| \geq 0.1$.
Neuronal output limits:
      $0.0 \leq y(t) \leq 1.0$.
Threshold:
      $\theta = 0.0$.
Amplitude of CS's:
      0.2, unless otherwise stated.
Amplitude of US:
      0.5, unless otherwise stated.
(Both CS's and US are called $x_i(t)$ for different i; see Figure 5.7. Do not forget to set any Δx_i to 0 if it becomes negative!) CS and US timing varies between simulation experiments.

(a) Simulate the acquisition of a conditioned response to a single CS. Let each trial (period in which both CS and US are presented) be 15 time units long; let the CS be on from time units 10 to 13 inclusive within each trial, and the US on only at time unit 14, within each trial, as shown in Figure 5.18.
      Simulate a sequence of 100 trials, and graph the CS-US synaptic weight $w_i(t)$ at the end of each trial. Notice that it has a sigmoid shape (initially positively accelerated, then negatively accelerated), as in Figure 3.6. Simulations of the earlier Rescorla-Wagner and Barto-Sutton models produce learning curves that are only negatively accelerated. Can you suggest a reason for this difference between the results of the models?

(b) Study the effects of varying both CS and US amplitudes in (a). (Klopf varied each amplitude between 1.0, 0.5, and 0.25, but you need not be bound by those values.)

(c) Simulate conditioned inhibition and extinction of conditioned inhibition, as follows:

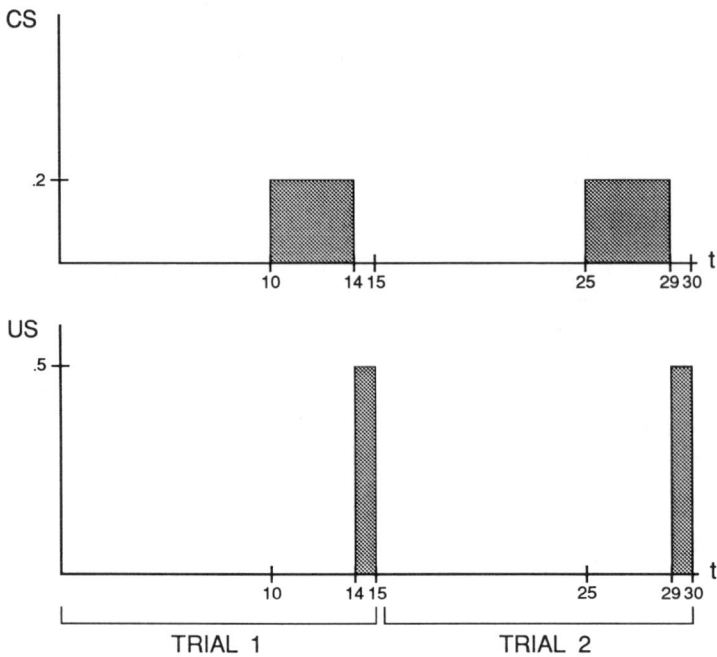

Figure 5.18. Time course of the stimuli for the drive-reinforcement network simulation of Exercise 2(a).

Trial lengths are now 25 time units instead of 15. On trials 1-70, present CS_1 at time units 10 to 12 of each trial, followed by the US at time units 13 to 15. On trials 71-200, do the same as on the earlier trials, but add presentation of both CS_1 and CS_2 at time units 20 to 22 of each trial. On trials 201-300, do the same as on trials 71-200 but leave off the US presentation. You should obtain the following results: for CS_1, the excitatory weight should increase up to trial 70, then decrease slightly and increase again up to trial 200, then decrease exponentially up to trial 300. For CS_2, the inhibitory weight should increase during trials 70 to 200, then remain the same through trial 300. In other words, the conditioned excitor CS_1 should extinguish, but not the conditioned inhibitor CS_2.

3. (a) Run a simulation of the Sutton-Barto equations (3.23) where the interstimulus interval is varied. Let the number of trials be 40 and each trial be 50 time units long. Let the CS duration be 3 time steps, the US duration 30 time steps. Use the parameters $\alpha = 0$, $c = .2$, $\beta = .9$, and the initial US associative strength $w_0 = .6$. Plot the asymptotic value of the CS associative strength as a function of interstimulus interval. Let CS and US both have amplitude 1.

(b) Run a simulation of a blocking paradigm using the Sutton-Barto equations and two conditioned stimuli, CS_1 and CS_2. Let CS durations be 5 time steps and US duration 10 time steps. For the first 10 trials, CS alone is presented and followed by US, starting immediately after CS termination. For the next 10 trials, CS_1 and CS_2 are presented simultaneously, followed immediately by the US. Use the parameters $\beta = 0$, $c = .5$, $\alpha = .6$, $w_0 = .6$.

*4. Simulate excitatory conditioning and extinction in the READ circuit, with the same parameters used by Grossberg and Schmajuk (1987), shown in Equations (5.9)-(5.20), with just a single CS. On each trial, present the CS for a duration of 40 time units, of which the US is also present for the last 15. Let the total trial length be 200 time units. The CS is paired with the US for 10 trials, and then presented in the absence of the US for the next 10 trials. The parameters in the differential equations (using the terminology of this book, not of Grossberg and Schmajuk) a=1, b=.005, c=.00125, e=20, h=20, k=20, m=.5, p=20, q=.005, r=.025. The strength of tonic arousal is 10, while the CS and US have amplitudes 1 and 200 respectively. For the functions $f(x)$ and $g(x)$, use .05x and x respectively if x is positive, and 0 if x is negative. Hint: this has tended to work better in simulations if on each time step, the equations for node activities are solved in the steady state, that is, with time derivatives equal to 0. Then the equations for connection strengths are integrated, using a differential equation solving routine, with the new values for node activities substituted in.

○5. Read in the psychological literature (see, *e.g.*, Parasuraman & Davies, 1984 or Rabbitt & Dornic, 1975) on some other attentional phenomenon not related to classical conditioning, and suggest a neural network model for the other type of attention based on competition, associative learning, opponent processing, and/or other principles discussed so far in this book.

○6. Read the article of Buonomano, Baxter, and Byrne (1990), in which the work of Gingrich and Byrne (1985, 1987) was extended to multineuronal networks in order to model classical conditioning. These modelers stated that "in our simulations, the threshold of the facilitatory neuron proved to be a critical parameter. With low thresholds, the network simulated second-order conditioning, but not blocking. Conversely, with a high threshold, the network simulated a small degree of partial blocking, but not second-order conditioning." Comment on why this is likely to be so, and how these authors solved this dilemma by the introduction of lateral inhibition. Compare the heuristics of their model with those of Grossberg and Levine (1987).

6

Coding and Categorization

The Nameless is the origin of heaven and earth; the Named is the mother of all things.

Lao Tzu

I hate definitions.

Benjamin Disraeli (*Vivian Gray*)

No two sensory events are exactly the same. We never see the same moon twice; even our wife's or husband's face exhibits subtle changes in expression from one viewing to another. Yet most of us manage to ascribe to the sensory world a fair degree of regularity. How do we decide when two sensory objects are similar enough to be listed in the same category and when they are not?

Of course, the rules for "similarity" vary tremendously with context. An apple and a radish are listed together if foods are classified by color, but not if foods are classified by taste (sweet, bitter, sour, or salt). How to model the capacity for multiple categorizations, in which the context determines which categorization is used, is a problem at the forefront of contemporary neural network theory (see the discussions of decision making and knowledge representation in Chapter 7). But much progress has been made in the last several years on the simpler problem of constructing neural networks that learn to encode a single categorization of sensory patterns, regardless of context.

Some neural network categorization models involve *supervised* learning; that is, certain output nodes are trained to respond to certain "exemplar" patterns, and the changes in connection weights due to learning cause those same nodes to respond to more general classes of patterns. Other models involve *unsupervised* learning; that is, input patterns are presented in some sequence and the network discovers through self-organization a "natural" categorization of the sensory world. The distinction between supervised and unsupervised has been made in the engineering literature for many years; see Duda and Hart (1973). Sometimes, in engineering applications, supervised categorization is called *classification*, whereas unsupervised categorization is called *clustering*. This distinction is not rigid because some "unsupervised" networks are actually supervised by an *internal* error signal (for discussion, see Dawes, 1991), but is a useful means of classifying models. Table 6.1 lists some of the best known neural networks for supervised and unsupervised categorization.

Neural network categorization algorithms have been quite diverse, but many of them (for example, Carpenter and Grossberg, 1987 a, b; Rumelhart and Zipser, 1985; Rumelhart *et al.*, 1986; Edelman, 1987) have some general points in common. All these networks include modifiable connections between one layer of nodes encoding features of the sensory environment, and another layer of nodes encoding categories of sensory patterns composed of those features (*cf.* the "generic" Figure 1.1 from Chapter 1). Typically, category nodes are initially random or neutral in their responses, but learn specific pattern categories by experience.

To prepare the way for understanding neural categorization, we first consider the slightly simpler issue of how a node in a neural network can learn to respond to particular patterns of activity at other groups of nodes. These patterns of activity, in turn, could represent combinations of sensory features. The next section deals with "coding" in that sense, not in the sense of how the primary representation of a sensory stimulus is actually formed in the nervous system. Network mechanisms for such higher-level coding, which are also discussed in Section 8.2 of Levine (1983b), have possible implications for biological organisms during development.

Supervised classification

Name of model	Article where model is described
Back propagation (BP)	Werbos (1974); LeCun (1985); Parker (1985); Rumelhart *et al.* (1986); McClelland and Rumelhart (1988)
Brain-state-in-a-box (BSB) with delta-rule error correction	Knapp and Anderson (1984); Anderson and Murphy (1986)
Reilly-Cooper-Elbaum and restricted coulomb energy (RCE)	Reilly, Cooper, and Elbaum (1982); Collins, Ghosh and Scofield (1988)

Unsupervised classification

Name of model	Articles where model is described
BSB without error correction	Anderson *et al.* (1977); Anderson and Mozer (1981)
Adaptive resonance for binary patterns (ART 1)	Grossberg (1976c); Carpenter and Grossberg (1987a)
Adaptive resonance for analog patterns (ART 2)	Carpenter and Grossberg (1987b, 1989)
Cognitron	Fukushima (1975)
Neocognitron	Fukushima (1980); Fukushima and Miyake (1982)
Darwin II	Reeke and Edelman (1984, 1987); Finkel and Edelman (1985); Edelman (1987)
Competitive learning (CL)	Rumelhart and Zipser (1985)

Table 6.1. Some of the more important neural network classification models to date.

6.1. INTERACTIONS BETWEEN SHORT- AND LONG-TERM MEMORY IN CODE DEVELOPMENT

Malsburg's Model with Synaptic Conservation

The study of neural networks for code development essentially began with a seminal article by Malsburg (1973) on the development and tuning of orientation-sensitive cells in the visual cortex. This model is discussed in some detail because its basic structure anticipates many of the more current models of coding and categorization.

Malsburg's model is based on unmodifiable recurrent excitation and inhibition between "cortical" nodes, combined with modifiable synapses to the "cortex" from an input ("retinal") layer of nodes. He was motivated to develop this model by a body of experimental results on the mammalian visual system. These results suggested that "The task of the cortex for the processing of visual information is different from that of the peripheral optical system. Whereas eye, retina and lateral geniculate body (LGB) transform the images in a 'photographic' way, *i.e.* preserving essentially the spatial arrangement of the retinal image, the cortex transforms this geometry into a space of concepts" (Malsburg, 1973, p. 85). (The lateral geniculate body is an area of the thalamus (*cf.* Appendix 1) through which most signals from the retina pass on the way to the visual cortex.)

In particular, Malsburg's model, and subsequent ones discussed in this chapter, drew their inspiration from physiological results on single-cell responses to line orientations. These models can explain findings that neurons in the cat or monkey visual cortex respond preferentially to lines of a particular orientation, and that cells responding to similar orientations are grouped close together anatomically, in columns (Hubel & Wiesel, 1962, 1963, 1965, 1968). These models also explain findings that preferred orientations of neurons are influenced by early visual experience (*e.g.*, Blakemore & Cooper, 1970; Hirsch & Spinelli, 1970). (More recent results on the various influences on orientation specificity, including the interplay between genetic and developmental factors, are summarized in Malsburg & Cowan, 1982 and Ferster & Koch, 1987).

Some models (*e.g.*, Perez, Glass, & Shlaer, 1974; Grossberg, 1976a, p. 152; Grossberg, 1976b, p. 131; Bienenstock et al, 1982, p. 1) also address evidence that there is a *critical period* in development of orientation detectors. That is, for a short period of time (in cats, age 23 days to four months), cortical orientation tuning is much more modifiable than it is either earlier or later.

Malsburg's simulated cortex is organized into two separate populations, excitatory and inhibitory nodes (he refers to them as "cells"). That is, Malsburg's model is *unlumped* in the terminology of Section 4.2. The variation of connection strengths with distance endows the simulated cortex with a crude form of the lateral inhibitory architecture shown in Figure 4.4: narrow-range excitation and broad-range inhibition. Malsburg's laws for lateral interaction between nodes are *additive* rather than *shunting* in the terminology of Sections 4.1 and 4.2.

The arrangement of excitatory-to-excitatory connections p_{ik}, excitatory-to-inhibitory connections r_{ik}, and inhibitory-to-excitatory connections q_{ik} in Malsburg's simulated cortex is shown in Figure 6.1. As that figure shows, excitatory and inhibitory nodes are organized into two parallel planes, each with a hexagonal arrangement of nodes. Excitatory nodes excite neighboring nodes, both excitatory and inhibitory ones, whereas inhibitory nodes inhibit excitatory nodes that are a distance of two away. The signal transmitted by each node is equal to the amount of the signaling node's activity that is above some threshold value. Equations relating all these variables, and the connection weights s_{ik} from simulated retinal afferents, are given in Section 6.5 below.

Of the connections in Malsburg's model, only the connections from retinal afferents to cortical nodes have modifiable weights. The rule for changing these weights combines an associative learning law with a synaptic conservation rule similar to the gamma-system learning law of Rosenblatt (1962) (*cf.* Section 2.1). Synaptic conservation was imposed to prevent the unbounded growth of synaptic strengths that would otherwise result from associative learning. This combination of laws is described as follows (Malsburg, 1973, p. 88):

> if there is a coincidence of activity in an afferent fibre i and a cortical E-cell k, then s_{ik}, the strength of connection between the two is increased to $s_{ik} + \Delta s$, Δs being proportional to the signal on the afferent fibre i and to the output signal of the E-cell k. Then all the s_{jk} leading to the same cortical cell k are renormalized to keep the sum $\sum_{j} s_{jk}$ constant.

Figure 6.2 shows the standard set of stimuli used on Malsburg's model retina. These stimuli correspond to bars of light at different orientations. As shown in Figure 6.3, orientation detectors, such as were found by Hubel and Wiesel (1962, 1963, 1968), develop spontaneously among Malsburg's simulated cortical cells. After 100 learning steps, the lateral excitatory and inhibitory interactions lead to self-organization of cortical nodes, whereby most nodes have preferred orientations and nodes of similar preferred orientations tend to be grouped together. Figure 6.4 shows a similar grouping of biological

be grouped together.   Figure 6.4 shows a similar grouping of biological
orientation detectors found by Hubel and Wiesel (1968) in the monkey striate,
or primary visual, cortex.

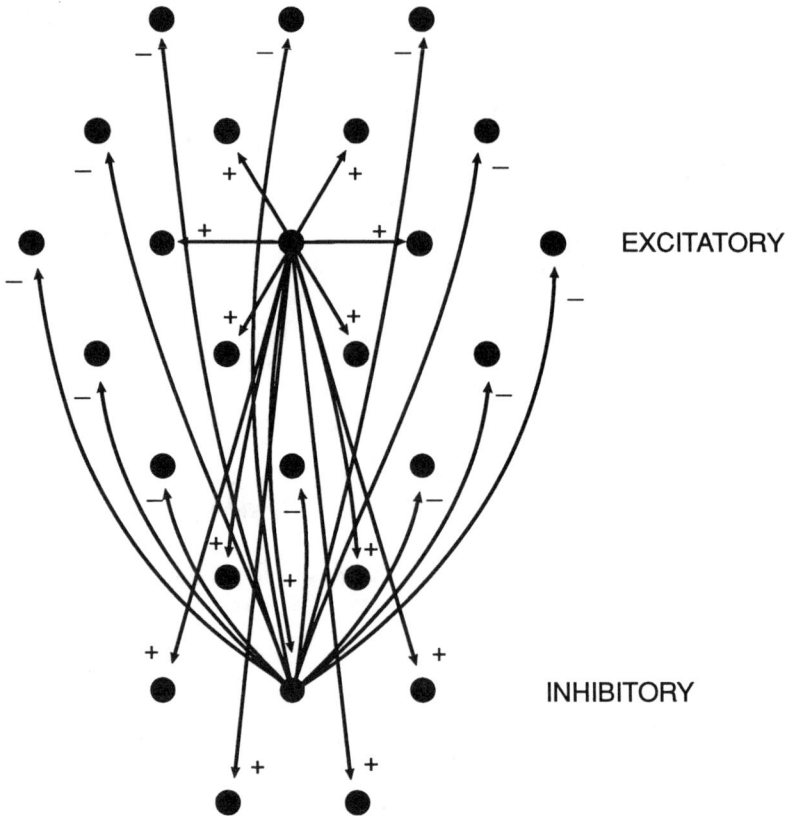

Figure 6.1. A small part of the simulated cortex of Malsburg
(1973), showing the arrangement of connections between
excitatory and inhibitory nodes.

The idea of synaptic conservation is intuitively based on the notion that
some chemical substance, whether a transmitter or second messenger (*cf.*
Section 3.1), is present in a fixed amount at postsynaptic sites and distributed

in variable fashion across impinging synapses. This mechanism is necessary for the effects in Malsburg (1973), and in two related models of the visual cortex by Perez *et al.* (1974) and Wilson (1975). Some more recent categorization models (*e.g.*, Rumelhart & Zipser, 1985; Carpenter & Grossberg, 1987a) also use learning laws whereby strengthening of some synapses weakens other synapses. Such laws are reminiscent of the learning scheme of Rescorla and Wagner (1972) (see Chapters 3 and 5), which includes an upper bound on the total associative strength of all stimuli with a given reinforcer.

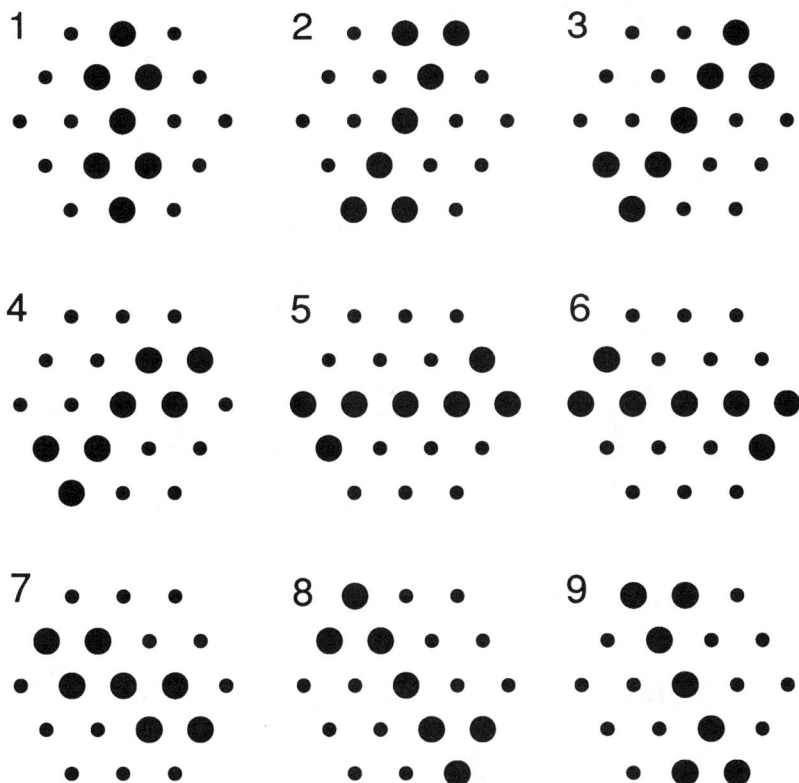

Figure 6.2. Standard set of stimuli used on the simulated retina. Larger dots denote locations of activated nodes. (Reprinted from Malsburg, 1973, with permission of Springer-Verlag.)

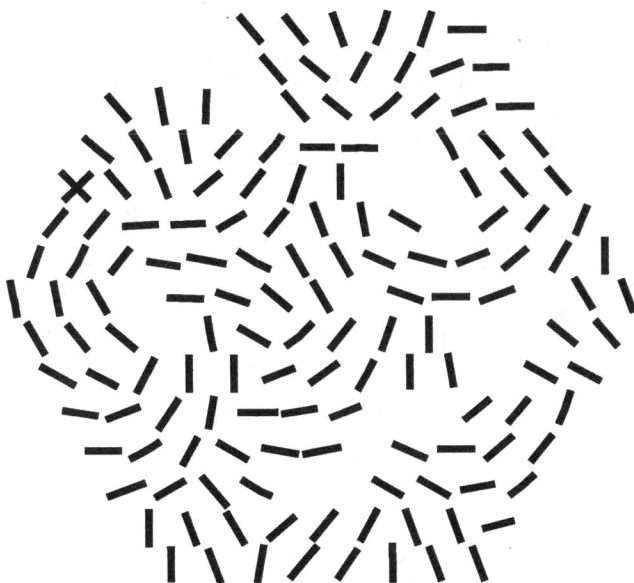

Figure 6.3. Simulated cortex after 100 time steps of learning. Each bar indicates the orientation to which the excitatory node at that location is most responsive. Blank spaces represent locations of nodes that never learn to react to any of the standard stimuli. (Adapted from Malsburg, 1973, with permission of Springer-Verlag.)

Grossberg's Model with Pattern Normalization

Grossberg (1976a, b) developed a model that has many principles in common with Malsburg's but does not use a synaptic conservation law for learning. He argued that such a conservation law is incompatible with classical conditioning. Moreover, while Malsburg used this law in order to keep synaptic strengths, and therefore total network activity, bounded, it is also possible to achieve boundedness by replacing additive lateral interactions with shunting interactions. These arguments will now be reviewed.

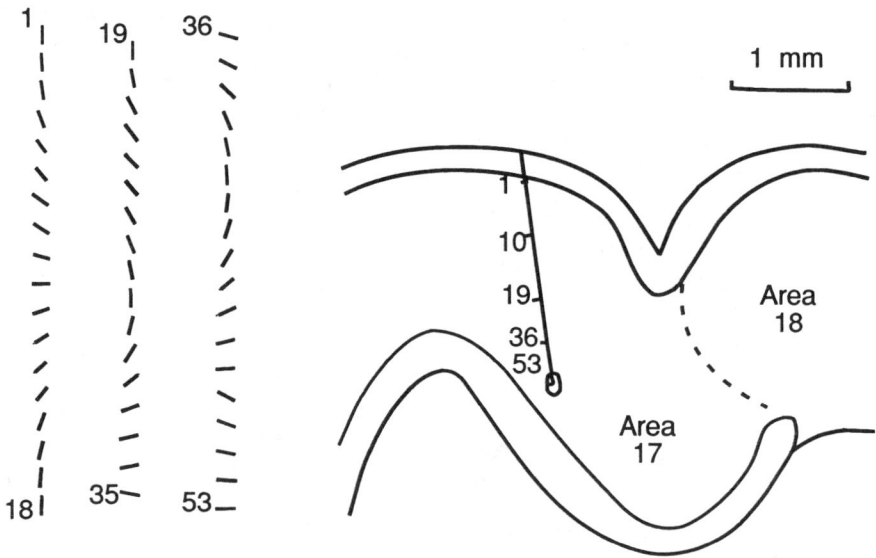

Figure 6.4. Reconstruction of an electrode penetration through an area of a monkey's primary visual cortex. Lines to the left indicate preferred orientations of the cells traversed. Areas numbered 17 and 18 are standard terms for subdivisions of the cortex. (Adapted from Hubel and Wiesel, 1968, with permission of Cambridge University Press.)

The argument that synaptic conservation is incompatible with conditioning was given in Grossberg (1976a, p. 149). Suppose a sensory cue S_1 elicits a response pattern R, and then another cue S_2 is paired with S_1 (see Figure 6.5). The pairing leads to a strengthening of the connection from the S_2 node to those nodes whose activation is strongest in the R response. But if, as in Malsburg's model, total strength of synapses impinging on a given node from the previous level is kept constant, strengthening of the connection S_2-to-R would force weakening of the connection S_1-to-R. In reality, secondary conditioning occurs (see Section 5.2), so that S_1 and S_2 can simultaneously be strongly connected to R. The argument that synaptic conservation is unnecessary for boundedness was based on the mathematical theory of shunting on-center off-surround networks (Grossberg, 1973; Ellias and Grossberg, 1975; Grossberg and Levine, 1975; see Section 4.2). This argument uses the nonrecurrent shunting equations (4.2): (recall the discussion of recurrent versus nonrecurrent inhibition in Section 4.1). If x_i is the activity of the ith node which receives input I_i, as shown in Figure 4.10, the equation for x_i is

$$\frac{dx_i}{dt} = -Ax_i + (B-x_i)I_i - x_i\sum_{k \neq i} I_k \qquad (4.2).$$

(See also Exercise 1 of Chapter 4). Suppose the inputs I_i have fixed relative sizes $I_i(t) = \theta_i I(t)$, as in outstar spatial patterns (see Section 3.2). Then the steady state value of x_i is obtained by setting $dx_i/dt = 0$ in (4.2), yielding

$$x_i = \theta_i\left(\frac{BI}{A+I}\right) \qquad (6.1).$$

Since the relative pattern weights θ_i add up to 1, (6.1) says that the sum of all the steady state activities equals $BI/(A+I)$, which is no larger than B regardless of the total input intensity I and the number of nodes in the network. Hence, total network activity never exceeds B.

Grossberg's argument is based on the factorization of spatial patterns into a product of relative pattern weights θ_i and total intensity I. This concept, sometimes called *factorization of pattern and energy*, was previously discussed in relation to outstar networks in Section 3.2. Factorization of pattern and energy will also play a role in models of motor control and of knowledge representation, to be discussed in Chapter 7.

A model for development and tuning of feature detectors, combining lateral inhibition for short-term memory with associative synaptic modification for long-term memory, is discussed in Grossberg (1976b). Figure 6.6 shows the minimal network of that article. This network, like that of Malsburg (1973), includes unidirectional modifiable synapses from an input layer F_1 to a "cortical" layer F_2, leading to coding of input patterns by cortical nodes. Grossberg (1976c) extended this model to include modifiable feedback from F_2 to F_1. For such "top-down" feedback, he coined the term *adaptive resonance*; this work ultimately led to the well-known adaptive resonance theory (ART) of Carpenter and Grossberg (1987 a, b).

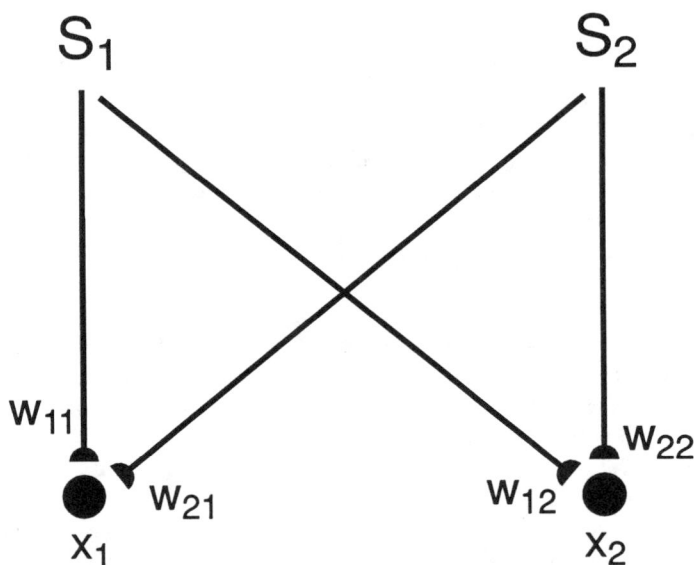

$$S_1 \qquad\qquad S_2$$

$$w_{11} \qquad w_{21} \qquad w_{12} \qquad w_{22}$$

$$x_1 \qquad\qquad x_2$$

$$w_{11} + w_{21} = w_{12} + w_{22}$$

$$w_{11} \gg w_{12}$$

$$w_{21} \ll w_{22}$$

Figure 6.5. Argument against conservation of synaptic strength. If w_{11} is much larger than w_{12}, the network shown here learns to respond to stimulus S_1 with a response pattern R, whereby x_1 is much more activated than x_2. Since the two sums are equal, w_{21} is then less than w_{22}, so the network cannot also learn to perform R in response to S_2. If S_2 is paired with S_1, as in secondary conditioning, this is contradictory. (Adapted from Grossberg, 1976a, with permission of Springer-Verlag.)

In the network of Figure 6.6, the input-receiving nodes x_{1i} are endowed with a nonrecurrent (feedforward) on-center off-surround anatomy, and the pattern-coding nodes x_{2i} with a recurrent on-center off-surround anatomy (*cf.* Figure 4.3). F_1 and F_2 represent successive layers in a hierarchical network. Grossberg suggested that variations on the same hierarchy could be repeated in different brain regions. In Malsburg (1973) and Perez et al (1974), F_1 wasinterpreted as either retina or thalamus, and F_2 as visual cortex (see Appendix 1). But F_1 might also be identified with a composite of early processing areas in the retina (*receptors, horizontal cells,* and *bipolar cells*) and F_2 with retinal areas closer to the optic nerve (*amacrine* and *ganglion* cells) (Grossberg, 1976b). Also, since the visual cortex itself contains several processing stages, identified with cell groups known as *simple, complex,* and *hypercomplex* cells (Hubel and Wiesel, 1962, 1963, 1965, 1968), F_1 and F_2 might be interpreted as different parts of cortex. Based on these cortical cell groups, Grossberg (1976 b, c) proposed more complex architectures, adding to Figure 6.6 another layer F_3 whose nodes code activity patterns across F_2. Nor are these architectures restricted to vision: Grossberg (1976b) described yet another interpretation, whereby F_1 is the olfactory bulb and F_2 is olfactory cortex.

In Grossberg (1976b), the input signals I_{2j} to the cortical nodes x_{2j} are linear combinations of the activities at the retinal nodes x_{1i}, weighted by the strengths of retinocortical synapses. The patterns viewed at the retina are assumed to be normalized so that their values θ_i add up to 1. (Recall the discussion in Section 4.1 of pattern normalization in networks with lateral inhibition.) For simplification, it is assumed that the x_{1i} activities represent the input pattern to the retina. Hence, the total signal at time t to a given cortical node x_{2j} due to the retinal pattern $\vec{\theta} = (\theta_1, \theta_2, ..., \theta_n)$ is

$$S_j(t) = \sum_{k=1}^{n} \theta_k w_{kj}(t) \tag{6.2}$$

where $w_{kj}(t)$ denotes the strength of the synapse from retinal node k to cortical node j. The linear combination in (6.2) also is incorporated in Equation (6.5) below, due to Malsburg (1973).

Grossberg (1976b) discussed several possible long-term memory laws for the synaptic strengths w_{kj} themselves. In one of these laws (shown in Equation (6.9) below), weights of connections to a node change only when short-term memory at that node is active.

Mathematical Results of Grossberg and Amari

Recall from Section 4.2 that a recurrent competitive network, such as F_2 in Figure 6.6, can either have exactly one or more than one node with positive asymptotic activity as time increases. Hence sometimes, but not always, competition is *winner-take-all*, that is, exactly one node at the competitive level gets its incoming signals stored in short-term memory (STM). From the viewpoint of categorization, the winner-take-all case is especially interesting. In particular, the "winning" node in the competition for short-term storage could be the one whose incoming signal S_j, as defined by (6.2), is the largest.

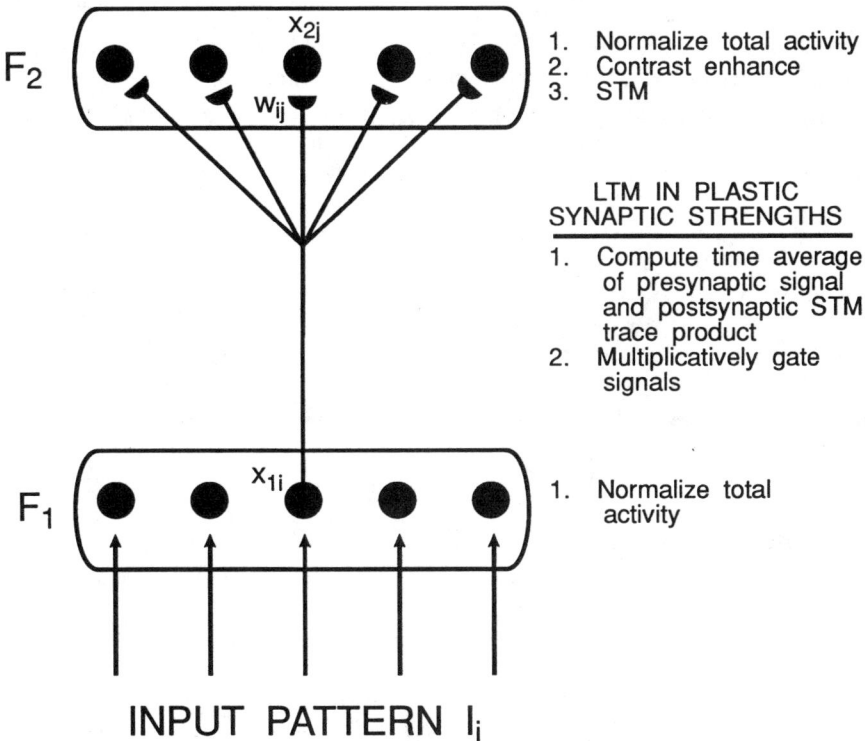

Figure 6.6. Minimal model of development and tuning of feature detectors using short-term memory (STM) and long-term memory (LTM) mechanisms. (Adapted from Grossberg, 1976b, with permission of Springer-Verlag.)

In Grossberg (1976b), the criterion of choosing the F_2 node with the largest incoming signal leads to a primitive scheme for categorizing patterns, each pattern defined as a vector of retinal node activities. This largest-linear-signal criterion is also the basis for categorization in Amari and Takeuchi (1978).

Grossberg (1976b) discovered, however, that such a categorization algorithm can lead to miscodings if the network is subjected to many different spatial patterns over time. For suppose that cortical node x_{21} is the most active in response to a particular retinal pattern, say $\vec{\theta}^1$. Then the vector $\vec{w}^1$ of synaptic weights from all the retinal nodes to that cortical node becomes closer to $\vec{\theta}^1$ as time increases, as is shown by Equation (6.9) below. But, as Figure 6.7 shows, bringing these weights closer to $\vec{\theta}^1$ could also bring them closer to a different spatial pattern $\vec{\theta}^2$. With enough repetition of this process, the node x_{21} can become unable to recognize the original pattern $\vec{\theta}^1$, forcing this pattern to be recoded by a different cortical node.

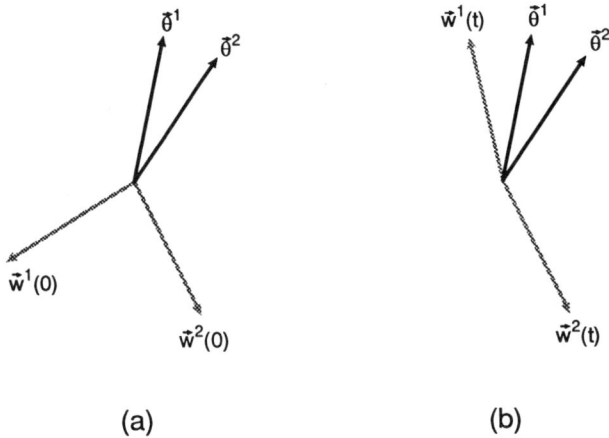

(a)   (b)

Figure 6.7. (a) Vectors of weights from F_1 to two different F_2 nodes at time 0, as compared with spatial patterns $\vec{\theta}^1$ and $\vec{\theta}^2$. (b) As node x_{21} of F_2 learns $\vec{\theta}^1$, its bottom-up weight vector gets closer to both $\vec{\theta}^1$ and $\vec{\theta}^2$ than the vector $\vec{w}^2$ of bottom-up weights to node x_{22}. (Adapted from Grossberg, 1976b, with permission of Springer-Verlag.)

To prevent presentation of one pattern from recoding other patterns, Grossberg (1976b) proposed adding to the network of Figure 6.6 some feedback connections from F_2 to F_1. If the node x_{21} perceives that $\vec{\theta}^2$ is sufficiently different from $\vec{\theta}^1$, this feedback can be used to suppress F_1-to-F_2 signals that would otherwise be generated by $\vec{\theta}^2$. This mismatch-detecting mechanism is at the heart of adaptive resonance theory, and is therefore discussed more fully in Section 6.3 below.

Grossberg (1976b, c) showed that every solution of the equations for the network of Figure 6.6 approaches some equilibrium point corresponding to a learned category. Amari (1977a, 1977c, 1980) proved an analogous result for a coding network similar to Grossberg's but with additive rather than shunting interactions. Amari and Takeuchi (1978) extended this result to a coding system which includes both excitatory and inhibitory modifiable interlevel synapses.

Feature Detection Models with Stochastic Elements

Bienenstock, Cooper, and Munro (1982) constructed a model of the development of orientation detectors in the visual cortex. Their model has much in common with Grossberg's but also includes stochastic elements. Bienenstock *et al.* added nonlinear interactions to the previous linear probabilistic model of Nass and Cooper (1975). (Recall from Section 3.2 that Nass and Cooper's model was in turn based on addition of decay terms to an early model of Anderson, 1973.) In addition to orientation preferences, units in the network of Bienenstock *et al.* can exhibit preferences for one or another eye; such ocularity preferences are influenced by the opening or closing of either eye during development (*cf.* Table 4.1).

In the model of Bienenstock *et al.* (1982), retinocortical connections are subject to a learning rule that includes the possibility of both synaptic increase (in the manner of Hebb, 1949) and decrease. If I_j is the jth retinal input, and w_j the strength of the synapse to a given cortical node with activity x, then the expression for the rate of change of w_j is of the form

$$- \epsilon w_j + \Phi(x(t))I_j(t) \qquad (6.3)$$

where the function Φ can either be positive or negative. In fact, Φ is positive for x above a certain threshold value, and negative for x below that threshold. That is, in current terminology, the learning law alternates between Hebbian with decay and *anti-Hebbian* with decay. (A more recent version of this type of learning law appears in Bear *et al.*, 1987.) This law embodies, in the learning equation itself, a form of contrast enhancement of significant inputs, or suppression of random noise, such as is often obtained with lateral inhibition (*cf.* Chapter 4). The network also, however, includes some actual lateral inhibition.

The inputs I_j of (6.3) are assumed to be random, with a probability distribution reflecting the distribution of patterns in a given sensory environment. Through computer simulation and mathematical analysis, the connection weights w_j were shown to converge to a steady state that is *selective* with respect to the distribution of the random inputs. That is, the response of the given cortical neuron, which equals the sum $\sum_j w_j(t)I_j(t)$, reaches its maximum possible value over a relatively small subset of the set of possible inputs. Bienenstock et al suggested that this kind of selectivity is analogous to the orientation selectivity of actual cortical neurons.

A series of articles by Linsker (1986a, b, c) carried the idea of multilayer structure further with a network of three (or more) layers with adaptable connection strengths, as shown in Figure 6.8. These layers were intended to be analogs of both the retina and the visual cortex, without mimicking the detailed anatomy of either area. From random inputs at the lowest layer, combined with Hebb-like learning at interlayer connections, emerge at higher layers some *spatial-opponent cells* (that is, nodes with on-center off-surround or off-center on-surround responses, like those of the retinal ganglion cells shown in Figure 4.2). At still higher layers, nodes emerge that are responsive to given orientations. Finally, the orientation nodes organize themselves into columns in much the same manner as occurs in the models of Malsburg, Grossberg, and Bienenstock *et al.*

Linsker's model differs from the other models discussed in this section in that prescribed orientations are *not* part of the pattern of its inputs. Rather, the orientation specificity emerges from mathematical properties of the Hebb-like learning laws at interlevel synapses. Possibly, Linsker's model could represent some aspects of visual pattern processing that are hard-wired, while the other models represent other aspects that are dependent on the organism's visual environment.

From Feature Coding to Categorization

The visual orientation detection networks discussed above perform a primitive form of pattern categorization. Hence, this type of network provides one of the bases for the more sophisticated categorizations (*e.g.*, "the shape of a dog," "the sound of a recorder") that we perform in daily life. As categorizations become more subtle, they depend heavily on the ability to notice common features in disparate, and sometimes noisy, inputs. For example, we classify both a cocker spaniel and a dachshund, with or without tail damage, as dogs.

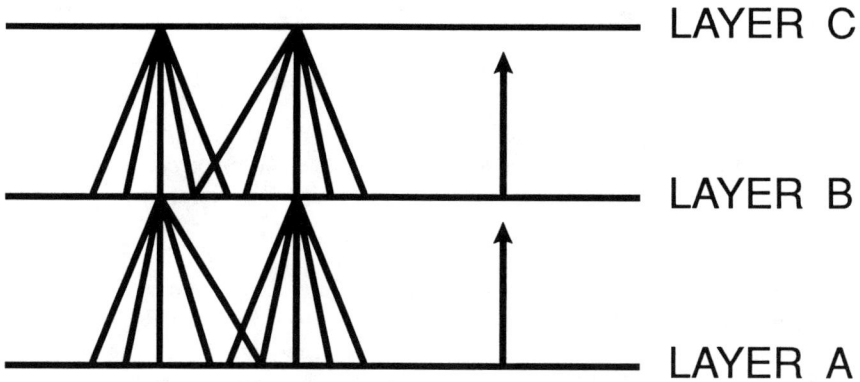

Figure 6.8. Three layers of a modular self-adaptive network. (Reprinted from Linsker, 1986a, with permission of the National Academy of Sciences.)

Recall from the start of this chapter the two general types of networks for pattern categorization or classification. These are networks with supervised and unsupervised learning. Both of these types of networks combine the coding of input patterns by internal layers of nodes with other functions. In supervised learning, the responses of an output layer to certain given patterns are compared with desired responses. Hence, in addition to a coding system, these networks typically require at some stage an error-correcting or delta learning rule (see Section 3.3). In unsupervised learning, the *input pattern* is compared with an internally generated prototype pattern. Hence, these networks typically require some architecture for measuring how similar a pattern is to previously encountered ones — in other words, for detecting familiarity or novelty. We now discuss these two types of networks in order.

6.2. SUPERVISED CLASSIFICATION MODELS

The use of neural networks for supervised learning of predetermined classifications dates back to the early work of Rosenblatt (*cf.* Chapter 2 of this book). His perceptron learning theorem (Rosenblatt, 1962, p. 596) states that:

> Given an α-perceptron, a stimulus world W, and any classification C(W) for which a solution exists; let all stimuli in W occur in any sequence, provided that each stimulus must reoccur in finite time; then beginning from an arbitrary initial state, an error correction procedure will always yield a solution to C(W) in finite time ... .

Since Rosenblatt, many other researchers have employed variants of the error correction or delta learning rule (for example, Widrow & Hoff, 1960; Widrow, 1962; Sutton & Barto, 1981; Barto & Anandan, 1985; LeCun, 1985; McClelland & Rumelhart, 1985; Parker, 1985; Anderson & Murphy, 1986). Rumelhart *et al.* (1986) showed how a multilayer network incorporating this rule can be applied to pattern classification.

While the perceptron, Widrow-Hoff, and back propagation rules are all variants of the error-correcting learning rule, there are important mathematical distinctions between them. A discussion of those distinctions, largely taken from Duda and Hart (1973), appears in Section 6.5.

The Back Propagation Network and its Variants

Recall the derivation in Chapter 3 of the back propagation learning algorithm, whereby the delta rule for changing connection weights to output units generalizes to a rule for changing weights to hidden units. Within such a feedforward scheme, this gradient descent method provides, in a sense, the most efficient weight changes for encoding the particular input-output mappings desired. (Indeed, the back propagation algorithm, before it was used for categorization, was employed in optimization problems of the type discussed in Section 7.1; see Werbos, 1974, 1988a.) At each iteration of this learning algorithm, the weights are set for a network of input, hidden, and output units with the generic architecture shown in Figure 6.9.

The types of nonlinear input-output relationships that can be learned by back propagation (BP) networks are essentially arbitrary. This is one of the main reasons for the broad appeal of such networks. Some relationships that

OUTPUT PATTERNS

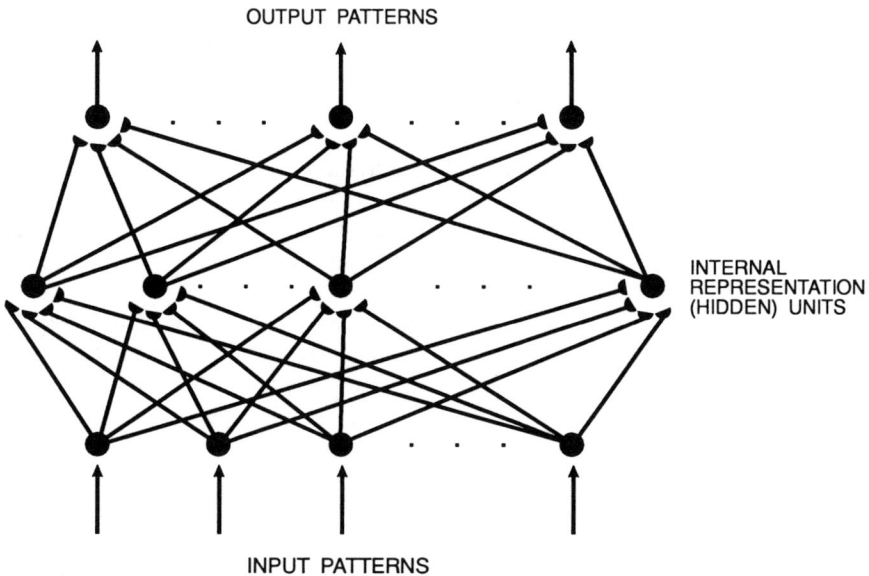

INTERNAL
REPRESENTATION
(HIDDEN) UNITS

INPUT PATTERNS

Figure 6.9. Generic architecture for a three-layer back propagation network. Error signals from output nodes, if their response to the input pattern is not the desired one, propagate backwards from hidden-to-output weights to input-to-hidden weights. In the process, the hidden units learn internal representations, that is, learn to encode certain classes of input patterns. (Adapted from Rumelhart *et al.*, 1986, with permission of MIT Press.)

can be taught to BP cannot be taught to networks without hidden units, because these relationships map dissimilar inputs into similar outputs or vice versa. Neither can they be taught to networks whose activation functions are all linear, in which case hidden units provide no advantage (*cf.* Exercise 1 of this chapter). The best known example of such a relationship is the *exclusive-or* (XOR) mapping of binary variables (see Table 6.2). (It is to be noted that XOR is also difficult for some nonlinear unsupervised categorization networks such as adaptive resonance networks; see Stork, 1989a).

The universality of the back propagation method has led to its use in a wide variety of applications. Perhaps the best known of its applications, or at least the first to be well known, is reading aloud. The NETtalk algorithm of Sejnowski and Rosenberg (1986) and Rosenberg and Sejnowski (1986) employed a BP network to associate written language to spoken sounds. Many researchers have also applied BP to character recognition. Early BP simulations often involved learning of a single letter discrimination, such as the discrimination between the letters "T" and "C" (regardless of rotation) that was taught to this network by Rumelhart *et al.* (1986) (see Exercise 3 of this chapter). More recently, this kind of simulation has been extended, using multiple output units, to the supervised learning of entire categorizations, such as teaching the network to discriminate between the ten possible digits in handprinted zip codes (*e.g.*, Kamangar & Cykana, 1988; Weidemann, Manry, & Yau, 1989; Gong & Manry, 1989).

But the back propagation method varies enormously in how many steps it requires to converge to the mapping it is supposed to learn. Typically, as shown in Figure 6.10, the convergence rate is strongly dependent on the number of hidden units, and that number must be decided separately for each application. Rumelhart (1988) discussed an algorithm for optimizing the number of hidden units, and criteria for selecting this number are also suggested by Theorem 3 of White (1987). Also, there is no universal guarantee of convergence, and the possibility exists that instead of converging to the desired (global minimum mean-square error) state, the system will be trapped in a local minimum of the error function (*cf.* Figure 4.13).

Input	Output
00	0
01	1
10	1
11	0

Table 6.2. The logical exclusive-or (XOR) relationship. Variables are assumed to be binary: "1" corresponds to "Yes" or "True," "0" to "No" or "False". Exclusive-or of two propositions is true whenever one of the propositions is true but *not both at once*.

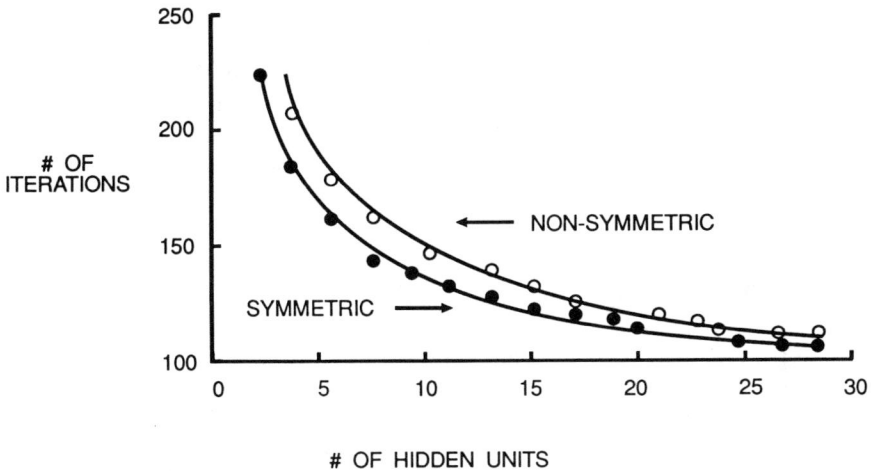

 # OF ITERATIONS

 # OF HIDDEN UNITS

Figure 6.10. Plot of number of iterations of the algorithm needed for convergence in the XOR problem as a function of number of hidden units, with two input units and one output unit. "Symmetric" means that node activities range from -.5 to .5, and "non-symmetric" means that they range from 0 to 1 as in typical back propagation networks. (Reprinted from Stornetta and Huberman, 1987, copyright © 1987 IEEE, by permission of the publisher.)

Conditions for convergence of the BP algorithm have been the object of much recent mathematical investigation. White (1987) placed the back propagation method in the context of other nonlinear regression methods. He gave some conditions that guarantee convergence to the desired global minimum, noting that his conditions (such as nonexistence of an alternative local minimum) are frequently not met in applied problems. Sethares (1988) suggested some slight modifications of the delta rule that could improve its convergence properties. Sontag and Sussmann (1989) proved that back propagation networks can at least learn all the types of discriminations that can be learned by the perceptrons of Rosenblatt (1962).

The setting of the learning rate is considered in Rumelhart *et al.* (1986) and McClelland and Rumelhart (1988, Ch. 6); the 1988 book chapter also discusses other BP implementation issues. There is a tradeoff in this setting: too small a learning rate can make convergence excessively slow, but too large a learning rate can cause oscillations that make convergence impossible. A method has been found for preventing such oscillations, thereby allowing a larger learning rate. This method involves adding to the weight change equations a *momentum* term which biases change in the same direction that the last previous change was made (see Section 6.5 for mathematical details). This reduces the likelihood of rapid oscillation between increases and decreases of the same weight.

Possible biological bases for the back propagation network have been suggested by many researchers. Grossberg (1987b) argued that BP is biologically unrealistic because the feedback in the network is not of electrical signals but of synaptic weights, and no mechanism is known for such weight transport in real nervous systems (see Figure 6.11). Stork (1989b), however, constructed a minimal collection of neurons and synapses implementing back propagation but not requiring actual weight transport (see Figure 6.12). His architecture is intricate and based on some unnatural assumptions, such as symmetry of connection weights (*cf.* Section 4.2) and lack of dependence of synaptic modifications on postsynaptic activity (*cf.* Sections 3.1 and 3.2). Whether a back propagation process can or does occur in actual nervous systems is an open question, but the answers may not be relevant to the utility of BP networks for industrial pattern processing applications.

The high degree of supervision required by the back propagation network makes it useful as a "neural network expert system" for classifying data into predetermined categories. The network architecture (number of hidden units, in particular) and the set of training inputs often have to be chosen carefully to allow the network to generalize to a correct classification of the desired set ofinputs. The issue of generalization in BP networks is discussed further in Huyser and Horowitz (1988) and Rumelhart (1988).

To illustrate use of BP in a specific problem domain, let us return to the case of teaching the network to discriminate between a "T" and a "C" regardless of position or orientation in the visual field (Rumelhart *et al.*, 1986). Figure 6.13 illustrates the different rotations of the T and C. The network for solving the T-versus-C problem is schematized in Figure 6.14. Each rotation has to be taught to the network separately. But translation invariance is achieved by adding an additional transformation to the rule for learning input-to-hidden-unit connections. To make the learning of a pattern independent of its location in the visual field, all hidden units are constrained to learn exactly the same pattern of weights. This is accomplished by adding together the weight changes dictated by the delta rule for each unit and then changing all weights by averages of those amounts. Other researchers (*e.g.*, Khotanzad & Lu, 1989) have added to the back propagation algorithm still more mathematical operations, involving Fourier or related transforms, to make the weight changes invariant under rotation as well as translation.

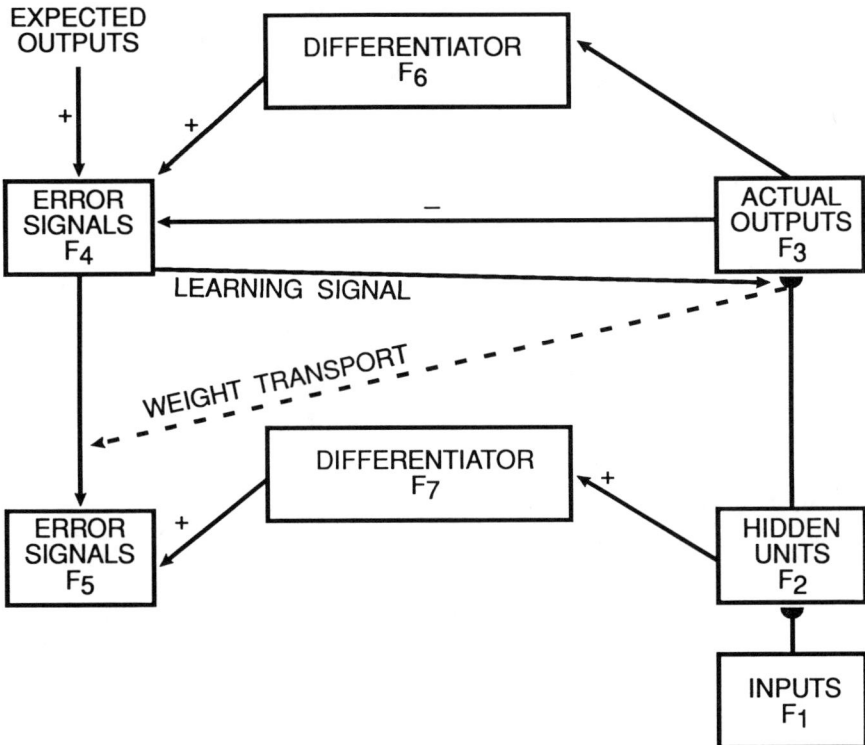

Figure 6.11. Possible circuit diagram of the back propagation model. Postulated interactions among node levels F_1, F_2, and F_3 suggest additional levels F_4, F_5, F_6, and F_7. Transport of weights from F_2-to-F_3 to F_4-to-F_5 pathways makes this circuit biologically implausible. (Reprinted from Grossberg, 1987b, with permission of Ablex Publishing Corporation.)

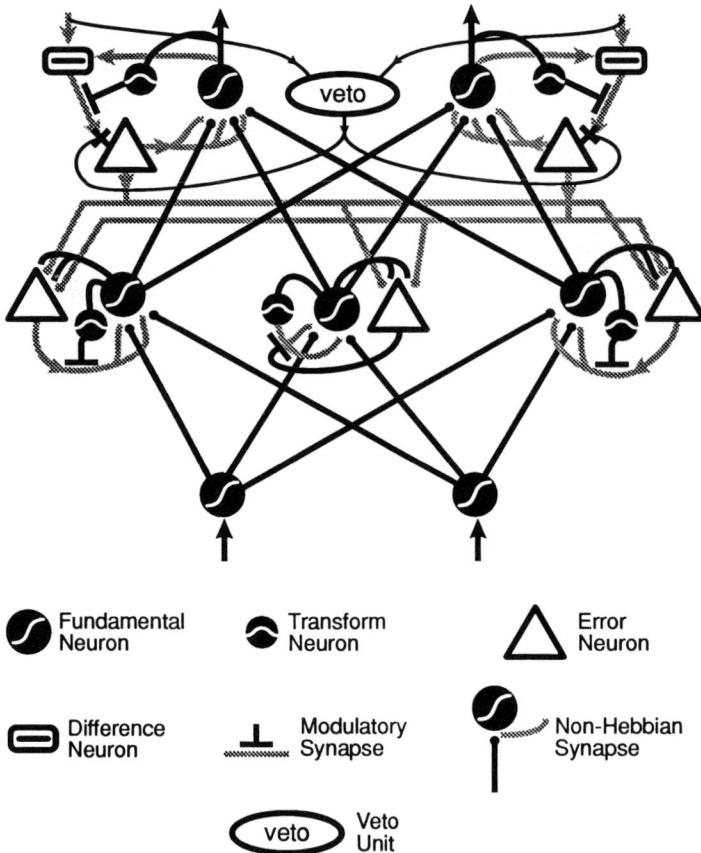

Figure 6.12. Another possible back propagation circuit. The types of nodes are defined as follows. *Fundamental neurons* and error neurons both emit a sigmoid function of the sum of inputs they receive. *Transform neurons* do a transformation of their inputs based on the derivative of that sigmoid. *Error neurons* also signals from *difference neurons* that emit the algebraic difference of their inputs. In *modulatory synapses*, the black activity gates the gray pathway's efficacy. In *non-Hebbian synapses*, the actual synapse is the black dot and the learning signal is the gray pathway. *Veto units* perform the logical NOR (firing when they receive no active inputs) and serve to prevent weight change during recall. (Adapted from Stork, 1989b, copyright © 1989 IEEE, by permission of the publisher.)

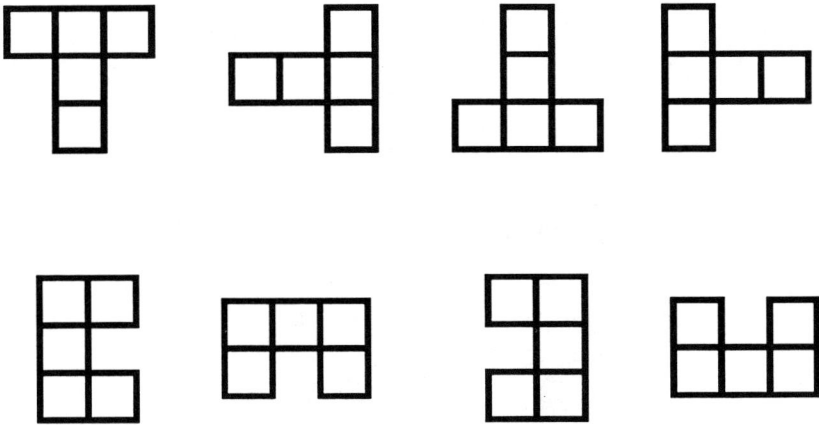

Figure 6.13. Stimulus set for the T-versus-C problem. The set consists of a block T and a block C in each of four orientations. One of the eight patterns is presented on each trial. (Reprinted from Rumelhart *et al.*, 1986, with permission of MIT Press.)

Averaging weight changes over the entire visual field is a *nonlocal* transformation, that is, a transformation in which a region is directly affected by an arbitrarily large region around it. Since locality is usually assumed in biological models, it might be desirable to achieve the same effect by averaging, at each connection, over some smaller diameter of the visual field. As is discussed in Section 6.4, the problem of translation- and rotation-invariance still poses a major challenge for theories of neural categorization.

Some Models from the Anderson-Cooper School

A class of categorization models, widely known by its nickname of "brain-state-in-a-box," has been studied since 1977 by Anderson and his colleagues. In early versions of these models (Anderson *et al.*, 1977; Anderson & Mozer, 1981), pattern classifications are unsupervised. In more recent versions (Anderson & Murphy, 1986), an error-correcting learning rule has been added to allow for learning of particular desired classifications. The brain-state-in-a-box (BSB) model was first derived by Anderson *et al.* (1977)

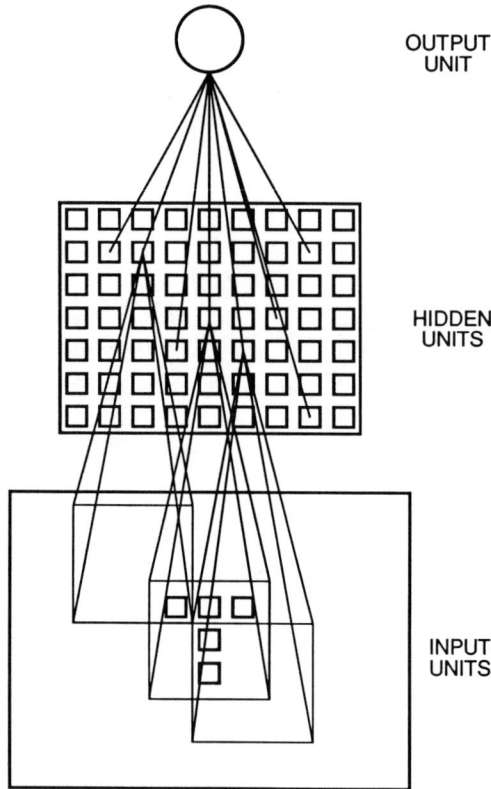

Figure 6.14. Schematic diagram of a network for solving the T-versus-C problem. Hidden units are organized into a two-dimensional grid with each unit receiving input from a square 3x3 region. The output unit is trained to take on the value 1 if the input is a T (at any position or orientation) and 0 if the input is a C. (Reprinted from Rumelhart *et al.*, 1986, with permission of MIT Press.)

as an offshoot of the linear model with saturation due to Nass and Cooper (1975) (see Section 3.2). Recall that saturation was imposed on a linear network with associative memory and positive feedback, in order to prevent activities of the nodes in the network from becoming unbounded. The "box" from which the model derives its name is shown in Figure 6.15. It is an abstract square, cube, or more generally, hypercube of possible n-dimensional activity vectors which constitute the numerical bounds for possible node activities.

The BSB model associates vector patterns of activities at a set of nodes with other patterns at the same nodes. The matrix consisting of the connection weights between nodes provides feedback that transforms the pattern, as will be developed below. The network then converges to one of the characteristic system states corresponding to corners of the "box" in Figure 6.15. Categorization of the original input pattern is based on which one of these corners is reached.

Recall from Section 3.4 the distinction between *autoassociative* and *heteroassociative* encoding (Kohonen, 1977, 1984). The connectivity matrix is designed to associate a vector pattern (input) $\vec{x}$ to another vector pattern (response) $\vec{y}$; this is called autoassociative if the two patterns are equal, and heteroassociative otherwise. The BSB model is applicable to both of these two types of encoding.

We shall now give a brief description of this algorithm; more mathematical detail appears in Section 6.5. As in Anderson's earlier models discussed in Chapter 3, the input pattern is interpreted as the initial state of a vector of node activities $\vec{x}(t) = (x_1(t), x_2(t), ..., x_n(t))$. These activities x_i vary between preassigned maximum and minimum values, which for mathematical simplicity are taken to be 1 and -1. The vector $\vec{x}$ at t=0 represents the input pattern to be classified; values of $\vec{x}$ at subsequent times are given by the rule

$$\vec{x}(t+1) = \vec{x}(t) + A\vec{x}(t) \qquad (6.4)$$

where A is the connectivity matrix of the system, or rectangular array of its connection weights. (For a good, accessible introduction to matrices as they relate to neural networks, see Jordan, 1986a).

Equation (6.4) represents positive feedback as it might occur in the brain, due to the past operation of a Hebbian associative learning law. This feedback has the desirable property of enhancing significant activities or stimuli, but often has an additional property that is undesirable. Repeated application of (6.4) to a pattern vector will drive the state of the system outside the "box" of Figure 6.15, that is, cause values of some or all of the x_i to become larger than 1 or smaller than -1. This is particularly true if the input pattern is one of the significant states of the system, which correspond to the vectors known as *eigenvectors* of the matrix A (see Section 6.5 for a definition of this term).

To prevent activities from becoming unbounded, Anderson *et al.* (1977) imposed an additional rule, whereby if any of the activities becomes greater than 1 as a result of (6.4), it is replaced by 1, and if that activity becomes less than -1, it is replaced by -1. The effect is to make the system converge to one of the corners of the box, which represent states in which all activities are 1 or -1.

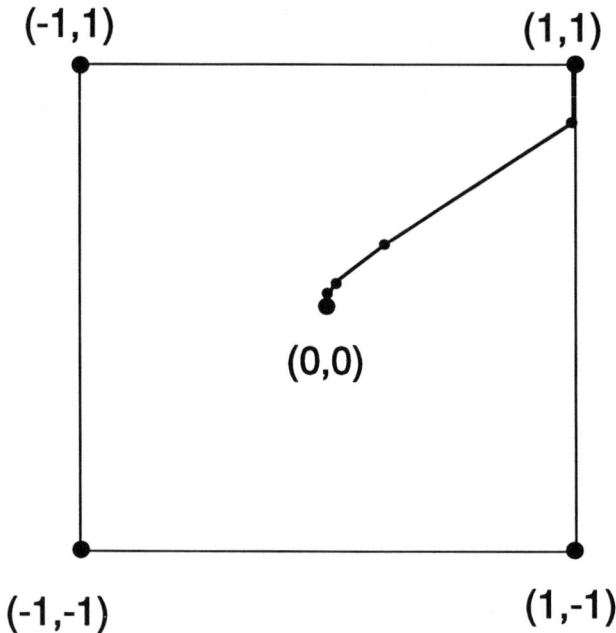

(-1,1)   (1,1)

(0,0)

(-1,-1)   (1,-1)

Figure 6.15. A two-dimensional example of the "box" from which the brain-state-in-a-box model (Anderson *et al.*, 1977) derives its name. The system state at any time is given by a vector of (in this case, two) numbers between the bounds for node activity (in this case, -1 and 1). Points along the curve drawn inside the box denote the system state at successive times. In this case, the points that are drawn arise if the initial state is (0, 0.05) and the connectivity matrix (see Equation (6.4)) is $\begin{bmatrix} 2 & 1 \\ 1 & 1 \end{bmatrix}$.

Hence, the BSB model, like all neural network models, includes a method for keeping activities within bounds, corresponding to the limits on possible firing frequencies of actual neurons. Grossberg (1978c) argued that this particular method (linearity plus "hard" saturation) is not biologically realistic and creates some inevitable distortions of pattern weights because distinctions among values close to 1 or -1 are lost. He argued that an inherently nonlinear mechanism is needed to avoid such distortion. (For the gist of his reasoning, see Exercise 1 of Chapter 4.) Anderson and Silverstein (1978) responded that linear transformations do occur in some regions of real nervous systems (as in the response of the horseshoe crab eye to visual stimuli). They added that a simple linear model illustrates some properties that are likely to be retained in more complex nonlinear models. There are certainly merits to both sides of this argument, but since 1978, both of those classes of models have developed further in ways driven more by the needs of application than by strict fidelity to neurobiology.

Further elaborations and simulations of the BSB categorization model were developed in Anderson and Mozer (1981), Anderson (1983), Knapp and Anderson (1984), and Anderson and Murphy (1986). Anderson and Mozer (1981, pp. 215-216) discuss the application of BSB to one form of autoassociation, namely, the recovery of all of a vector pattern from part of it. The other articles discuss its application to heteroassociation.

The network of Knapp and Anderson (1984) is particularly interesting from the standpoint of prototype formation. That network models some psychological data indicating that humans tend to have a mental prototype of any category they encode, and that their reaction times to members (*exemplars*) of that category are shortest for exemplars that are close to the prototype (*e.g.*, Mervis & Rosch, 1981). For example, a sparrow is recognized as a "bird" sooner than an ostrich is. Moreover, if a subject is taught exemplars that are random variations on a general pattern of dots, the prototype formed will be close to the average of these variations. The subject will then learn the prototype faster than any of the actual exemplars even without having seen the prototype (*e.g.*, Posner & Keele, 1970). Knapp and Anderson simulated some of these random-dot data.

The prototype model, however, is not universally accepted as the basis for all categorization (see Homa, 1984 for a review). Models based on exemplars or features have also been suggested for various experimental paradigms, and even proponents of the prototype theory (*e.g.*, Posner & Keele, 1970) do not claim that the prototype is the *sole* memory representation of a category.

Anderson and Murphy (1986) combined BSB with the delta or error-correction learning rule (in a form used previously by McClelland & Rumelhart, 1985) to deal with the problem of noisy encoding. Previous BSB networks succeeded in distinguishing mutually orthogonal (perpendicular) input vectors, but did not always succeed in distinguishing inputs whose vector directions were relatively close together. For this reason and others, the connectivity matrix learned from previous pattern associations sometimes failed to yield the desired encoding if the data were somewhat noisy.

In the Anderson-Murphy simulations, the connectivity matrix **A** is initially chosen at random. Then the system is fed a desired association of a vector $\vec{x}$ to another vector $\vec{y}$. On each time step, the matrix **A** is changed by a correction term based on the difference between $\vec{y}$ and the product $\mathbf{A}\vec{x}$ (see Section 6.5 for details). Once the desired matrix has been found, the system goes through the saturating-linear BSB algorithm as in Anderson *et al.* (1977).

Anderson and Murphy's supervised version of the BSB model has been applied to processing linguistic inputs that are converted to vectors of 1's and -1's by means of ASCII codes. This model has reproduced the disambiguation by context of words with more than one meaning; for example, the word "bat" by itself could mean a flying animal, "ball" could mean a dance, and "diamond" could mean a jewel, but all three words together must refer to baseball (see Table 6.3). Other applications of this categorization system have included medical diagnosis (Anderson, 1986; see Table 6.4) and radar signal classification (Anderson, Penz, Gately, & Collins, 1988), neither yet at the practical implementation stage. (The same radar problem has also been treated with an unsupervised adaptive resonance network; *cf.* Levine & Penz, 1990.)

A different approach to supervised category learning, also based on earlier work of Anderson, Cooper, and their colleagues, was developed by Reilly, Cooper, and Elbaum (1982). These authors set out to construct a network that can learn an arbitrary classification of geometric vectors, one where the boundary between members of different categories might be oddly shaped (see Figure 6.16). Their network was also designed to allow different prototype patterns to activate the same classifier node: in one example, small "a" and capital "A" activate the same sound representation even though they are quite different in appearance.

The algorithm of Reilly *et al.*, like that of Anderson and Murphy (1986), includes a matrix that maps input vectors to prototype vectors. But after the prototype level, as shown in Figure 6.17, comes a classification level that is also influenced by "external instruction." Both transitions, input to prototype and prototype to classification, are trainable. In the training stage, if the pattern at the input nodes does not cause the "correct" classifying node to fire, a new prototype node is committed to the current input pattern, and the synapse between the current prototype node and the appropriate classifying node is given the maximum strength. Also, synapses from input to prototype levels are changed so as to facilitate future responses of the newly committed prototype node to that input pattern or similar ones. If another, "incorrect" classification node fires, the instruction signal causes "negative reinforcement," which diminishes the strength of synapses from input nodes to currently active prototype nodes associated with the wrong response.

1. Single ambiguous words

Input	Output	Time Steps
--------Bat ---------	Vampire MythBat NiteDracu	81
-----------Ball-----	Tennis GameCortBallRackt	105
----------------Diamd	GeoShapeTwoDCrclSqreDiamd	134
----Game-------------	Poker GameBeerTablCards	68

2. Pairs of ambiguous words

--------Bat Ball-----	BaseballGameBat BallDiamd	30
--------Bat-----Diamd	BaseballGameBat BallDiamd	28
-----------BallDiamd	BaseballGameBat BallDiamd	27
----GameBat ---------	BaseballGameBat BallDiamd	28

3. Pairs of words, one of them ambiguous

--------Bat Nite-----	Vampire MythBat NiteDracu	23
--------Bat Wing-----	Animal LiveBat WingFlyng	25
---Shape--------Diamd	GeoShapeTwoDCrclSqreDiamd	22

4. Three or more ambiguous words

--------Bat BallDiamd	BaseballGameBat BallDiamd	18
----GameBat ----Diamd	BaseballGameBat BallDiamd	18
----GameBat Ball-----	BaseballGameBat BallDiamd	25
----GameBat BallDiamd	BaseballGameBat BallDiamd	14

Table 6.3. Response of a BSB network to various word inputs. Any two of the words "bat," "ball," and "diamond," or "bat" and "game," evoked a baseball association, whereas each word alone evoked a different association. Adding words reduced the network's "reaction time". (Adapted by permission of the publisher from Anderson & Murphy, *Physica D* **22**, 318-336. Copyright 1986 by Elsevier Science Publishing Co., Inc.)

Output	Time Steps
--------+bacMening-------	80
--------+bacMening-------	90
--------+bacMening-------	102
-o------+bacMening------m	125
-o------+bacMening-e----m	133
Co-----'+bacMening-en--im	139
Co-----'+bacMening-en--im	143
Co-----'+bacMening-en--i-	145
Co-----'+bacMening-en--i-	152
Co-----'+bacMening-eni-i-	156
Co-----'+bacMening-enici-	160
Co-----'+bacMening-enici-	168
Co-----'+bacMeningPenici-	174
Co-y-e-'+bacMeningPenicil	178
Co-y-e-'+bacMeningPenicil	181

Table 6.4. Output of an error-corrected BSB network for generalization about medical data. The system lacks information about meningitis caused by Gram positive bacilli, but has the information that other conditions caused by such bacilli are treated with penicillin. Thus, it "guesses" that penicillin is the treatment of choice for that form of meningitis (possibly wrongly, and taking a long time to converge to an answer). (Adapted from Anderson, 1986, with permission of the author).

···· REAL BORDERS
—— LEARNED BORDERS

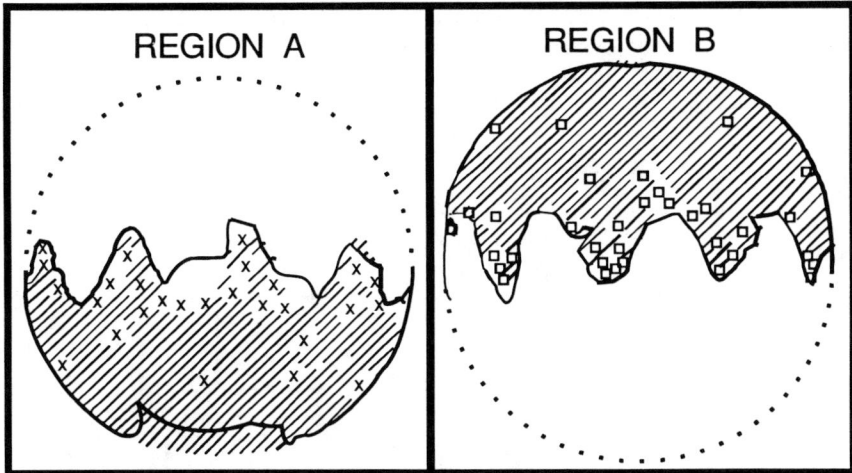

Figure 6.16. Example of supervised learning of a discrimination between two regions on the surface of a sphere. In this example, the boundary between regions is sinusoidal. (Adapted from Reilly *et al.*, 1982, with permission of Springer-Verlag.)

This ability to assign multiple prototypes to the same classifier has been the basis for more recent network architectures that include multiple learning modules (*e.g.*, Reilly, Scofield, Elbaum, & Cooper, 1987). Some of these architectures have been applied to the study of knowledge representation; one example, a simulation of an insurance underwriter's decisions (Collins, Ghosh, & Scofield, 1988), is discussed in Section 7.2.

Classification algorithms that are tightly supervised have some distinct advantages in speed and reliability for applications in which particular, known outputs are desired. Supervised learning, however, is probably not a realistic model of biological categorization processes. Living organisms are internally supervised by reinforcement and drive systems that function as *critics* (*cf.* Grossberg, 1971; Barto *et al.*, 1983; Section 5.2 of this book) but do not dominate classification decisions.

Hence, living neural systems seem to include learning modules that find without direct supervision the naturally recurring input classes in the environment, but then are subject to attentional control. This attentional control causes finer distinctions to be made among those inputs that are most important to the system's goals. Some of the important unsupervised categorization algorithms are discussed in the next section. Early work on attentional control of such categorization systems is discussed in Chapter 7.

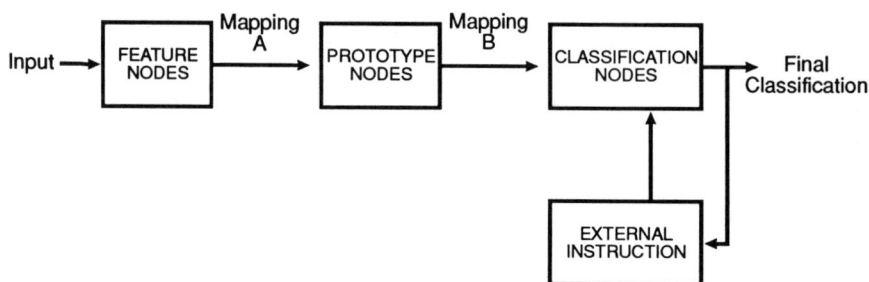

Figure 6.17. Architecture of a classification network. Prototype and classification levels are separated to allow different prototype patterns (*e.g.*, "a" and "A") to be classified together. (Adapted from Reilly *et al.*, 1982, with permission of Springer-Verlag; in that article, feature nodes were called "coding cells".)

6.3. UNSUPERVISED CLASSIFICATION MODELS

As stated above, early versions of the brain-state-in-a-box model of Anderson and his colleagues are unsupervised. Other important unsupervised categorization algorithms include competitive learning (CL) (Rumelhart & Zipser, 1985); adaptive resonance theory (ART) (Carpenter & Grossberg, 1987a, b); and Darwin II (Edelman, 1987; Reeke & Edelman, 1987).

The Rumelhart-Zipser Competitive Learning Algorithm

The term *competitive learning* is in general usage for multilevel networks that combine associative and competitive principles, including most of the networks discussed in this chapter (see Grossberg, 1987b for a discussion). The same term is sometimes used more specifically for a subclass of this type of network developed by Rumelhart and Zipser (1985). These researchers studied a simple system capable of detecting pattern regularity and illustrating some basic competitive learning principles.

The Rumelhart-Zipser model is based on the multilayer architecture shown in Figure 6.18. The lower level consists of input units (feature detectors), and inputs are treated as binary patterns activating some of these nodes. Nodes in succeeding layers group into clusters, and there is winner-take-all competition within a cluster. Winning nodes then send signals up to the next layer. There is no feedback from higher to lower layers.

The weights of connections from lower to higher layers change according to a rule similar to the synaptic conservation rule of Malsburg (1973). Only connections to winning units in clusters are modified. The connection weights w_{ij} to a given unit j, from units at the next layer below, add up to a fixed amount, and a proportion of the weight shifts from inactive to active pathways; the equations for this process are given in Section 6.5 below.

Each cluster of nodes classifies the stimulus set into as many groups as there are units in the cluster. If the arriving inputs fall into "natural clusters," the clusters of network nodes tend to find them. If there are no "natural clusters," responses of those nodes can at times become oscillatory or chaotic rather than converging to an equilibrium.

Rumelhart and Zipser (1985) applied the competitive learning algorithm to a variety of binary patterns — letters, horizontal and vertical lines, and "dipoles," which activated exactly two neighboring input units[1]. In one experiment, the stimuli are letter pairs drawn from the set {AA, AB, BA, BB}. Due to randomness in the initial weights, the units in higher layers develop activation patterns that correspond either to A or B in the first serial position, or else to A or B in the second position. Moreover, on any given run, only one of the two positions is preferentially detected.

---

[1] Rumelhart and Zipser's usage of the term "dipole" is entirely unrelated to the gated dipole introduced in Section 3.3.

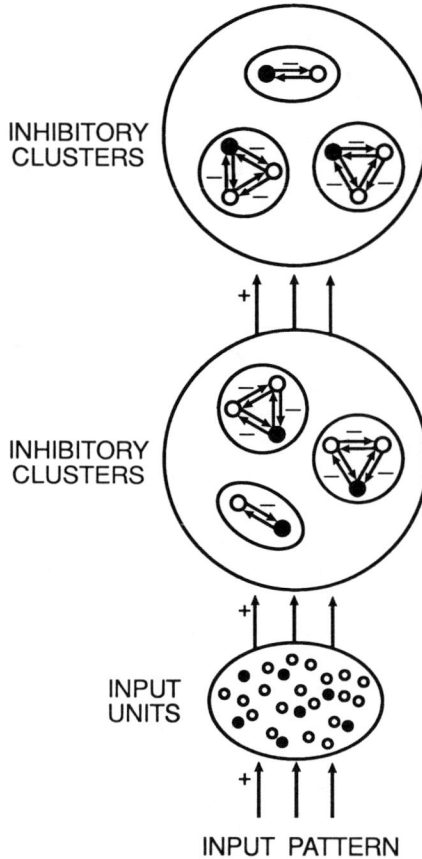

Figure 6.18. Competitive learning architecture. Active units are denoted by filled circles, inactive ones by open circles. Mutual inhibition between clusters at the top two layers results in exactly one unit per cluster being active. Architecture can be extended to arbitrarily many layers. (Adapted from Rumelhart & Zipser, 1985, with permission of Ablex Publishing Corporation.)

Another experiment done by Rumelhart and Zipser draws letter pairs from the set {AA, BA, SB, EB}. In this case, as shown in Figure 6.19, all higher-level units become detectors of the letter in the second position; that is, they learn to respond to either the class {AA, BA} or the class {SB, EB}. This is striking because "A" and "E" are similar in their dot patterns as represented, as are "B" and "S." But the network completely ignores these similarities in favor of identity in the second position. In other experiments where there is no repetition of letters at either position, the responses of higher-level units do not reach an equilibrium. This model is discussed again in Section 7.2, where it is compared with other models of letter and word recognition by Rumelhart and McClelland.

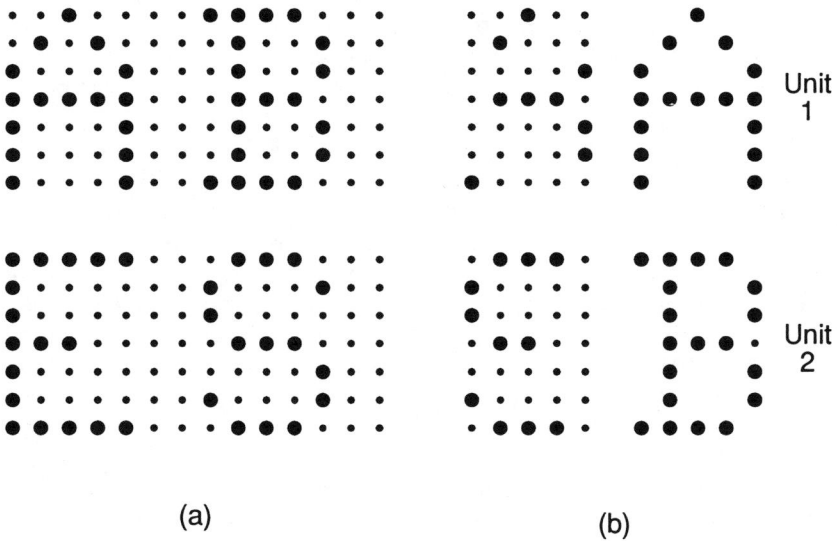

(a)   (b)

Figure 6.19. (a) Inputs used in simulated letter and word experiments. Note that "E" is similar to "A" and "S" to "B". (b) Pattern of weights from input layer to two second-layer units in the experiment with letter pairs AA, BA, SB, and EB. Note that each unit became a position-specific letter detector. Similarity of E to A and S to B had little effect. (Reprinted from Rumelhart & Zipser, 1985, with permission of Ablex Publishing Corporation.)

Adaptive Resonance Theory

Adaptive resonance theory (ART) is best discussed in two articles, Carpenter and Grossberg (1987a) which describes the ART 1 model for classifying binary inputs, and Carpenter and Grossberg (1987b) which describes the ART 2 model for classifying analog inputs. Further refinements, in the model known as ART 3, are developed in Carpenter and Grossberg (1990). The architectures of these networks were based on the idea of adaptive resonant feedback between two layers of nodes as developed in Grossberg (1976c).

Figure 6.20 illustrates the basic structures of ART 1. Equations governing this network will be given in Section 6.5 below. The field F_1 is assumed to consist of nodes that respond to input features, analogous to cell groups in a sensory area of the cerebral cortex. (The detailed structure of how inputs are processed at F_1 is disregarded in this network but is considered in ART 2.) The field F_2 is assumed to consist of nodes that respond to categories of F_1 node activity patterns. Synaptic connections between the two fields are modifiable in both directions, according to two different learning laws.

The F_1 nodes do not directly interact with each other, but the F_2 nodes are connected in a recurrent competitive on-center off-surround network (*cf.* Figure 4.3). As discussed in Chapter 4, such competition is a common device in neural networks, inspired by visual neurophysiology, for making choices in short-term memory. In this version, the simplest form of choice (winner-take-all) is made: only the F_2 node receiving the largest signal from F_1 becomes active. To compute the signal received by a given F_2 node, the activity of each F_1 node in response to the input pattern is weighted by the strength of the bottom-up synapses from that F_1 node to the given F_2 node, and all these weighted activities are added.

Inhibition from the F_2 field to the F_1 field (via one of the gain control nodes, shown in the figure as filled-in circles) serves two related purposes. First, it prevents F_2 activity from always exciting F_1, thereby preventing "hallucinations" from occurring when a category node is active. Second, it shuts off most neural activity at F_1 if there is mismatch between the input pattern and the active category's prototype. Only with a sufficiently large match are enough of the same F_1 nodes excited by both the input and the active F_2 category node, which is needed to overcome nonspecific inhibition from F_2.

If match occurs, as shown in Figure 6.20, then F_1 activity is large because many nodes are simultaneously excited by input and prototype. Then F_1 inhibits the activity of the node A representing the orienting subsystem. This stabilizes the categorization of the given input pattern in the given F_2 node. If mismatch occurs, by contrast, F_1 activity is not sufficient to inhibit A, which thereby becomes active. The A node activity leads to F_2 reset which shuts off

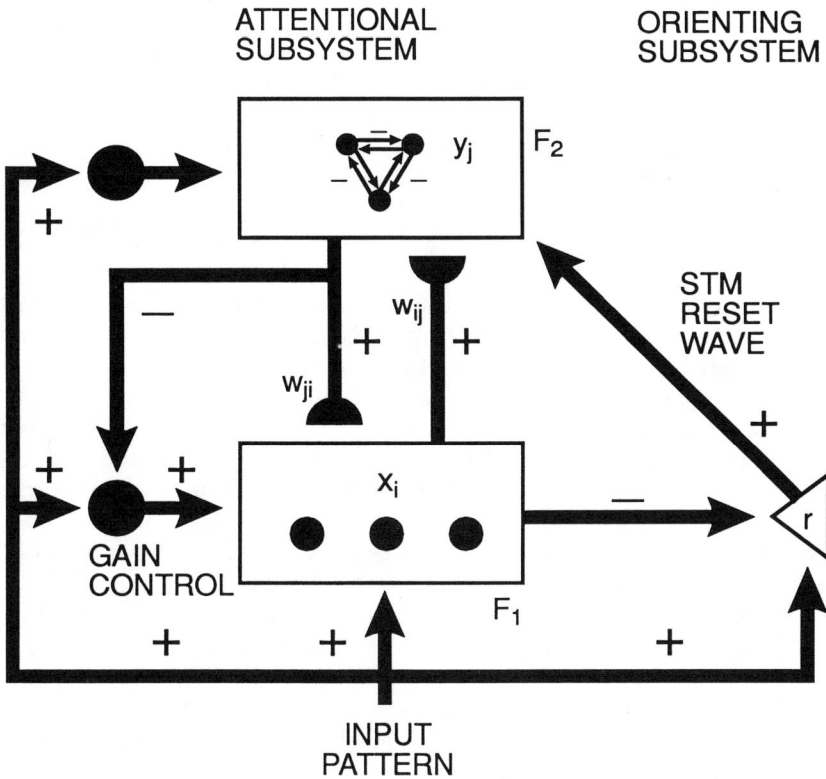

Figure 6.20. ART 1 architecture. Short-term memory at the feature level F_1 and category level F_2, and bottom-up and top-down interlevel long-term memory traces, are modulated by other nodes. The orienting system generates a reset wave to F_2 when bottom-up and top-down patterns mismatch at F_1, *i.e.*, when the ratio of F_1 activity to input activity is less than a *vigilance level* r. This wave tends to inhibit recently active F_2 nodes. Functions of gain control nodes are described in the text. (Adapted from Carpenter & Grossberg, 1987a, with permission of Academic Press.)

the active category node as long as the current input is present. The F_2 node receiving the next largest F_1 signal is then tested, and the process repeated. The exact criterion for mismatch is that the ratio of F_1 activity to total input intensity be less than some prescribed parameter. That is, if [I] is the number of active pixels in the (binary) input pattern, and [X] the number of F_1 nodes active after input and prototype presentation, then mismatch is said to occur if [X]/[I] < r for some positive constant r, which is called the *vigilance* of the network.

The short-term memory (STM) and long-term memory (LTM) equations for this network, shown in Section 6.5, incorporate all these effects along with two additional rules. The *2/3 Rule* says that an F_1 node must be activated by at least two out of three signal sources if it is to generate suprathreshold output signals. These sources are outside inputs, gain control (activated by inputs but inhibited by F_2), and the top-down signal from the active category node. The *Weber Law Rule* says that LTM size should vary inversely with input pattern scale. This rule is designed to prevent a category node that has learned to code a particular binary pattern (1's in particular locations) from also coding every superset pattern (a pattern that has 1's in those same locations and some others). Figure 6.21 shows an example of coding by ART 1; note the importance of the 2/3 Rule for code stability.

The ART 2 network of Carpenter and Grossberg (1987b), shown in Figure 6.22, is designed to categorize analog (or continuous-valued, or gray-scale) input patterns. Its architecture builds on the ideas of ART 1 with two layers and modifiable synapses in both directions, but adds several sets of processing nodes at the F_1 layer. These extra nodes are designed to contrast-enhance significant parts of the pattern and suppress noise, according to general principles developed in Grossberg (1973). Figure 6.23 illustrates the type of contrast enhancement performed by ART 2.

While the preprocessing is more complex in ART 2 than in ART 1, the learning laws are simpler. Since supersets are not an issue with analog patterns, the Weber Law Rule is dispensed with and the top-down and bottom-up LTM equations are the same (although top-down weights are initially 0, while bottom-up weights are initially random and positive). The matching criterion for ART 2 is also different, since [X] and [I] no longer make sense. In ART 2, the quantity to be compared with the vigilance value is the cosine of the angle between input and prototype vectors, a measure used in some older, non-neural-network clustering algorithms (*e.g.*, Duda & Hart, 1973, Chapter 6; Kohonen, 1984).

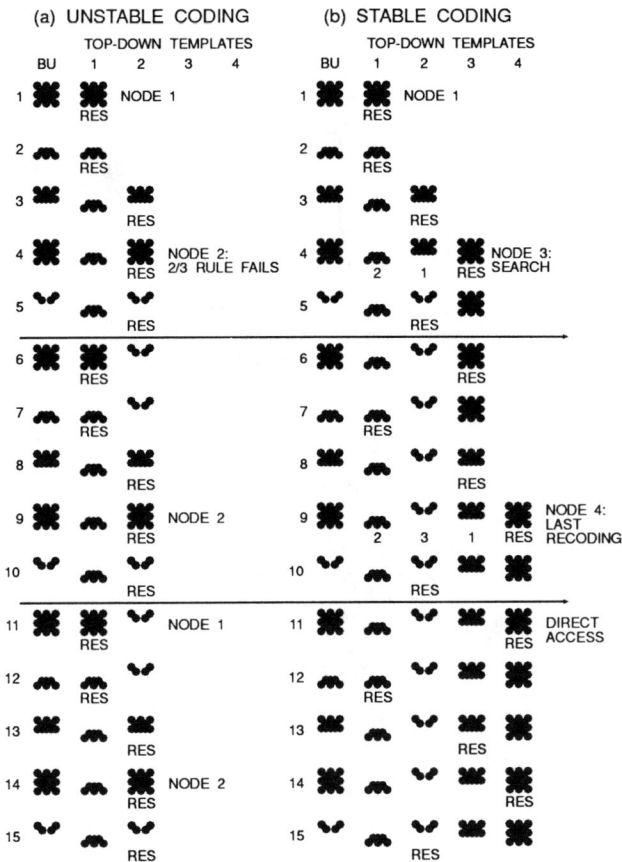

Figure 6.21. Categorization of binary patterns by ART 1. The same input sequence, four patterns A, B, C, and D in the order ABCAD, is presented repeatedly in both (a) and (b). In (a), top-down inhibitory gain control (see Figure 6.20) is weak, and the 2/3 rule is violated. This leads to ceaseless recoding of A. In (b), after some initial recoding, all patterns resonate (as shown by the symbol "RES") in distinct stable categories. (Reprinted from Carpenter and Grossberg, 1987a, with permission of Academic Press.)

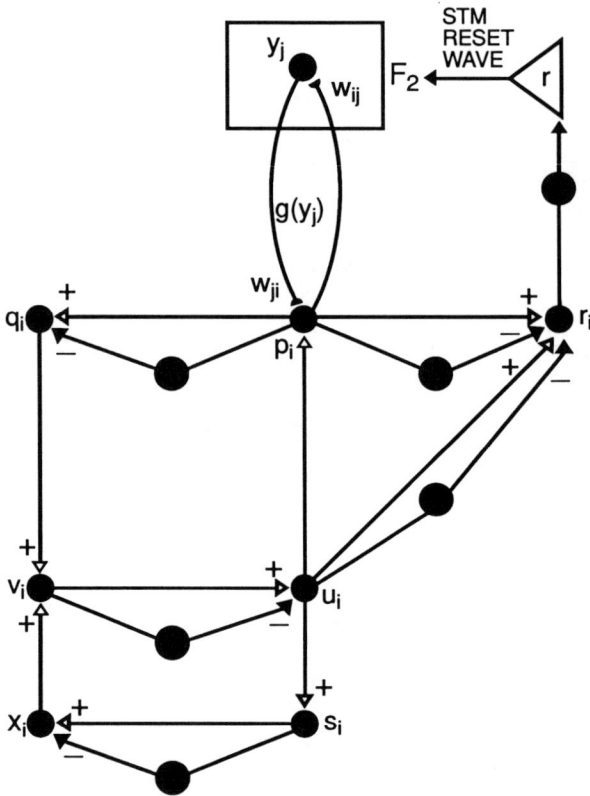

Figure 6.22. ART 2 architecture. F_1 is replaced by seven fields of nodes, p_i, q_i, r_i, u_i, v_i, s_i, and x_i, designed to enhance contrast and suppress noise. Open arrows indicate specific patterned inputs, filled arrows nonspecific gain control inputs. Gain control nodes (unlabeled filled circles) inhibit other nodes in proportion to total STM activity in their fields. When F_2 makes a choice, $g(y_J)>0$ if the Jth F_2 node is active and $g(y_j)=0$ otherwise. (Adapted from Carpenter & Grossberg, *Applied Optics* **26**, 4919-4930, 1987, with permission of the Optical Society of America.)

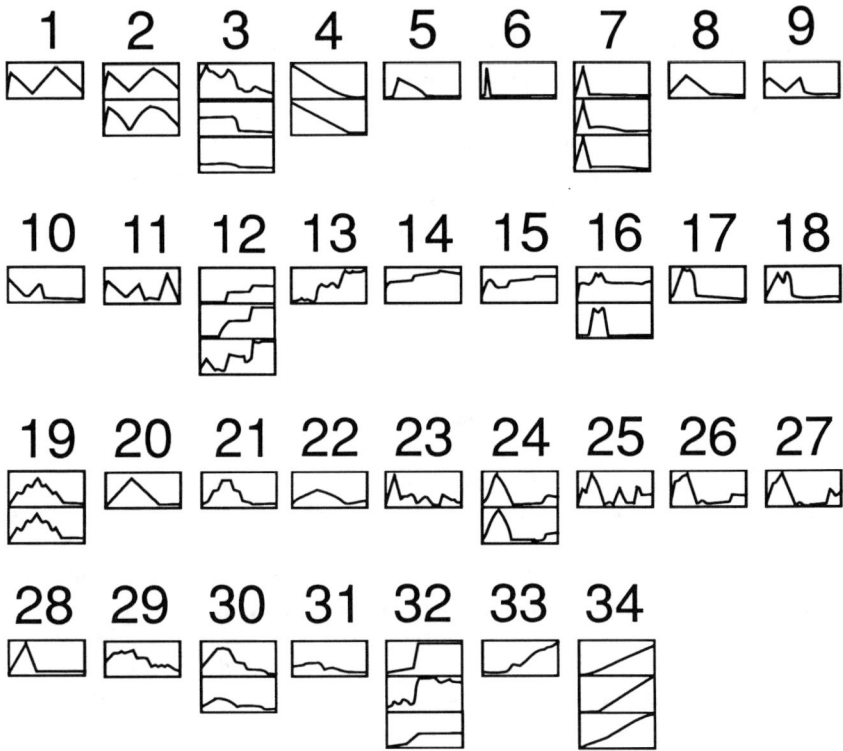

Figure 6.23. Example of categorization of analog patterns by ART 2. Fifty patterns, drawn as graphs of functions, are grouped into 34 recognition categories. (Reprinted from Carpenter & Grossberg, *Applied Optics* **26**, 4919-4930, 1987, with permission of the Optical Society of America.)

ART has several advantages over competing neural network classification theories. First, it exhibits considerable stability. Theorems proved in Carpenter and Grossberg (1987a) show that (in ART 1 at least) after a given collection of input patterns has been learned, the categorization of those patterns is not perturbed by an arbitrary barrage of new inputs. Second, the category prototypes against which an input is tested change over time to reflect the type of patterns that are most frequently observed in the environment. Third, the model allows for influences on the feature and category layers from other subsystems external to these layers, such as the attentional and orienting systems. Hence, it is part of an interconnected set of theories of many other cognitive processes (see, *e.g.*, Grossberg, 1982c, 1987c, 1988; Levine, 1983b).

Yet the current form of the ART model also has several limitations. Some other modelers, largely motivated by specific technological applications, have built networks with variations of the ART architecture designed to overcome some of these limitations. These modelers include Ryan and Winter (1987), Ryan (1988), Levine (1989a), and Weingard (1990).

Ryan and Winter (1987) noted that ART 1 networks had not entirely solved the code stability problem that motivated their original design (*cf.* Grossberg, 1976b). For example, suppose that a previously unclassified pattern perturbs the F_1 field and is placed in a new category, say the one coded by node J at F_2. Then suppose that, during the early training stage before asymptotic weights have been established, F_1 receives a number of variations on that input pattern, gradually shifted in a given direction in pattern space. Each variation would then match previous exemplars closely enough to be placed in the same category. This would cause a gradual shift in the weights to node J from F_1, ultimately a large enough shift that node J would not recognize the original pattern.

Ryan and Winter (1987) prevented recoding due to novel inputs by including thresholds, which they termed *symmetric adaptive thresholding*. They also included in their network a collection of "reset nodes" whose dynamics were influenced by those same thresholds. Carpenter and Grossberg (1987a, b) had previously assumed that mismatch-activated reset at F_2 is mediated by a field of interacting gated dipoles (*cf.* Grossberg, 1980; Levine & Prueitt, 1989; Sections 3.3 and 5.2 of this book) but had not explicitly included such dipoles in their simulations. Ryan (1988) made further extensions of these architectures, whose effect is to reduce the number of categories that need to be searched after input presentation.

Levine (1989a) noted that ART 1 and ART 2 networks always come to a decision about where to categorize an input. The amount of ambiguity in the input, that is, how close it is to a boundary between categories, influences the speed of decision but is not permanently coded. By contrast, the competitive learning (Rumelhart & Zipser, 1985) and back propagation (Rumelhart *et al.*, 1986) networks tend to register the degree of "confusion" in inputs coming from the environment. Hence, while ART's stability is usually an advantage, it can at times obliterate useful information. Ambiguity detection is useful not only in human psychology but also in automatic control systems. An automatic pilot or threat-detection system, for example, could be designed to run by itself if arriving inputs are unambiguous, but transfer control to a human operator, or to a specialized subsystem, if inputs are ambiguous.

Levine (1989a) found a method for recording, rather than immediately resolving, ambiguity in an ART-type network. The method is not presented here but is left to the reader as Exercise 9 of this chapter. Levine and Penz (1990) showed that signal classification can be improved by making category assignments for both ambiguous and unambiguous inputs, but changing synaptic weights only for unambiguous inputs. In this way, patterns close to category boundaries do not cause errors by shifting future classifications.

ART's inability to solve the exclusive-or problem, or more generally to make classifications where the shapes of category boundaries are irregular (*cf.* Stork, 1989a), was mentioned in Section 6.2. This limitation is related to the fact that ART's vigilance is uniform across both features and categories. A dramatic example of the consequences of this uniformity is shown in Figure 6.24. In classification of binary "pixel" patterns as letters of the alphabet, the same noise that changes an "O" into a "Q" simply changes an "I" into a noisy "I." Hence, any vigilance setting that separates O and Q would also separate I and noisy I. A more realistic model of biological categorization would allow some features to be more attended to than others, and some categories to have more rigid membership standards than others.

Overcoming the problems caused by uniform vigilance would involve adding to ART some "supervision" based on specific knowledge of the current data set. This issue has been addressed by Weingard (1990). His "self-organizing analog field" model adds attentional modulation to an ART network, modulation that allows for dynamic setting of the vigilance parameter. Attentional modulation of ART also appears in a model of frontal lobe function by Levine, Leven, and Prueitt (1991), which is discussed in Section 7.2.

Carpenter and Grossberg (1989) made different extensions of the ART architecture. Their extension includes multiple levels of nodes, denoting increasing levels of abstraction in the network's categorizations. Also, it includes an explicit mechanism for category search and reset, which is absent in their 1987 simulations. The reset mechanism is based on some known qualitative properties of chemical transmitter storage, utilization, and release by neurons. This work is further extended in the "ART 3" model of Carpenter and Grossberg (1990).

These recent articles illustrate that the influences on feedback between ART's F_1 and F_2 layers can be widely varied to reflect the action of other important neural subsystems. Hence, ART, like BP and BSB, is a general structure that is amenable to modification to overcome some of its limitations.

Edelman and Neural Darwinism

Edelman (1987), using a theoretical development based partly on analogies with immunology, proposed that the nervous system performs a selection between pattern encodings in which the "fittest" encodings survive. (This is the meaning of his book's title, *Neural Darwinism*.) Reeke and Edelman (1987) describe simulations of a categorization network incorporating this selection idea.

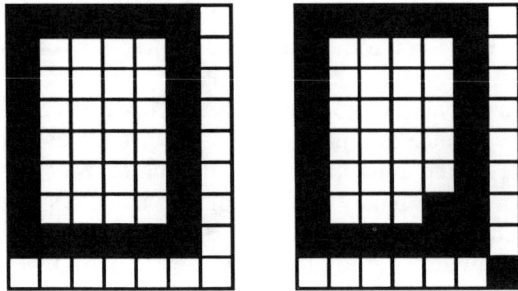

(a)

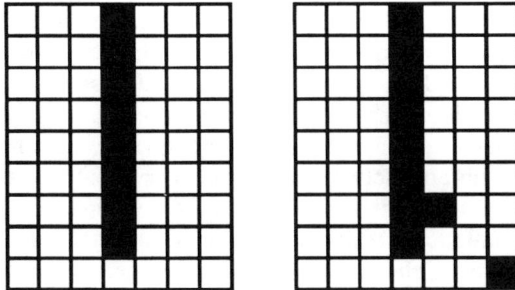

(b)

Figure 6.24. The same two extra pixels which change an "O" into a "Q" in (a) change an "I" into a noisy "I" in (b). (Reprinted from Stork, *Journal of Neural Network Computing* 1, 26-42 (New York: Auerbach Publishers). Copyright © 1989 Warren, Gorham, & Lamont, Inc. Used with permission.)

While Edelman's pattern selection idea is sometimes regarded as a new view of the brain (Rosenfield, 1988), I have argued elsewhere (Levine, 1988) that Edelman's philosophical approach is not essentially different from the approaches of many other neural modelers. Darwinian selection among encodings is a striking metaphor for the much older idea of competition between neurons and neuron groups. Moreover, modifiability of visual or somatosensory maps, which forms the biological basis of Edelman's arguments, is also the basis for modifiable interlayer synapses proposed by Malsburg, Grossberg, Bienenstock *et al.*, Linsker, and others.

Edelman and his colleagues may have made a greater contribution in proposing a theory about which functional groups of neurons in living animals correspond to "nodes" in a neural network. The basic idea of this theory (*cf.* Edelman, 1987 and Levine, 1988 for details) is that chemical markers, called *cell adhesion molecules* (CAM's), determine boundaries between groups of neurons. Sensory inputs during development alter the distribution of CAM's, in a way that has not been fully described. In adult life, there is less shifting of cell group boundaries, and the main mechanism for change, in this theory, is synaptic modification via a non-Hebbian associative learning law (*cf.* Edelman & Reeke, 1982; Finkel & Edelman, 1985). In this law, modification of a synapse depends not only on activities of the two neurons connected by that synapse, but on activities of all neurons in a group.

An example of the categorization network based on Edelman's scheme is shown in Figure 6.25. It includes two subnetworks, called "Darwin" and "Wallace," which perform different processing stages. This network is discussed further in next section, since it bears on the problem of translation invariance.

6.4. TRANSLATION AND SCALE INVARIANCE

Thus far, we have mostly discussed the categorization of patterns that are vectors of input activities at defined locations. But in reality, we see the letter "A," for example, as an "A" regardless of its location in the visual field. Minsky and Papert (1969) and many others have noted that this problem of *translation invariance* adds considerable difficulty to pattern categorization.

In Section 6.2 we mentioned "fixes" to categorization algorithms, some requiring extra nodes, that partly solve the problem of translation invariance (and, in some cases, rotation and scale invariance as well). One example is the averaging of weight changes that enables the back propagation network to distinguish a "T" from a "C" regardless of position or rotation. Another example is the mechanism of Reilly *et al.* (1982) whereby multiple prototypes can map into one classification node. These types of modifications are easier to add to supervised than to unsupervised algorithms.

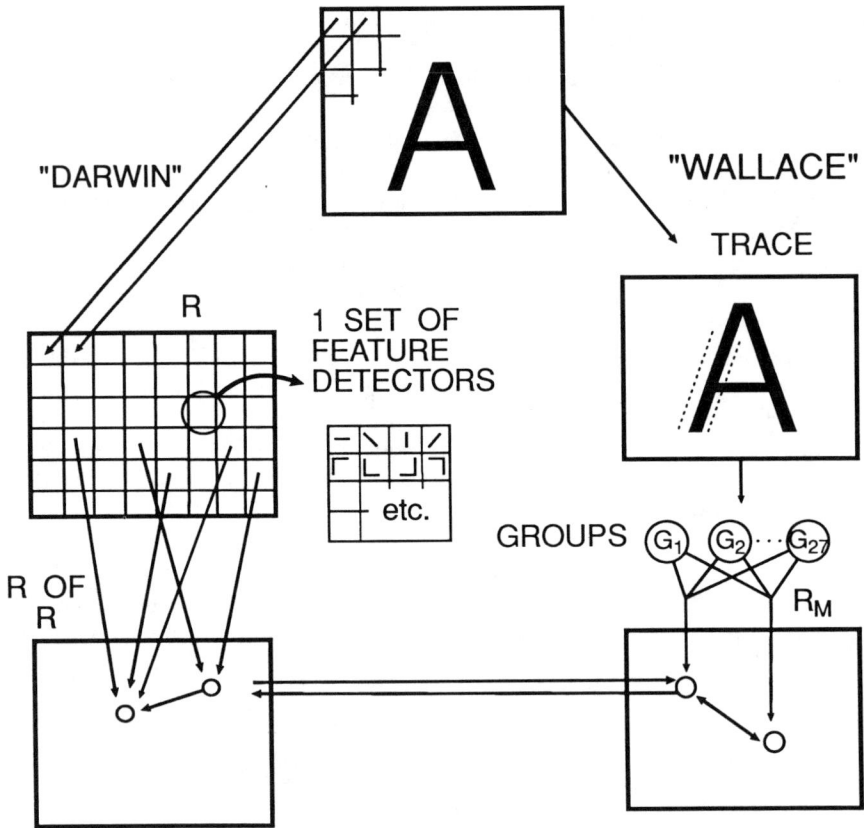

Figure 6.25. Simplified Darwin II architecture. The "R" on the "Darwin" side stands for "recognizers" or feature detectors, an example of which is shown in the inset to the right of the R array. R_M is an abstracting network that responds to patterns of activity but is insensitive to translation or rotation. (Adapted from Neural Darwinism, by Gerald Edelman. Copyright © 1987 by Basic Books, Inc. Reprinted by permission of Basic Books, Inc., Publishers, New York.)

The Neocognitron of Fukushima (1980) and Fukushima and Miyake (1982) also achieves translation invariance by taking a previously developed multilevel hierarchical classifier that is position-dependent (the Cognitron of Fukushima, 1975) and adding further levels. Fukushima drew his inspiration from the physiological findings of Hubel and Wiesel (1962, 1965), discussed in Section 6.1, that the cat or monkey visual cortex contains a hierarchy of cell types from *simple* to *complex* to *hypercomplex*. In real animals, as one ascends this hierarchy, the highest layers tend to respond most selectively to complicated pattern features but least selectively to location.

Fukushima's model mimics the hierarchy developed by Hubel and Wiesel, and extends it beyond hypercomplex cells to cells in association areas of the cortex (see Appendix 1). As shown in Figure 6.26, this model includes modifiable connections between successive hierarchical layers, as do most of the models of this chapter. The cells in each layer are organized so as to have the same receptive field structure but at different positions.

A different approach to translation and rotation invariance (and scale invariance, as well) is seen in the "Wallace" part of the network of Reeke and Edelman (1987), shown in Figure 6.25. Reeke and Edelman (1984, p. 188) describe this approach as "probabilistic matching." More specifically (p. 189): "Wallace begins with a tracing mechanism designed to scan the input array, detecting object contours and tracing along them to give *correlations of features* (emphasis mine) that ... respond to some of their characteristics, such as junctions of various types between lines."

6.5. EQUATIONS FOR VARIOUS CODING AND CATEGORIZATION MODELS

Malsburg's and Grossberg's Development of Feature Detectors

The equations of Malsburg (1973) for excitatory and inhibitory node activities $x_{E,k}$ and $x_{I,k}$ are of the form

$$\frac{dx_{E,k}}{dt} = -\alpha_k x_{E,k}(t) + \sum_i p_{ik} x_{E,i}^*(t) + \sum_i s_{ik} A_i^*(t) - \sum_i q_{ik} x_{I,i}^*(t)$$

$$\frac{dx_{I,k}}{dt} = -\alpha_k x_{I,k}(t) + \sum_i r_{ik} x_{E,i}^*(t)$$

$$(6.5)$$

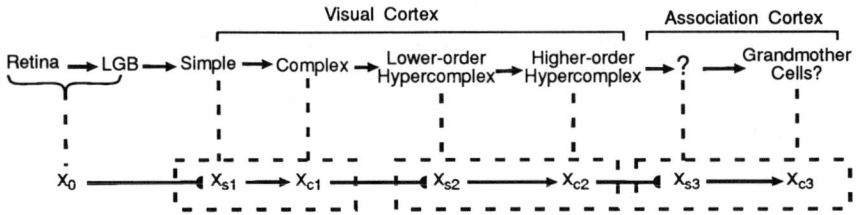

(a)

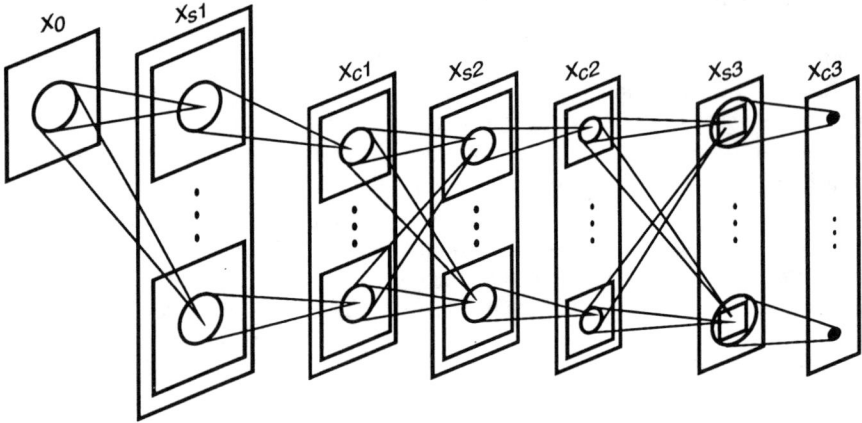

(b)

Figure 6.26. Neocognitron architecture. (a) Schematic of layers in the network's hierarchy and their rough correspondence to the actual biological hierarchy of Hubel and Wiesel. "LGB" stands for the lateral geniculate body of the thalamus, a processing station between the retina and the visual cortex. "Grandmother cell" is a colloquial term for a neuron that fires only in response to a very specific pattern. (b) More detailed diagram of the network's interconnections and their locations within layers. (Adapted from Fukushima, 1980, with permission of Springer-Verlag.)

where the "*'s" denote signals from other nodes (excitatory nodes in the case of E_i^*, inhibitory nodes in the case of I_i^*, and retinal afferents in the case of A_i^*). These equations had previously been used, for a different purpose, by Grossberg (1972d).

Malsburg's associative law is

$$s_{ik}(t+1) = s_{ik}(t) + hA_i^* E_k^* \qquad (6.6),$$

that is, retinal-to-cortical connection weights grow with the cross-correlation of presynaptic retinal afferent activity and postsynaptic excitatory cortical node activity. The other connection weights, p_{ik}, q_{ik}, and r_{ik}, do not change over time.

In the model of Grossberg (1976b), the activities x_{1i} of the V_i obey the nonrecurrent STM equations

$$\frac{dx_{1i}}{dt} = -Ax_{1i} + (B-x_{1i})I_i - x_{1i}\sum_{k \neq i} I_k \qquad (6.7)$$

where I_i are the inputs to the "retinal" nodes F_{1i}. Note that equations (6.7) are identical to (4.10) with x_i replaced by x_{1i}. To store patterns of retinal node activity in cortical STM, the activities x_{2j} of the F_{2j} are governed by the recurrent equations

$$\frac{dx_{2j}}{dt} = -Ax_{2j} + (B-x_{2j})[f(x_{2j})+I_{2j}] - x_{2j}\sum_{k \neq j} f(x_{2k}) \qquad (6.8)$$

where f is a signal function (typically sigmoid) and I_{2j} represents the excitatory input to F_{2j} from the F_1 level. Equations (6.8) are a subcase of Equations (4.1), which was developed in Grossberg (1973).

Grossberg (1976b) listed many possible laws for long-term memory (LTM) and for how the LTM traces affect the interlevel signals I_{2j} of (6.8). In general, I_{2j} is a weighted sum of the form $\sum_i \theta_i w_{ij}$, where the θ_i are the *normalized* activities at the F_1 level, and the w_{ij} are the weights of the corresponding connections from F_1 to F_2. The w_{ij} in turn obey equations such as

$$\frac{dw_{ij}}{dt} = (-w_{ij} + \theta_i) x_{2j} \qquad (6.9).$$

Equations (6.9) say that the weights of connections to cortical node j change only when short-term memory at that node is active, so that x_{2j} is nonzero.

This equation bears a close resemblance to the outstar learning law (3.14), except that in the outstar, weights of connections *from* a given node change only when short-term memory at the node is active. Indeed, the subnetwork of Figure 6.6 consisting only of a single cortical node and its connections behaves like a reverse outstar, and is frequently called an *instar*.

If a constant spatial pattern is presented through time to the network defined by Equations (6.8) and (6.9), a theorem in Grossberg (1976b) shows that this pattern is learned by the vector of connection weights to some F_2 node from the F_1 field. More precisely, recall from Section 6.1 that the total signal at time t to a given cortical node x_{2j} due to the retinal pattern $\vec{\theta} = (\theta_1, \theta_2, ..., \theta_n)$ is

$$S_j(t) = \sum_{k=1}^{n} \theta_k w_{kj}(t) \qquad (6.2).$$

Grossberg showed that if j is such that the value of S_j as defined by (6.2) is the largest, then the angle between the vector $\vec{w}_j = (w_{1j}, w_{2j}, ..., w_{nj})$ of weights to the jth node and the input vector θ is decreasing for all time and approaches 0. The result of Amari and Takeuchi (1978) also states that asymptotic weights to an F_2 node have a "winning dot product" with the original pattern. (Recall from Section 3.2 that the dot product of two vectors, of the same number of components, is found by multiplying the two vectors componentwise and then summing the products.)

Error-correcting Rules and Linear Classification Algorithms

All classification algorithms can be treated as rules for decision among n-dimensional vectors. Classical decision theory for vectors involves a Bayesian algorithm, which assumes that the probabilistic structure of the inputs is known. But in most neural network applications, the probabilistic structure of the inputs is unknown; in such cases, the Bayesian algorithm is replaced by one that minimizes some function of the vectors, called an *objective function*. Typically, the objective function also depends on a set of weights. These weights are coefficients of some linear combination of the vector components, which is called a *linear discriminant* function because its values determine the classification of vectors. This subsection indicates how the perceptron learning algorithm, the Widrow-Hoff algorithm, and the back propagation algorithm fit into linear discriminant analysis. Much of the discussion is taken from Duda and Hart (1973, Chapter 5).

In general, a linear discriminant function, for an n-dimensional vector $\vec{x} = (x_1, x_2, ..., x_n)$, is a real-valued function of the form

$$g(\vec{x}) = (\vec{w} \cdot \vec{x}) + \Gamma \qquad (6.10)$$

where $\vec{w} = (w_1, w_2, ..., w_n)$ is a weight vector and Γ is the negative of a threshold. If there are two categories, the rule is that $\vec{x}$ is assigned to Category 1 if $g(\vec{x}) > 0$, that is, if the dot product of weight and input vectors exceeds the threshold $-\Gamma$; to Category 2 if $g(\vec{x}) < 0$; and to neither category if $g(\vec{x}) = 0$. (The set of points satisfying the latter equation is called a *decision surface*.) If there are n categories, $n > 2$, then (6.10) is replaced by the n linear discriminant functions

$$g_i(\vec{x}) = (\vec{w}_i \cdot \vec{x}) + \Gamma_i \qquad (6.11).$$

For the discriminant functions defined by (6.11), $\vec{x}$ is assigned to Category i if $g_i(\vec{x}) > g_j(\vec{x})$ for all $j \neq i$. (Note that the networks of Grossberg, 1976b and Amari & Takeuchi, 1978, though unsupervised, use similar linear classification rules.)

Now consider the decision scheme for elementary perceptrons as developed in Rosenblatt (1962) (*cf.* Chapter 2). For simplicity, consider one half of the classification: let $\vec{y}_i$, $i = 1$ to n, be a set of vectors that we want to be members of some class, and define membership by the rule $g(\vec{y}) = \vec{w} \cdot \vec{y} > 0$. Then the objective function to be minimized is

$$J_p(\vec{w}) = \sum_{\vec{y} \in Y} (-\vec{w} \cdot \vec{y}) \qquad (6.12)$$

where the set Y consists of those vectors that are *misclassified* by the given set of weights $\vec{w}$. The function J_p is always nonnegative because for misclassified vectors, $\vec{w} \cdot \vec{y} < 0$.

If the given classification problem is *linearly separable*, that is, if there exists a set of weights leading to the correct classification for all vectors, it has been shown that those weights can be found by applying a *steepest descent* (or *gradient descent*) method to the function J_p. For any objective function J, the steepest descent method increments the current weight vector by some negative constant multiple of the gradient of J at the weight vector. (Recall from Section 3.5 that the gradient is the vector of partial derivatives with respect to the variables on which J depends; in this case, the variables are the weights w_i.) That is, at time t,

$$\vec{w}(t+1) = \vec{w}(t) - r(t) \nabla J(\vec{w}(t)) \qquad (6.13)$$

where r(t) is a "learning rate" (possibly different at each iteration). This means that on the next time step, the weight vector will move in the direction along which J decreases most sharply.

For the perceptron objective function J_p defined by (6.12), the jth component of the gradient ∇J_p is $\dfrac{\partial J_p}{\partial w_j}$, which equals $\sum_{\vec{y} \in Y} -y_j$. Hence the gradient vector equals $\sum_{\vec{y} \in Y} -\vec{y}$, and the gradient descent algorithm (6.13) becomes

$$\vec{w}(t+1) = \vec{w}(t) + r(t)\sum_{\vec{y} \in Y} \vec{y}$$
.

The procedure of Widrow and Hoff (1960) was originally designed for iterative solution of a system of linear equations. If A is an n-by-n matrix and $\vec{b}$ is a vector with n components, the goal is to find another vector $\vec{x}$ with n components such that $A\vec{x} = \vec{b}$; since A may not be invertible, that equation may have only an approximate solution. The objective function to be minimized here is the *mean-square error*

$$J_s(x) = \|A\vec{x}-\vec{b}\|^2 = (A\vec{x}-\vec{b})\cdot(A\vec{x}-\vec{b}) \qquad (6.14).$$

It can be shown (see Exercise 8 of this chapter) that the gradient of the function J_s defined by (6.14) is the vector $2A^t(A\vec{x} - \vec{b})$, where the superscript "t" denotes the transpose of a matrix (reversing rows and columns). Since the Widrow-Hoff procedure is sometimes used to simultaneously approximate the solutions to many equations with variable A and $\vec{b}$, this leads to a gradient descent algorithm whereby

$$\vec{x}(t+1) = \vec{x}(t) + r(t)(\vec{b}(t)-A^t(t)\vec{x}(t))\cdot\vec{x}(t)$$

(see Table 6.5). In a variant of this procedure (*e.g.*, Anderson & Murphy, 1986), the *matrix* is incremented iteratively as more vector pairs are presented.

Finally, the back propagation algorithm, like the Widrow-Hoff algorithm, uses a mean-square-error objective function, in this case $\|\vec{y}_p-\vec{t}_p\|^2$, where $\vec{y}_p$ is the actual output vector (for the pth pattern) and $\vec{t}_p$ the target vector. The learning procedure based on the gradient of this function was derived in Section 3.5. Table 6.5 summarizes the objective functions and the gradient descent algorithms for all of the rules discussed in this subsection.

	Perceptron	Widrow-Hoff	Back propagation
Objective function	$J_p(\vec{w}) = \sum_{\vec{y} \in Y} -\vec{w} \cdot \vec{y}$	$J_s(\vec{x}) = \|A\vec{x} - \vec{b}\|^2$	$\|\vec{y}_p - \vec{t}_p\|^2$
Descent algorithm	$\vec{w}(t+1) = \vec{w}(t)$ $+ r(t) \sum_{\vec{y} \in Y} \vec{y}$	$\vec{x}(t+1) = \vec{x}(t)$ $+ r(t)[\vec{e}(t) \cdot \vec{x}(t)],$ $\vec{e}(t) = \vec{b}(t) - A'(t)\vec{x}(t)$	$\vec{w}(t+1) = \vec{w}(t)$ $+ r(\vec{y}_p - \vec{t}_p) \cdot \vec{x}(t)$

Table 6.5. Summary of descent procedures for obtaining linear discriminant functions. (Adapted from Duda & Hart, *Pattern Classification and Scene Analysis*, copyright © 1973 John Wiley & Sons, by permission of the publisher.)

Some Implementation Issues for Back Propagation Equations

Recall from Sections 3.3 and 3.5 that the back propagation network of Rumelhart *et al.* (1986) uses nonlinear (typically sigmoid) activation or input-output functions f. The output of unit j (output or hidden) is

$$y_{pj} = f(net_{pj}) = f(\sum_i w_{ij} y_{pi}) \qquad (3.24)$$

with the sum taken over units i from the previous layer.

In most of the simulations done by Rumelhart *et al.*, a "bias term" θ_j is added to net_{pj}, the total signal received by unit j. This bias, which can be positive or negative, is interpreted as the spontaneous activation level of the jth unit, regardless of what inputs it is or is not receiving. Also, these simulations use a specific form of the activation function, namely, the *logistic function*, f(x) = 1/ (1 + exp (-x)). Hence, the special case of (3.24) used therein is

$$y_{pj} = \frac{1}{1 + \exp(-(\sum_i w_{ij} y_{pi} + \theta_j))} \qquad (6.15).$$

The logistic function has the convenient property that its derivative equals

$$f'(x) = f(x)(1 - f(x)) \qquad\qquad (6.16).$$

By (6.16), the change in f is largest when f is 1/2, and 0 when f is 0 or 1. In other words, output units that have not "made up their mind" whether to respond (with a 1) or not respond (with a 0) are subject to the greatest weight changes. In applied problems, as Rumelhart *et al.* (1986, p. 329) point out, values above .9 or below .1 are usually taken to be "decisions."

The logistic function also simplifies the equations given in Section 3.5 for changes of weights to both output and hidden units. Recall that the expression for the error signal at an output unit, derived from the chain rule for derivatives, is

$$\delta_{pj} = f'(net_{pj})(t_{pj} - y_{pj}) \qquad\qquad (3.10a)$$

where y_{pj} is the actual value of the jth output unit's activity and t_{pj} its desired value. Combining (3.26) with (6.16) and $y_{pj} = f(net_{pj})$, we obtain

$$\delta_{pj} = (t_{pj} - y_{pj})y_{pj}(1 - y_{pj}) \qquad\qquad (6.17).$$

From Section 3.5, error signals at hidden units are related to error signals at output units by

$$\delta_{pj} = f'_j(net_{pj})\sum_k \delta_{pk}w_{jk} \qquad\qquad (3.10b)$$

where the summation is over all output units k that receive inputs from the jth hidden unit. Equation (3.10b), combined with (3.26), (6.16), and $y_{pj} = f(net_{pj})$, yields

$$\delta_{pj} = y_{pj}(1 - y_{pj})\sum_k \delta_{pk}w_{jk} \qquad\qquad (6.18).$$

Rumelhart *et al.* originally made the change in the weight w_{ij} proportional to the product of activation level y_{pi} and error term δ_{pj}, the latter defined by (6.17) or (6.18). But it was found that with such a rule, too low a constant of proportionality (learning rate) makes learning much too slow, whereas too large a learning rate can lead to wild oscillations. They prevented oscillations at high learning rates by including a *momentum* term that makes the direction of present weight changes partly dependent on the direction of recent past changes. Hence the rule they actually used is

$$\Delta w_{ij}(t+1) = \beta(\delta_{pj}y_{pi}) + \alpha\Delta w_{ij}(t) \tag{6.19}$$

where ß (the learning rate) and α (the momentum) are two separate positive constants.

Brain-State-in-a-Box Equations

Section 6.2 listed the equation of Anderson *et al.* (1977) for linear transformation of the node activity vector (initially an input vector) over time. This equation is reproduced here:

$$\vec{x}(t+1) = \vec{x}(t) + A\vec{x}(t) \tag{6.4}$$

where $\vec{x} = (x_1, x_2, ..., x_n)$ is the vector of node activities at any given time, and A is the connectivity matrix of the system. But it was noted that Equation (6.4) can create instability through positive feedback; that is, it can drive the state vector of the system outside the hypercube (box) of Figure 6.15. To prevent such instability, Anderson *et al.* added a rule whereby any node activity driven above 1 by (6.4) is replaced by 1, and symmetrically, any activity driven below -1 is replaced by -1. Hence, if the subscript "$_i$" below any vector denotes its ith component, i between 1 and n, (6.4) is replaced by the equation

$$\vec{x}(t+1)_i = \max(-1, \min(1, [(I+A)\vec{x}(t)]_i)),$$
$$i = 1,...,n \tag{6.20}$$

where I denotes the identity matrix (1 along the main diagonal, 0 elsewhere).

Equation (6.20) is the one that Anderson et al (1977) actually used in their simulations. The matrix A is constrained to be *symmetric*: that is, for each i and j, the entry a_{ij} in the ith row and j^{th} column is identical to a_{ji}. Symmetry simplifies a matrix's mathematical properties. For example, a symmetric n-by-n matrix always has n mutually orthogonal *eigenvectors*, that is, n-dimensional vectors $\vec{x}_i$ such that $A\vec{x}_i$ is a constant multiple of $\vec{x}_i$. The constant that is multiplied is called an *eigenvalue* of the matrix A (see Jordan, 1986 for discussion). In order to simulate the positive feedback process involved in classification, Anderson *et al.* used a matrix for which all eigenvalues are positive, and all eigenvectors are corners of the box in Figure 6.15. (For examples, see Exercise 6 of this chapter.)

Anderson and Murphy (1986) use a variant of (6.20) that incorporates the initial state (*i.e.*, input) vector $\vec{x}(0)$, in order to keep input information present. Their equation for the change in the state vector $\vec{x}$ over time is

$$(\vec{x}(t+1))_i = \max(-1, \min(1, [\alpha A\vec{x}(t) + \Gamma\vec{x}(t) + \delta\vec{x}(0)]_i)),$$
$$i = 1,...,n$$

where α, Γ, and δ are different positive constants.

In addition, Anderson and Murphy (1986) include a scheme for learning the correct matrix A. The optimal A for forming a heteroassociative connection $\vec{x}$-to-$\vec{y}$, with $\vec{x}$ and $\vec{y}$ both vectors, is one for which

$$A\vec{x} = \vec{y} \tag{6.21}$$

(see the discussion in Section 6.2). Anderson and Murphy's error-correcting scheme for incrementing the matrix, based on the rule of Widrow and Hoff (1960), was designed to move A closer to satisfying (6.21). Their equation for the change in A is

$$\Delta A = \beta[(\vec{y} - A\vec{x}) \cdot \vec{x}] \tag{6.22}.$$

The learning rate β in (6.22) can be either fixed or adjustable. The matrix learning scheme defined by (6.22) can also, the authors pointed out, be used in autoassociative encoding by letting $\vec{y} = \vec{x}$. In that case, the learning scheme becomes an algorithm for making $\vec{x}$ an eigenvector of A with an eigenvalue equal to 1.

Rumelhart and Zipser's Competitive Learning Equations

The competitive learning algorithm of Rumelhart and Zipser (1985) is defined by changes in the interlevel weights w_{ij} of Figure 6.18. At each of Layers 2 and 3 in that figure, node j receives at each time step a signal equal to $\sum_i x_i w_{ij}$, the sum taken over nodes i in the next lower layer, where x_i is 1 if node i is active and 0 otherwise. (The algorithm can be extended to arbitarily many layers). If the lower layer is Layer 1 (input units), activity or inactivity is determined by which nodes are excited by the input pattern. If the lower layer is Layer 2, activity or inactivity is determined by whether the ith node won or lost its intra-cluster competition at the last time step. Within each cluster of the top two layers in turn, the node receiving the largest (linearly weighted) signal wins and the others lose.

The w_{ij} in turn change at each time step according to a rule similar to the gamma system rule of Rosenblatt (1962). No change occurs in connections to losing nodes; hence $\Delta w_{ij} = 0$ if unit j loses. For connections to winning nodes, weight is shifted from inactive to active pathways. Each of the input pathways to a winning node gives up some proportion k (between 0 and 1) of its weight, and that weight is then distributed equally among the active input pathways (*cf.* Malsburg, 1973). Hence

$$\Delta w_{ij} = k[\frac{c_i}{n} - w_{ij}] \;\; if \; unit \; j \; wins \qquad (6.23)$$

where n is the total number of active units at the next lower layer in the current pattern, and c_i is 1 if unit i at the lower layer is active and 0 otherwise. Note that

$$n = \sum_i c_i \qquad (6.24).$$

Equations (6.23) and (6.24) together imply that $\sum_i w_{ij}$ remains constant for a given j. In the simulations done by the authors, that sum is kept equal to 1.

Adaptive Resonance Equations

In the ART 1 network, the STM activity of the ith F_1 node is denoted by x_i and the STM activity of the jth F_2 node by x_j. The convention that the subscript "i" relates to F_1 and "j" to F_2 is observed throughout; hence, w_{ij} values represent bottom-up LTM strengths (synaptic weights) and w_{ji} values represent top-down LTM strengths.

The STM traces at F_1 are assumed to change quickly under the influence of shunting (multiplicative) excitation from outside inputs and from top-down signals, and shunting inhibition from F_2 (mediated by one of the gain control nodes, shown in Figure 6.20 as dark circles). Hence

$$\epsilon \frac{dx_i}{dt} = -x_i + (1-A_1 x_i)(I_i + D_1 \sum_j f(x_j) w_{ji})$$
$$- (B_1 + C_1 x_i) \sum_j f(x_j) \qquad (6.25)$$

where ϵ is small (.05 or .1). The function f of Equation (6.25) is defined by f = 1 if node j of F_2 is active and 0 if node j is inactive. Only one node of F_2 is active at a time. After an input comes in, the jth F_2 node receives a bottom-up signal equal to $T_j = D_2 \sum_i h(x_i) w_{ij}$, where D_2 is a positive constant and h is the Heaviside (unit step) function. The category chosen to be active (that is, to be tested for match or mismatch between bottom-up and top-down patterns) is the one for which T_j is the largest.

The activities x_j of the F_2 nodes are in turn governed by the equations

$$e \frac{dx_j}{dt} = -x_j + (1 - A_2 x_j)(g(x_j) + T_j) - (B_2 + C_2 x_j) \sum_{k \neq j} g(x_k) \qquad (6.26)$$

where g is a sigmoid function and T_j is as defined above. The summation in Equation (6.26) indicates lateral inhibition in an on-center off-surround field.

The LTM trace of the top-down pathway from w_j to w_i obeys the learning equation

$$\frac{dw_{ji}}{dt} = f(x_j)[- w_{ji} + h(x_i)] \qquad (6.27).$$

where h is again the unit step function. The bottom-up pathway obeys an equation similar to (6.27) except that the synaptic decay term embodies the Weber Law Rule. As discussed in Section 6.3, this rule is designed to prevent access to a category by supersets of the category prototype; this is achieved by selectively decreasing bottom-up signals from input patterns that activate large numbers of F_1 nodes. The design is achieved through competition between LTM traces, resulting in equations of the form

$$\frac{dw_{ij}}{dt} = Kf(x_j)[- E_{ij} w_{ij} + h(x_i)] \qquad (6.28)$$

where $E_{ij} = h(x_i) + (1/L) \sum_{k \neq i} h(x_k)$ with K a constant, L > 1, and the sum taken over all F_1 indices k not equal to i. It can be shown that the other rule described above, the 2/3 Rule, is satisfied by choosing the parameters of (6.25) such that max $(1, D_1) > B_1 > D_1$. Equations (6.27) and (6.28) guarantee that learning only occurs at synapses to or from active category nodes.

The ART 2 equations, which are not given here, also involve shunting excitation and inhibition in the manner of equations (6.25) and (6.26). The shunting terms in the F_1 STM equations reflect influences from six extra sets of input processing nodes shown in Figure 6.22, designed to suppress noise that is characteristic of analog patterns. The LTM laws are simpler than in ART 1, because superset encoding is irrelevant when patterns consist of continuous values rather than just 1's and 0's. Hence the Weber Law Rule is not used and top-down and bottom-up LTM equations are nearly the same.

EXERCISES FOR CHAPTER 6

*1. Simulate Malsburg's equations (6.5) and (6.6), using the parameters used in his original article. In each equation, let the starred values represent the values of the received signal minus the threshold 1; for example, $x_{e,i}^* = $ max $(x_{e,i}-1, 0)$. Let the "retina" consist of 19 nodes, that is, the main diameter is 5 nodes long, as shown in Figure 6.2. Let the "cortex" consist of 169 excitatory and 169 inhibitory nodes, that is, the main diameter of each layer is 15 nodes long, as shown in Figure 6.3.

The values of the connection strengths p_{ik} are all equal to .4, the q_{ik} to .3, and the r_{ik} to .286. For each excitatory cortical node k, the s_{ik} are first chosen randomly with uniform distribution over the set [0.0.25] and then normalized in order to sum to the mean value of possible sums, which is $19(0.25)/2 = 2.375$. That is, for a fixed k, each s_{ik} is replaced by

$$2.375 \ \frac{s_{ik}}{\sum_{i=1}^{19} s_{ik}}$$

The learning rate h of Equation (6.6) is 0.05. The s_{ik} are renormalized after each learning time step.

Run 100 time steps of these equations, on each one presenting a random stimulus chosen from the ones shown in Figure 6.2. In Equation (6.6), the ith retinal node sends a signal of $A_i^* = 1$ if the node is darkened in that figure, and $A_i^* = 0$ otherwise. After 100 time steps, look at the optimal orientations for all the cortical nodes, and compare with Figure 6.3. (Because of the random inputs, exact replication of the figure is not expected. A smaller hexagon, with fewer than 169 nodes, could be used to produce the same effects; if there are too few nodes, however, the boundary might need to "wrap around" to correct for edge effects in the network.)

**2. Show that if a classification problem is unsolvable with a single layer linear machine (that is, one that uses linear activation functions), then it is unsolvable for linear machines with an arbitrary number of layers. Does the same result apply to nonlinear machines? If yes, give a proof; if no, give a counterexample.

**3. Consider the competitive learning model of Grossberg (1976b) defined by Figure 6.6. Let the activities x_{1i} of the F_1 layer encode the relative input intensities, *i.e.*,

$$x_{1i} = \frac{I_i}{\sum_j I_j}$$

Let $x_{2j} = 1$ if $S_j > \max (\tau, S_k: k{\neq}j)$,
        0 if $S_j < \max (\tau, S_k: k{\neq}j)$,

where for $j = 1, 2, 3, S_j = \sum_i \theta_i w_{ij}$, $i = 1, 2$. For each j, let the vector of synaptic weights to node j of F_2 be $\vec{w}_j = (w_{1j}, w_{2j})$, let $\vec{\theta} = (\theta_1, \theta_2)$, and let $\vec{w}_j$ obey the equation

$$\frac{d\vec{w}_j}{dt} = x_2(- \vec{w}_j + \vec{\theta})$$

with initial conditions $w_0(0) = (-1,0)$, $w_1(0) = (0,1)$, and $w_2(0) = (1,0)$.

Give an example of a spatial pattern $\vec{\theta}$ for which a stable coding can be achieved, and an example for which a stable coding cannot be achieved. Prove your claims.

4. Simulate a back propagation network with 0 biases. Let the input units form a 5x5 grid, and each hidden unit respond to a 3x3 subgrid, as described in Rumelhart et al (1986) and shown in Figure 6.27.

Use initial weights of .5 for all connections. Teach the network the "T-C" discrimination, so that it should respond with a 1 to T, 0 to C, in any rotation but centered at (3,3) in the visual field, as shown in Figure 6.28.

Use the sigmoid function $1/(1 + \exp(-x))$. Do not correct for edge effects. Use a positive momentum term (you can experiment with that). Look at the final response patterns of the hidden units after several thousand iterations.

○5. Design a classification problem involving written or spoken text, and a back propagation network to solve it.

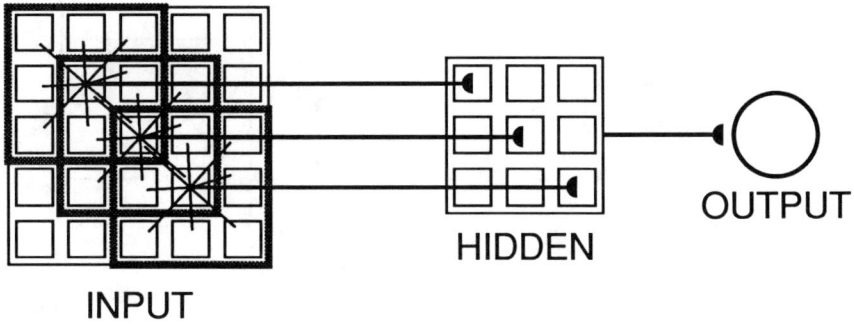

Figure 6.27. Example of the three layer back propagation network used in the simulation of Exercise 4.

6.   Simulate the brain-state-in-a-box equations due to Anderson et al (1977):

$$\vec{x}(t+1)_i = \max(-1, \min(1, [(I+A)\vec{x}(t)]_i)),$$
$$i = 1, \ldots, n$$

(6.20).

(a) Choose n = 4, and choose the matrix **A** to be

$$
\begin{bmatrix}
\frac{3}{2} & 0 & 0 & \frac{1}{2} \\
0 & \frac{3}{2} & \frac{1}{2} & 0 \\
0 & \frac{1}{2} & \frac{3}{2} & 0 \\
\frac{1}{2} & 0 & 0 & \frac{3}{2}
\end{bmatrix}
$$

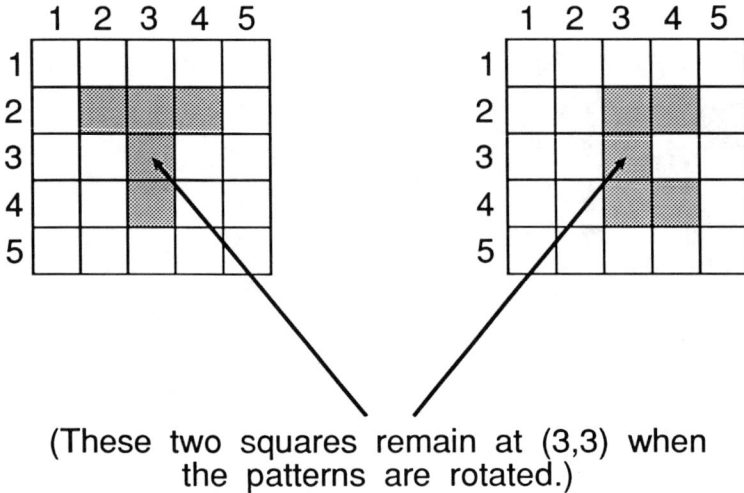

(These two squares remain at (3,3) when the patterns are rotated.)

Figure 6.28. Basic "T" and "C" input patterns used by Rumelhart *et al.*(1986). The network of Figure 6.27 can be trained to classify all of the four rotations of the "T" in one category and all of the four rotations of the "C" in another category.

This matrix has the two 4-dimensional hypercube corners $\vec{P} = (1, 1, -1, -1)$ and $\vec{Q} = (1, -1, 1, -1)$ as eigenvectors, both with eigenvalue 1. Define the 16 starting points $\vec{Q}_i$, i = 0 to 15, by $\vec{Q}_0 = \vec{P}$, $\vec{Q}_{15} = \vec{Q}$, and $\vec{Q}_i = (1, r_i, s_i, -1)$, where $r_i = \cos\theta_i + \sin\theta_i$, $s_i = -\cos\theta_i + \sin\theta_i$, $\theta_i = 12$ i degrees. Thus the $\vec{Q}_i$ have their first and last components fixed at 1 and -1, their middle two at equal spaces along the arc of the circle between (1, -1) and (-1, 1).

Test the "categorizations" made by the network based on the $\vec{Q}_i$ as starting positions, that is, whether the final state of the network is $\vec{P}$ or $\vec{Q}$. With no noise added to the $\vec{Q}_i$, the final state should always be $\vec{P}$ for i = 0 to 7, $\vec{Q}$ for i = 8 to 15. Then test the categorizations with Gaussian noise of various standard deviations[2] added to each component of each $\vec{Q}_i$.

(b) Do the same as in part (a) except with the matrix A equal to

$$
\begin{bmatrix}
\dfrac{3}{2} & 0 & \dfrac{1}{2} & 0 \\[2ex]
0 & \dfrac{3}{2} & 0 & \dfrac{1}{2} \\[2ex]
0 & \dfrac{1}{2} & 0 & \dfrac{3}{2} \\[2ex]
\dfrac{1}{2} & 0 & \dfrac{3}{2} & 0
\end{bmatrix}
$$

This matrix has $\vec{P}$ and $\vec{Q}$ as eigenvectors with eigenvalues 1 and 2, respectively, giving the network a bias in favor of going to the corner $\vec{Q}$.

*7. The BSB model can be mathematically described by the equations

$$
x_i(t+1) = S\left(x_i(t) + \sum_{j=1}^{n} w_{ij} x_j(t)\right) \tag{6.29}
$$

where the weight matrix W is symmetric ($w_{ij} = w_{ji}$), and the function S is linear with hard saturation, i.e.,

$$
\begin{aligned}
S(a) = & \ -1 \text{ if } a < -1 \\
& \ a \text{ if } -1 \le a \le 1 \\
& \ 1 \text{ if } a > 1
\end{aligned}
$$

---

[2] Some computer systems have access to a package that generates Gaussian noise, that is, generates a random variable with a normal ("bell-shaped") distribution. If you do not have access to such a program, the following procedure (Box & Muller, 1958) can be used for this purpose. First let u_1 and u_2 be two random variables *uniformly* distributed between 0 and 1, each obtained, for example, by the algorithm in Chapter 2, Exercise 3. Then $x_1 = [-2 \ln u_1]^{1/2} \sin(2\pi u_2)$, $x_2 = [-2 \ln u_1]^{1/2} \cos(2\pi u_2)$ are two independent Gaussian variables of mean 0 and standard deviation 1; to get a variable of standard deviation s, multiply x_1 or x_2 by s.

The matrix W is defined to be *diagonally dominant* (DD) if each of its diagonal entries is nonnegative and at least equal to the sum of absolute values of other entries in the same column, *i.e.*,

$$w_{jj} \geq \sum_{i \neq j} |w_{ij}|, \, j = 1,2,...,n$$

If W is DD, prove that every extreme point of the hypercube of possible states, that is, every n-dimensional vector with 1 and -1 as components, is an equilibrium point of the system defined by (6.29).

**8. Verify the gradient given in Section 6.5 for the Widrow-Hoff objective function defined by (6.19).

○9. How could the ART 1 network of Carpenter and Grossberg be modified to include a statement of the degree of certainty in the choice made? In particular, if a pattern just barely passes the vigilance criterion for more than one category at a time, how would "ambiguity detection" be built in? Note: *Recording* ambiguity is not the same as *resolving* ambiguity.

*10. Run a simulation of the ART 1 network, using Equations (6.25) to (6.28). For the ART 1 equations discussed above, choose any parameter settings that obey the following inequalities from Table 1 of Carpenter and Grossberg (1987a):

$A_1 \geq 0$, $C_1 \geq 0$, max $(1, D_1) < B_1 < 1 + D_1$, $0 < \varepsilon \ll 1$ (where "$\ll$" is an imprecise term meaning "much smaller than"), K is close to 1, L > 1, $0 < z_{ij}(0) < L/(L-1+M)$, $1 \geq z_{ji}(0) > (B_1-1)/D_1$.

Let F_1 consist of a 5-by-5 "pixel" array, and let the patterns A, B, C, and D be as shown in Figure 6.29.

(a) Following the simulations shown in Figure 6 of Carpenter and Grossberg (1987a), present these patterns repeatedly in the order ABCAD, with vigilance level r=.8. Show that after several repetitions, the patterns A, B, C, and D are eventually coded in separate, stable categories at F_2.

(b) Lower the value B_1 to a value below max $(1, D_1)$, which means that the 2/3 Rule is violated. Show that periodic recoding of pattern A can occur.

(c) Progressively lower the vigilance level r by increments of .1 until two or more of the patterns are coded in the same category, then until all four are coded in the same category.

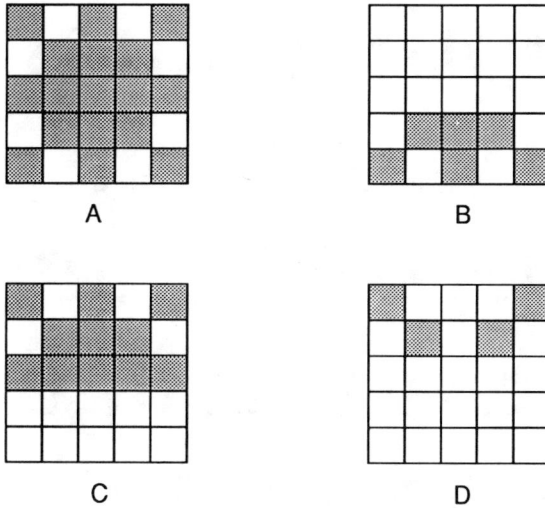

A

B

C

D

Figure 6.29. Input patterns used in the ART 1 simulation of Exercise 10.

7

Optimization, Control, Decision, and Knowledge Representation

Thinking in its lower grades is comparable to paper money, and in its higher forms it is a kind of poetry.

Havelock Ellis (*The Dance of Life*)

We may affirm absolutely that nothing great in the world has been accomplished without passion.

Georg Hegel (*Philosophy of History*)

The preceding chapters covered the areas of neural network research that have been the most well developed to date. This chapter takes us into the farther reaches of neural and cognitive modeling, problems that are largely unsolved and profoundly important.   Potential applications of the work discussed include the development of robots with capacities for planning, the building of "intuitive" capabilities in intelligent machines, and the treatment of mental illness.

Some topics covered in this chapter (optimization, motor control, speech synthesis and recognition) have been the object of much recent work, which is discussed in a condensed manner for space reasons.   Good sources on these topics include the relevant sessions in the proceedings of the various International Conferences and International Joint Conferences on Neural Networks, sponsored by IEEE and the International Neural Network Society (INNS), and in Volume 1, Supplement 1 of the journal *Neural Networks*.   Other topics covered in this chapter (decision making, higher-order control, concept relationships, and inference) are still largely *terra incognita*.   This chapter gives a good picture of the state of the art in those areas.

Some cognitive scientists believe that high-level cognitive problems pose difficulties for neural network theory.   These researchers (*e.g.*, Chandrasekharan, Goel, & Allemang, 1988; Fodor & Pylyshyn, 1988) assert that current neural network models are mainly useful for pattern recognition and classification, and inherently unsuitable for more complex functions such as concept formation, semantic information processing, and inference.   But the human capacity for such higher-order functions must have *some* biological basis.   The principle of parsimony, operative in other areas of biology, suggests that brain areas involved in higher-order processes should contain some of the same neural architectures used for lower-order processes, albeit in different arrangements.

Hence we must ask: can the fundamental ideas of earlier chapters — associative learning, competition and cooperation, sigmoid signal functions, shunting interactions, opponent processing, modifiable interlevel feedback, etc. — play a role in modeling complex mental functions?   There are enough preliminary results to give us optimism on this score.

In the frontier areas of neural networks covered in this chapter, any way of organizing the discussion must be somewhat arbitrary.   As a starting point for our exposition, we use the principle of optimization, or reinforcement learning.   We begin with a discussion of neural networks for solving optimization and control problems.   Then we investigate the inadequacies of optimization theory with respect to understanding certain biological behavioral phenomena.   This brings us to a discussion of decision systems with multiple, and highly context-dependent, decision criteria.

7.1. OPTIMIZATION AND CONTROL

Several different types of problems can be grouped together under the rubric of optimization and control in neural networks. One type of problem relates to conditioning, already discussed extensively in Chapter 5. Through selective attention (as indicated by experimental paradigms such as blocking), the organism attempts to predict the environment so as to maximize (optimize) positive reinforcement and minimize negative reinforcement (*e.g.*, Sutton & Barto, 1981, 1991; Grossberg & Levine, 1987; Klopf, 1982, 1988). This kind of prediction of the environment is sometimes called *reinforcement learning* (Hinton, 1987; Werbos, 1988a), and has been related to artificial networks designed for other forms of prediction such as economic forecasting (Werbos, 1988b).

A second type of problem relates to motor control, with applications both to controlling robots and to understanding biological movement. A target position for a muscle is calculated and the current position is continuously updated by comparison with the target (*e.g.*, Grossberg & Kuperstein, 1986; Kawato, Furukawa, & Suzuki, 1987; Bullock & Grossberg, 1988). This comparison fits within the rubric of supervised learning as discussed in Section 6.3.

A third type of problem relates to what is sometimes called classical optimization. In this kind of problem, artificial neural networks are used to converge to the best solution for a particular task (*e.g.*, Kirkpatrick *et al.*, 1983; Hopfield & Tank, 1985; Hinton & Sejnowski, 1986; Szu, 1986). Two examples of such optimization problems are stated by Hopfield and Tank (1985, p. 141): "Given a map and the problem of driving between two points, which is the best route? Given a circuit board on which to put chips, what is the best way to locate the chips for a good wiring layout?" The solutions obtained may not resemble the ways in which humans perform these tasks, but use "connectionist" techniques from neural network theory, rather than symbolic processing techniques from traditional artificial intelligence.

Robotics and classical optimization are both areas in which neural network principles are not well established. It is possible, but uncertain, that similar principles will prove valuable for both types of problems. In our discussion, we start with classical optimization problems for two reasons. First, such problems provide a fairly simplified framework for discussing some network ideas that may have wider utility. Second, the differences between classical optimization situations and more realistic biological situations will be instructive for later sections of this chapter.

Hopfield, Tank, and the Traveling Salesman Problem

One of the longest-standing optimization problems is the Traveling Salesman Problem (TSP), discussed by Hopfield and Tank (1985, 1986). The TSP is the problem of how to cover all of a given number n of cities while traveling the minimum total distance. After covering all of them, the salesperson is assumed to return to the starting city. Traditional computing techniques have encountered difficulties in obtaining the best solution to this problem, because the solution to the TSP is computed in a time that grows exponentially with the number of cities (Garey & Johnson, 1979).

The equations used by Hopfield and Tank (1985) in their TSP simulation is given in Section 7.3. They are extensions of the analog (continuous-time) Equations (4.15) from other articles by the same authors (cf. Chapter 4). The innovation of the 1985 article is to make the network two-dimensional: each node has two indices, one for a city and the other for a possible position within the tour. The network interactions are reminiscent of some of the two-dimensional visual networks discussed in Section 4.3 (Amari & Arbib, 1977, where the dimensions were position and disparity, and Grossberg & Mingolla, 1985 a, b, where the dimensions were position and orientation).

As in previous Hopfield-Tank articles, the system used in the TSP simulation always converges to a steady state. As before, this steady state corresponds to one of the local minimum states of a certain global energy function for the system (cf. Figure 4.11). But in this case the steady states have a particular meaning, as illustrated in Figure 7.1. Corresponding to each city, there is a vector of node activities corresponding to the city's possible order of arrival on the salesperson's tour. In a steady state, that vector will have a 1 in one position and 0 in all the others; for example, if the city in question is third on the tour, the 1 will be in the third position of the vector.

Hence, steady states of the network of Hopfield and Tank (1985) correspond to possible tours between the cities. But what guarantees that the network will converge to the *optimal* (*i.e.*, shortest-distance) tour? In fact, it may not, but a term is added to the differential equations for the system (and therefore to the system's energy function) that weights the interactions by the distances between cities, so that solutions associated with longer-distance tours are penalized. This means that the system, while not guaranteed to converge to the optimal steady state, is biased in favor of doing so.

Hopfield and Tank's article has inspired a variety of other neural network schemes for solving the TSP (*e.g.*, Aarts & Korst, 1987; Durbin & Willshaw, 1987; Angeniol, DeLaCroixVaubois, & LeTexier, 1988). In particular, Angeniol *et al.* (1988) used an idea of Kohonen (1984), called the *self-organizing feature map*, to improve on Hopfield and Tank's computation time (proportional to the square of the number of cities). In their article, n nodes (as many as there are cities) are placed at random locations on a two-dimensional map, and at each iteration, each node moves toward the city that it is nearest to.

As discussed in Section 6.2, even highly supervised networks like the back propagation network of Rumelhart *et al.* (1986) are not guaranteed to converge to a particular, desired equilibrium. This insight motivated the development of various schemes for perturbing a network that is approaching a suboptimal steady state, thereby increasing the probability that the network will ultimately approach another (optimal) steady state. We now turn to some of these schemes, starting with the notion of simulated annealing.

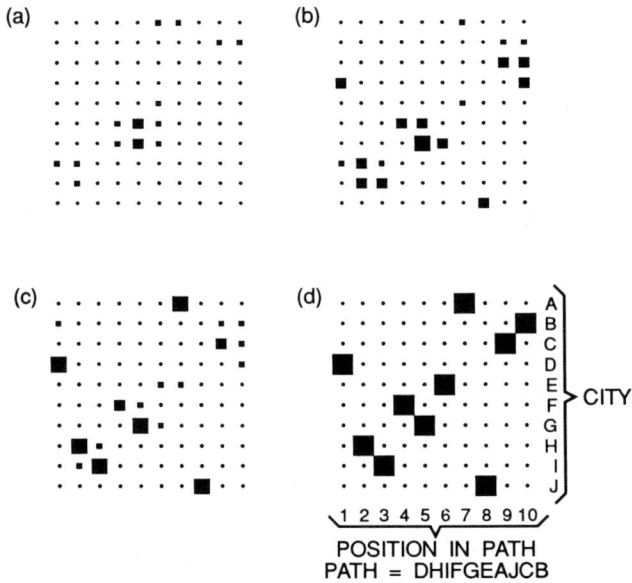

Figure 7.1. Convergence of a 10-city 10-position neural network to a tour in the TSP. The size of each square is proportional to the activity of the node x_{ij} for the ith city and the jth position. Indices in (d) illustrate how the final state is decoded in a tour. (Reprinted from Hopfield & Tank, 1985, with permission of Springer-Verlag.)

Simulated Annealing and Boltzmann Machines

The term *simulated annealing* was used by Kirkpatrick *et al.* (1983) to describe a formal analogy between their network and a certain physical system. Annealing is the process whereby a substance is first melted at a very high temperature, and then slowly cooled to a desired shape (for example, a crystal structure of a solid).

Kirkpatrick *et al.* adapted an algorithm from statistical mechanics for converging to one of many possible "cooled" (low energy) states. In this algorithm, energies obey a probability distribution called the Boltzmann distribution, in which the probability of a given energy E is an exponentially decaying function of E. As temperature increases, the decay rate becomes slower, which in turn increases the probabilities of high-energy states. The "temperature" in the network of Kirkpatrick *et al.* is not a literal temperature, but a control parameter, set outside the network, determining the relative probabilities of different system energy levels. As in the networks of Hopfield (1982, 1984) and Hopfield and Tank (1985, 1986) (*cf.* Chapter 4 and the last subsection of this chapter), the system converges to a state at which the total system energy has a local minimum. A small random displacement of one unit (atom in the case of a solid, node in the case of a neural network) occurs at given time intervals. If the change ΔE in energy caused by this perturbation is negative or zero, the system displacement is "accepted" (*i.e.*, allowed to reset the network) and carries over to the next time step. If, however, ΔE is positive, the displacement is accepted with probability $P(\Delta E)$, where P is an exponentially decaying (*i.e.*, Boltzmann-type) function. Otherwise, the system reverts to its unperturbed state. Thus the system can sometimes be perturbed out of equilibrium states; the network is designed so that perturbation be most likely from states that are highest in energy, that is, furthest from globally optimal (*cf.* Figure 4.12).

In the optimization problems discussed by Kirkpatrick *et al.*, the energy function is replaced by a cost function. The simulated annealing process is superimposed on the perturbation algorithm described above. As shown in Figure 7.2, the function P is such that the higher the temperature, the greater the likelihood of a perturbation from an equilibrium state. Hence, the algorithm calls for raising the temperature when the system is at or near a nonoptimal steady state, and lowering it when the system is at or near an optimal steady state.

After the article of Kirkpatrick *et al.*, Geman and Geman (1984) proved that the simulated annealing algorithm converges for a wide range of conditions. Also, there has been extensive work on incorporating the simulated annealing algorithm into neural network structures. Perhaps the best known network of this sort is the *Boltzmann machine* (Hinton, Sejnowski, & Ackley, 1984; Ackley, Hinton, & Sejnowski, 1985; Hinton & Sejnowski, 1986) which combines simulated annealing with a network structure of the Hopfield-Tank type.

An outline of the Boltzmann machine's learning algorithm is given here; more mathematical details appear in Section 7.3. The network is organized into *visible* and *hidden* units, but unlike the back propagation network of Rumelhart *et al.* (1986), it lacks output nodes. Its visible units are all input nodes, and its "output" consists of the classifications learned by hidden units. The algorithm incorporates a form of supervised learning. The correction is done not by a target output (see the discussion in Section 6.2) but by the probability distribution of the input patterns in the environment.

In the Boltzmann machine, external inputs reconfigure the states (which are "on" or "off" as in Hopfield, 1982) of visible units but not of hidden units. The states, combined with the connection weights, determine the system energy, which has the same functional form used by Hopfield (*cf.* Section 4.5). The energy level in turn, via the Boltzmann distribution, determines the probability associated with the current state.

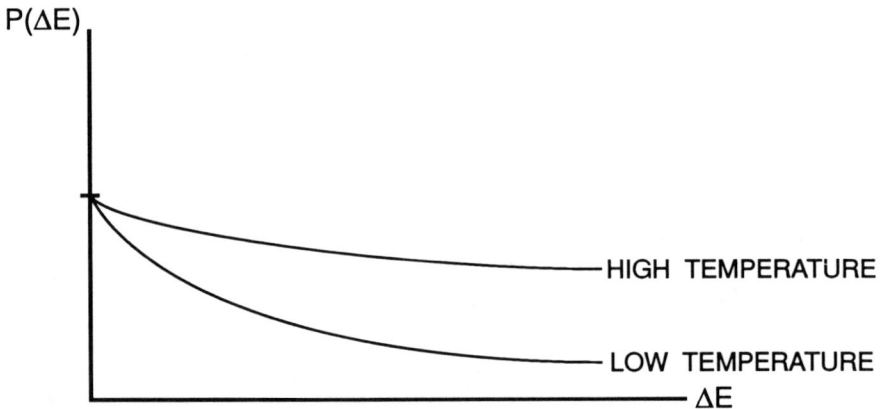

Figure 7.2. Probability of transition to a new system state as a function of increase in system energy, in a simulated annealing algorithm.

Finally, these state probabilities determine the changes in weights over time. Specifically, the change in the weight w_{ij} between the ith and j^{th} units (whether visible or hidden) is proportional to the influence of the current input on the probability of both those units being simultaneously in the "on" state. This influence, in the earliest versions of this network, is measured as p_{ij}-p'_{ij}, where p_{ij} is the probability of both units being on in the presence of the current input, and p'_{ij} is the same probability in the absence of this input.

The energy function in the Boltzmann machine has some formal similarity to the *harmony* function developed by Smolensky (1986). Since Smolensky's harmony is a measure of the coherence of an input or a proposed action with existing knowledge representations, his network is discussed in the subsection on knowledge processing in Section 7.2.

The networks discussed in this section have all achieved some success in using physical analogies to solve difficult problems efficiently. There have been many recent variations on these networks, some using different probability distributions than the Boltzmann (*e.g.*, Levy & Adams, 1987; Scheff & Szu, 1987). And the Boltzmann machine itself has been subject to many mathematical variations (*e.g.*, Aarts & Korst, 1989). But physical analogies have some limitations, based on the homogeneity of the ensuing network structures (see Mingolla & Bullock, 1989 for further discussion). Networks of this sort tend to lack functional subnetworks that correspond either to brain regions or to cognitive subprocesses. This lack of definable subnetworks does not prevent these models from being useful in computing applications, but significantly limits their ability to explain human or animal behavior. We now proceed to some neural models of motor control systems, many of which have been directly inspired by biological data.

Motor Control: the Example of Eye Movements

Some general issues involved in a mechanistic understanding of motor control were discussed by Kuperstein (1988b, p. 1308): "The human brain develops accurate sensorimotor coordination despite many unforeseen changes in the dimensions of the body, strength of the muscles, and placements of the organs. This is accomplished for the most part without a teacher." Two other issues are the ability to learn an invariant movement regardless of its velocity, and the synchronization of different muscles into an overall coordinated movement (*cf.* Bullock & Grossberg, 1988).

Neural network models of motor control have tended to focus on specific biological systems, such as eye movements, arm movements, and speech production. Of these, eye movements provide the simplest laboratory for studying complex kinds of control. Eye movements only require three pairs of agonist-antagonist muscles, and do not involve variable load as do arm or leg movements. Yet (and this is a somewhat daunting fact) the *control* of eye movements (by a combination of intention and visual inputs) is complex enough to be the subject of a book of over three hundred pages filled with large circuits (Grossberg & Kuperstein, 1989)!

The eyes move either by slowly tracking objects or by jumping abruptly from one fixation to another. The models we discuss concentrate on the latter type, known as *ballistic* or *saccadic* movements. In these movements, one or both eyes move within the head so as to bring an object, noticed either through peripheral vision or through another sense such as audition, into perception by the *fovea* or central part of the retina; this is called *foveating* the object. Once the saccadic movement (*saccade*) is initiated, it is fairly insensitive to feedback until after it is performed, unlike arm movements, which can be interrupted by a change in the environment. Yet, the simple saccades are subject to control by the retina and many brain areas (superior colliculus, parietal cortex, cerebellum, peripontine reticular formation, visual (occipital) cortex, frontal cortex, and oculomotor nuclei; *cf.* Appendix 1).

Robinson (1981) noted that a saccade to a visual target does not always immediately perform an accurate foveation. He therefore proposed an error-correcting mechanism for saccade generation, based largely on ideas from traditional control theory as used in engineering. Robinson's theory is basically linear but does recognize the necessity of nonlinear transformations from neural activity to muscle activity.

The model of Grossberg and Kuperstein (1989) uses much the same error-correcting mechanism as does Robinson's, but also incorporates the necessity of translating visual information into motor control information. The visual information leads to a *target position* code, and the current locations of the eyes to a *present position* code. The difference between these two positions generates an error signal that effects a movement command as shown in Figure 7.3.

One challenge that Grossberg and Kuperstein faced was to transform visual coordinates, represented by a two-dimensional vector denoting retinal location, into oculomotor coordinates, represented by a six-dimensional vector denoting activities of the six external eye muscles. (These muscles are organized into three antagonistic pairs, for movement out and in, up and down, and forward and back.) The mechanism for organizing such compensatory eye movement should also, if possible, organize the eye movements in the *vestibulo-ocular reflex* (VOR). The VOR is the reflex whereby if the head moves, vestibular signals move the eyes in the opposite direction to maintain fixation of a visual target.

The control of both saccades and the VOR involves both a learned and an unlearned component, as shown in Figure 7.3. The learned component appears to be identifiable with the cerebellum (see Optican & Robinson, 1980, and Ito,

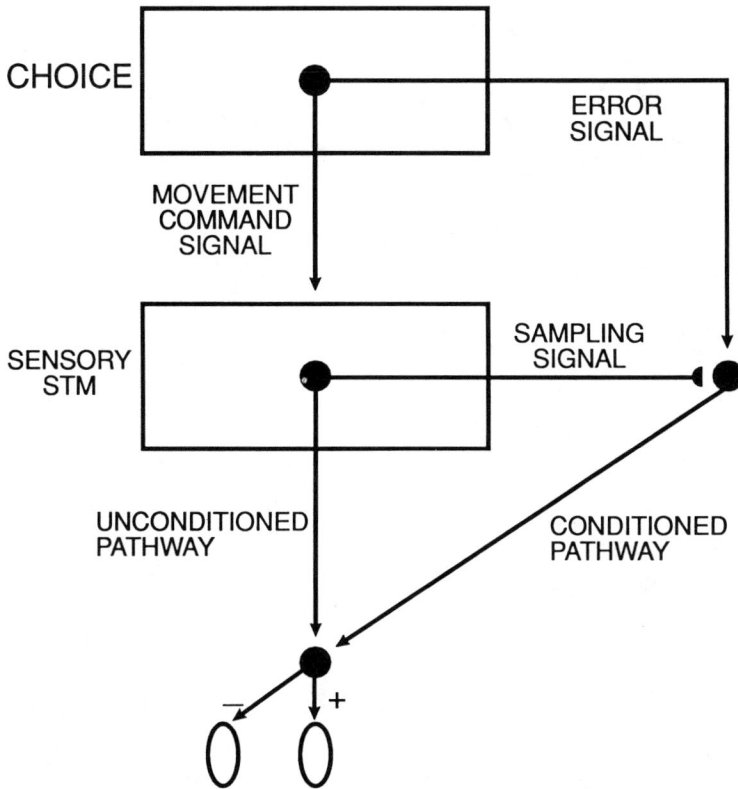

Figure 7.3. A visual target chosen to be foveated gives rise to an unconditioned and a conditioned movement signal. The unconditioned signal causes movements and the conditioned signal corrects these movements. The conditioned signal's strength can be altered by error signals mediated by a second visual target. Vertical ellipses at the bottom represent antagonistic pairs of eye muscles. (Adapted with permission from Grossberg & Kuperstein, *Neural Dynamics of Adaptive Sensory-motor Control*. Copyright 1989 Pergamon Press.)

1982; also see Houk, 1987 for a data-driven network model of the cerebellum's role in controlling other kinds of movements). Also, the first saccade to a visual target does not always immediately bring the target to the fovea; it sometimes overshoots or undershoots, and must then be followed by a smaller corrective saccade.

These considerations led Grossberg and Kuperstein to construct the co-ordinate map shown in Figure 7.4. The visual space on the retina was divided into *sectors* denoting the six eye muscles (labeled α^+, α^-, β^+, β^-, Γ^+, and Γ^-). The locations in this figure indicate what type of corrective movement by which muscle is needed to foveate the desired visual target.

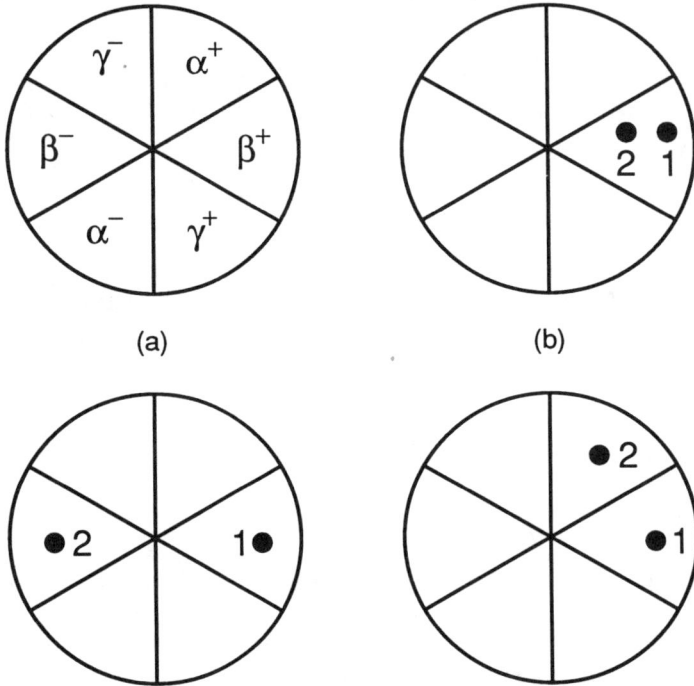

(a) (b)

Figure 7.4. (a) Sectors of the retina of one eye. If light falls on the given sector, the indicated muscle must move to foveate the light. (α^+, α^-), (β^+, β^-), and (Γ^+, Γ^-) are antagonistic pairs. (b) A saccadic undershoot error. "1" locates the retinal position of the first light, "2" of the second light. (c) A saccadic overshoot error. (d) A skewed undershoot error. (Reprinted with permission from Grossberg & Kuperstein, *Neural Dynamics of Adaptive Sensory-motor Control.* Copyright 1989 Pergamon Press.)

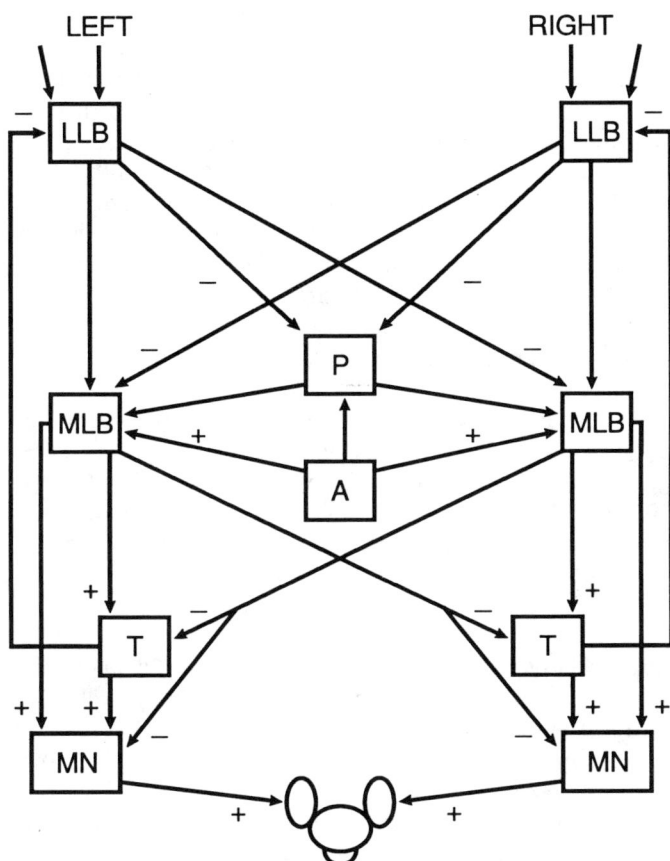

Figure 7.5. Saccade generator circuit controlling a pair of mutually antagonistic (left and right) muscles. Nodes correspond to groups of cells whose abbreviations denote their firing properties (LLB=long lead burster, MLB=medium lead burster, P=pauser, T=tonic or steadily firing cell) or functions (A=arousal, MN=motor neuron). Actions of these nodes depend on eye position maps. The circuit helps the eye to maintain its final position and terminate the saccade when that position is reached. (Reprinted with permission from Grossberg & Kuperstein, *Neural Dynamics of Adaptive Sensory-motor Control.* Copyright 1989 Pergamon Press.)

The later parts of Grossberg and Kuperstein's book are largely devoted to the details of the control structures in Figures 7.3 and 7.4, including correspondences with the physiology of the brain areas mentioned above. For instance, Figure 7.5 illustrates details of the bottom two pathways of Figure 7.3.These networks obey equations discussed in Section 7.3. The network of Figure 7.5 is designed to shut off the saccade output even in the presence of the (visual) input that originally generated that output, once the saccade has been performed and the error corrected. The mechanism to accomplish this task uses feedback inhibition and agonist-antagonist (gated dipole) interactions, with the addition of special types of cells with desired firing properties (tonic cells, burst cells, pause cells, etc.). The equations describing this network were shown to fit a variety of saccade-related data, ranging from single-neuron responses in the brain stem and cerebellum to effects of arousal on movements.

The final concern of Grossberg and Kuperstein's work is the formation of target position maps that are invariant in spite of movement of the organism. This idea was inspired by the notion of sensory-motor feedback loop or *circular reaction* (Piaget, 1952) and was further developed in Kuperstein (1988b). Figure 7.6 shows a network for such invariant target recognition.

Motor Control: Arm Movements

Bullock and Grossberg (1988) modeled a variety of data on the invariances of planned arm movements. This includes, for example, the "bell-shaped" velocity profile shown in Figure 7.7, based on data of Atkeson and Hollerbach (1985): the velocity of movement as a function of time has the same qualitative shape over a wide range of movement sizes and speeds.

There are at least two possible ways to model such invariances. One of them is a network based on optimization ideas (*e.g.*, Flash & Hogan, 1984; Hogan, 1984) which includes a high-level stage of nodes that explicitly calculates the invariant trajectory. The other is a network in which globally invariant properties are not explicitly programmed but emerge from events distributed across many interacting sensory, neural, and muscular loci (*e.g.*, Bullock & Grossberg, 1988). The latter class of models includes error correction (reminiscent of the circular reaction of Piaget, 1952) but no explicit optimization. Both types of models have been suggested as bases for robotic control.

Bullock and Grossberg (1988) were led to a model without explicit trajectory representation by both the variable-speed and synchronization issues mentioned in the last subsection. The same muscles could be parts of more than one synergy (for example, muscles of the shoulder, elbow, wrist, and fingers are all involved in a reaching movement but only the wrist and fingers are involved in a grasping movement). Moreover, many synergies we perform are learned and not hard-wired, such as those involved in playing a musical instrument. Movement from a fixed initial position to a fixed target position can also be made at different velocities, and therefore over different durations.

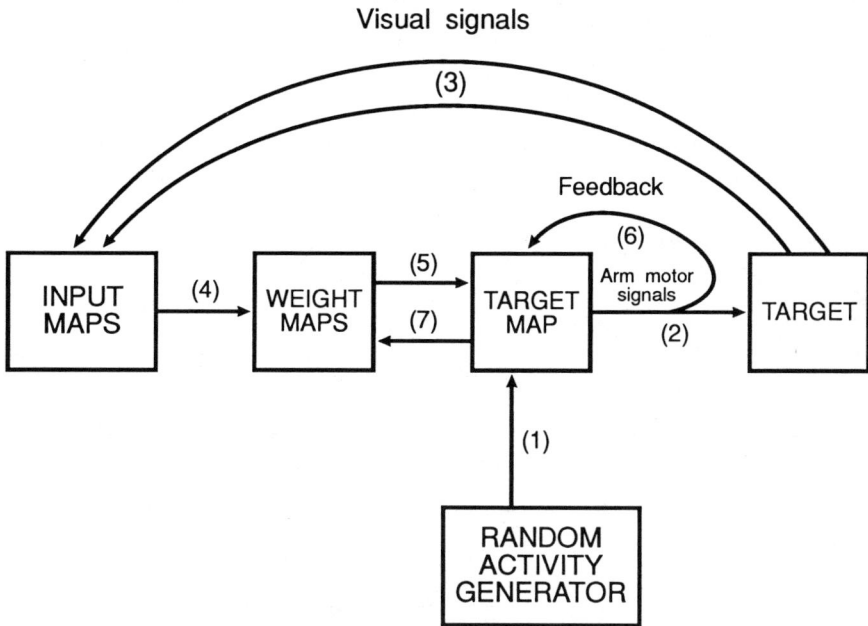

Figure 7.6. Network for circular reaction. Self-produced motor signals that manipulate an object target are correlated with target sensation signals. Numbers in parentheses denote the order of training, except that (5) and (6) are trained together; the order of performance is (3), (4), (5), (2). (Reprinted from Kuperstein, *Science*, **239**, 1308-1311, 1988b, with permission of the American Association for the Advancement of Science.)

The issue of variable velocity also led Bullock and Grossberg to reject models based on the physical analogy of a mass and a spring (*e.g.*, Kelso & Holt, 1980). (Recall the discussion earlier in this chapter on the limitations of physical system analogies.) In spring models, if an arm is at a particular position and a command occurs to move it to another position, an incremental move toward the new position creates a force imbalance. This imbalance in turn causes the arm to spring in the direction of the larger force at a rate proportional to the force difference. But Bullock and Grossberg argued that such a model would not allow for independent control over distance and speed. Also, they argued, in such a model it would be hard to terminate the movement at an intermediate position if contingencies changed.

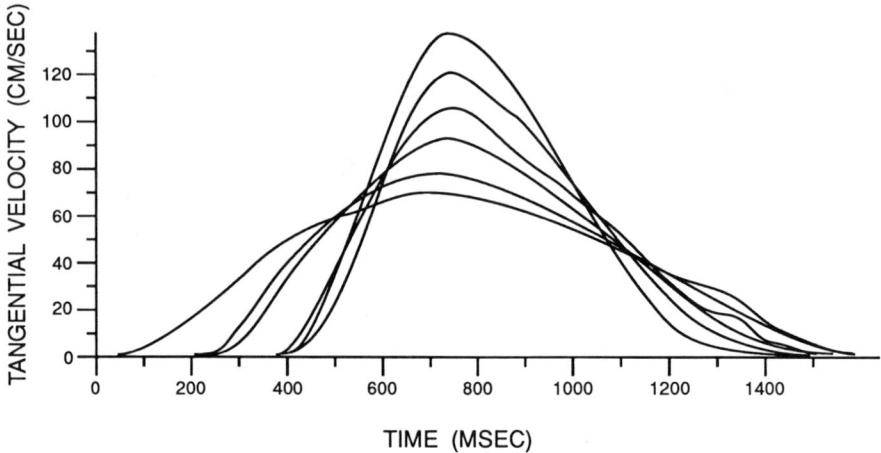

Figure 7.7. A set of typical tangential velocity profiles recorded from the wrist joint during an unrestrained vertical arm movement between two targets. These targets were light-emitting diodes, to enable visual tracking of movements. (Reprinted from Atkeson & Hollerbach, 1985, by permission of the *Journal of Neuroscience*.)

Bullock and Grossberg (1988) developed a model called *vector integration to endpoint* (VITE) where a given movement can be performed at variable velocities depending on the activity of a "GO" signal (see Figure 7.8). The "GO" activity is multiplied by the computed vector of muscle activities. Such *factorization* of a neural activity vector into a product of energy (total intensity) and pattern (relative strengths) has been a theme of Grossberg's work over the years, in perceptual as well as motor contexts. For example, this theme appears in the studies of relative weights in an outstar (*cf.* Section 3.2) and of sensory pattern normalization (*cf.* Section 4.1).

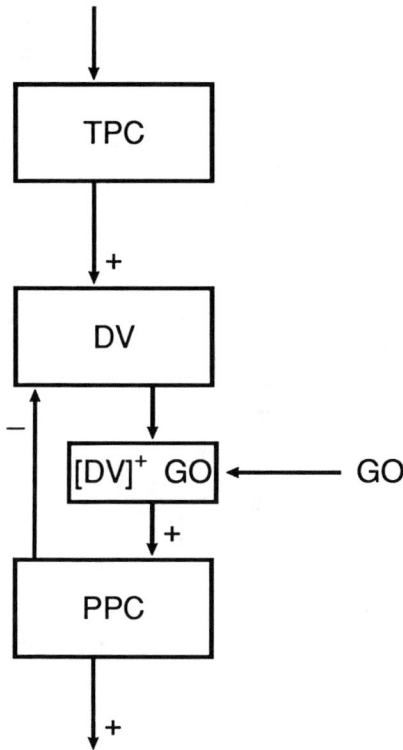

Figure 7.8. Main variables of the VITE motor control circuit. TPC=target position command, PPC=present position command, DV=difference vector (error), GO=GO signal which is multiplied by the difference vector if that vector is positive. The full circuit also includes interactions between DV and PPC stages of agonist and antagonist muscle commands. (Adapted by permission of the publisher from Bullock & Grossberg, in W. A. Hershberger (ed.), *Volitional Action*, pp. 253-298. Copyright 1989 by Elsevier Science Publishing Co., Inc.).

In the VITE model, as shown in Figure 7.8, the present position is compared with the a target position command (TPC) to form a difference vector (DV). The GO command (identified with output from the globus pallidus of the basal ganglia; *cf.* Figure 9.8 of Appendix 1) interacts with the DV. The present position command is gradually updated (see Figure 7.9) by integrating the multiplied vector, that is summing it over (continuous) time, hence the model's name. The effect of the present position command on motoneurons is organized through agonist-antagonist (dipole) pairs of muscles. Cells analogous to difference vector cells have, in fact, been located in arm zones of the premotor, motor, and parietal areas of the cerebral cortex (*e.g.*, Georgopoulos, Kalaska, Caminiti, & Massey, 1984). Equations for the VITE model, with the addition of update of position by passive movements, will be given in Section 7.3 below.

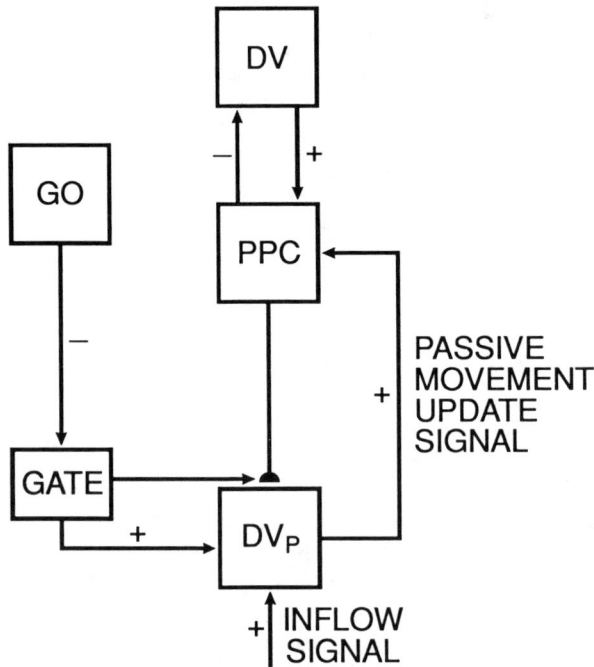

Figure 7.9. Passive update of position circuit. The subscript "P" on the second DV refers to "posture"; hence, the network is sensitive both to active and passive movement. During passive movements, the output from GO is zero; hence, DV_P is disinhibited and can update the PPC as does the active DV. (Reprinted by permission of the publisher from Bullock & Grossberg, in W. A. Hershberger (ed.), *Volitional Action*, pp. 253-298. Copyright 1989 by Elsevier Science Publishing Co., Inc.).

The factorization idea also plays a role in a complementary model (Bullock & Grossberg, 1989) designed to explain how the same posture can be held at different intended rigidities. The latter model is called *factorization of length and tension* (FLETE). In the authors' words (p. 255): "It (FLETE) models a part of the spino-muscular system that cooperates with the VITE network to generate the forces needed to ensure that the limb follows the trajectory commanded by the VITE circuit."

Transformation from visual ("task-oriented") co-ordinates, previously discussed in the context of eye movements by Grossberg and Kuperstein (1989), was also the basis for a series of models of voluntary movement by Kawato *et al.* (1987) and Kawato, Isobe, Maeda, and Suzuki (1988). Their 1987 article developed and simulated a control circuit driven by sensory signals and inspired by known anatomy and physiology of several brain areas. This circuit is shown in Figure 7.10. Like the Bullock-Grossberg network, the networks of Kawato *et al.* can learn a movement at one speed and then perform the same movement at a different speed.

An alternative approach to modeling arm movements was developed by Massone and Bizzi (1989). Their model is based on the *sequential network*, previously developed by Jordan (1986b). The sequential network, shown in Figure 7.11, is a standard back propagation network with the addition of some feedback and some *plan units* activated by external stimuli. The Massone-Bizzi network is simpler in its control structure than the Bullock-Grossberg or Kawato networks, but notable in its incorporation of more realistic arm-joint geometry into the network. (The simulations of Bullock and Grossberg, 1988, 1989 have thus far dealt with single joints, but extensions of their simulations to multijoint networks are currently underway.)

Speech Recognition and Synthesis

Cohen, Grossberg, and Stork (1987) noted that many aspects of the above models of arm movement could be adapted to other kinds of movement, such as the articulatory movements of speech. Hence, their approach to speech production included an auditory-articulatory "circular reaction" including target position command (TPC), present position command (PPC), and difference vector (DV) as in the VITE model of Bullock and Grossberg (1988). In the model of Cohen *et al.*, different articulatory subsystems, each with its own TPC-PPC-DV combination, are needed for different types of characteristic sounds.

Figure 7.12 shows the overall architecture developed by Cohen *et al.* (1987) for feedback between the auditory system and the speech motor system. The organization of this network on the auditory side is designed for the purpose of *unitization*, that is the grouping of heard speech into significant "chunks." First, "items" are identified, and then the items are grouped into sequences or "lists."

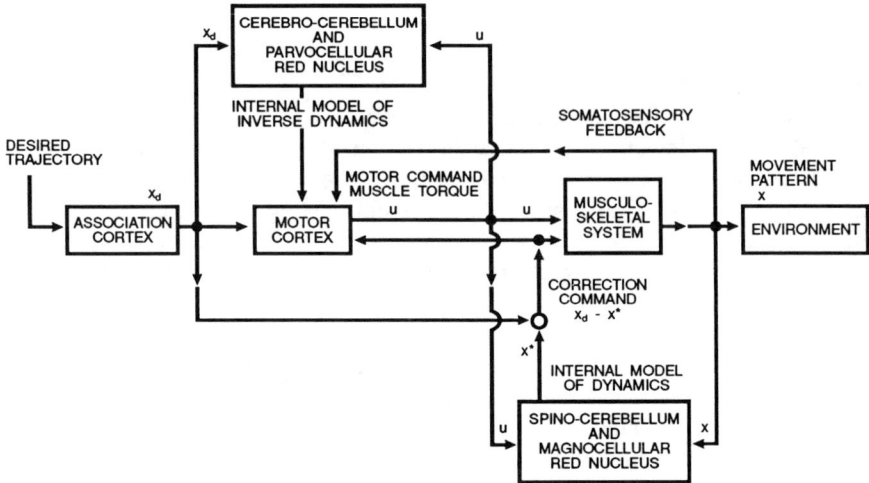

Figure 7.10. Network model for movement control and learning. The model includes: (1) A pathway from cortex to muscles, and a cortical loop, denoted by heavy lines; (2) A system, in the spinal cord, cerebellum, and magnocellular red nucleus, that changes a motor command into a predicted movement x^*; (3) A system, in the cerebrum, cerebellum, and parvocellular red nucleus, which changes a desired trajectory into a motor command. (Adapted from Kawato *et al.*, 1987, with permission of Springer-Verlag.)

The grouping of items into lists is based on a network architecture called the *masking field*, developed by Cohen and Grossberg (1986, 1987) and shown in Figure 7.13. The masking field is a competitive on-center off-surround network of representations, not of individual phonemes but of sequences of phonemes, or more generally, sequences of sensory stimuli. Built into this field is a bias favoring longer over shorter time sequences, in order to give an advantage to list nodes pooling information from a larger number of items. For example, if there is a node for the sound "MYSELF," it should tend to inhibit the "MY," "SELF," and "ELF" nodes. The masking field and its associated context-dependent item-list interactions are discussed twice in Section 7.2: in relation to the modeling of frontal lobe function, and of word recognition.

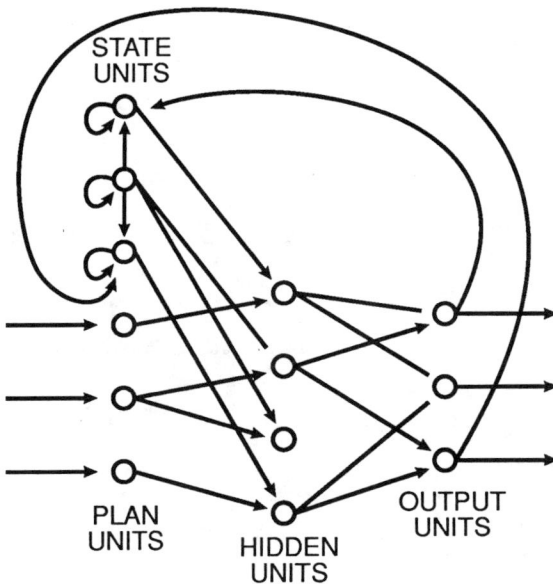

Figure 7.11. Basic architecture of Jordan's sequential network. (Reprinted from Massone and Bizzi, 1989, copyright © 1989 IEEE, by permission of the publishers.)

There has been a variety of other artificial neural networks designed for speech recognition and synthesis, many of which have not used specific structures of auditory information. Some, in fact, have used visual information, most notably the NETtalk algorithm, based on a back propagation method, that converts written English text to phonemes (Rosenberg & Sejnowski, 1986; Sejnowski & Rosenberg, 1986). NETtalk uses a completely supervised learning paradigm, without access to semantic information, or to contextual information other than groups of five letters in a row. It has achieved 95% success in

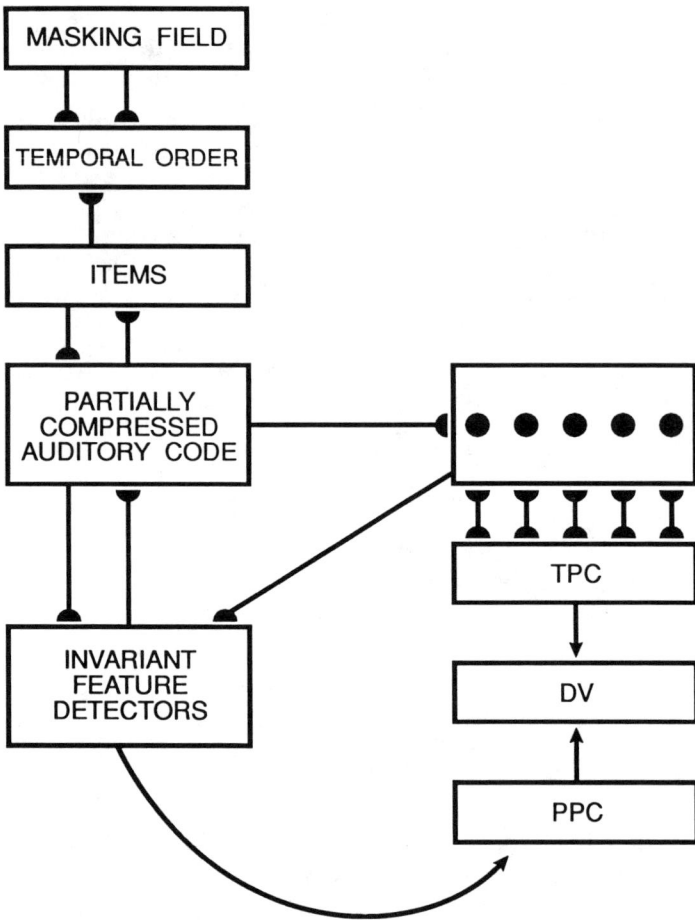

Figure 7.12. Large circuit linking auditory processing areas (on left) and speech articulatory centers (on right). TPC, PPC, and DV have the same meaning as in the arm movement architectures of Figures 7.8 and 7.9. The masking field is discussed further in the caption for Figure 7.13. (Adapted from Cohen *et al.*, 1987, copyright © 1987 IEEE; reprinted by permission of the publishers.)

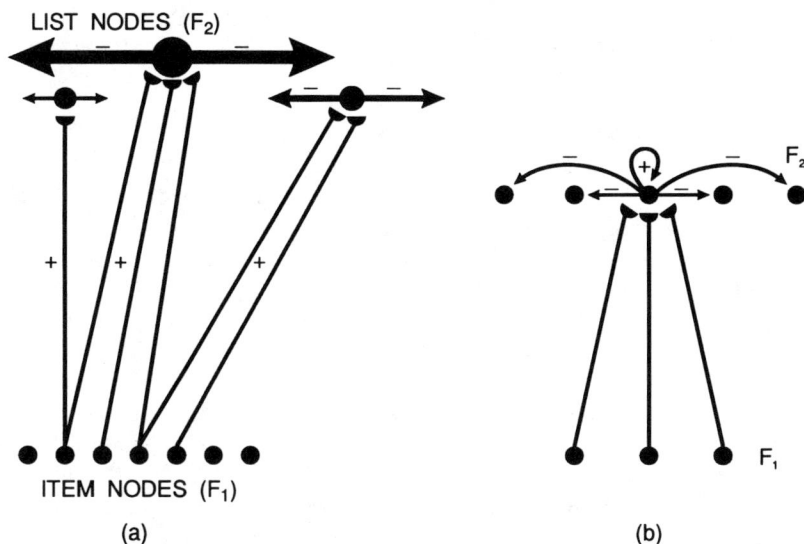

LIST NODES (F₂)

ITEM NODES (F₁)

(a)

(b)

Figure 7.13. Masking field interactions. F_1 and F_2 are as in ART (*cf.* Figure 6.22); here, nodes in F_1 code items and nodes in F_2 (the *masking field*) code lists. (a) F_1-to-F_2 connections grow randomly. F_2 nodes grow so that larger item groupings, up to an optimal size, activate nodes with stronger interactions. (b) Interactions within F_2 are on-center off-surround (*cf.* Chapter 4). F_1-to-F_2 long-term memory traces amplify F_2 reactions to item groupings which have previously excited F_2 nodes. (Modified from Cohen & Grossberg, *Applied Optics* **26**, 1866-1891, 1987, with permission of the Optical Society of America.)

pronouncing English words, and is able to reproduce certain observed effects of the spacing of word presentations.

The success of the NETtalk algorithm, while it is not a model of biological language comprehension, has spurred other efforts in the area of language processing, such as the work of Kohonen (1986, 1987) on understanding both Finnish and Japanese words. Some of these models are discussed in Chapter 8, in the context of technological advances.

Robotic Control

The models discussed in the last section, especially those of Kuperstein and Kawato, have found application to the building and control of robots. There is also a large literature on adaptive control neural networks whose biological inspiration is somewhat less direct than that of the networks discussed earlier. Some of this literature is discussed in the book edited by Miller, Sutton, and Werbos (1990), and particularly in that book's survey article by Barto (1990).

Barto's article places neural network control algorithms in the context of classical control methods for other types of engineering systems. These methods have often relied on widely known mathematical results in the control of linear and nonlinear dynamical systems.

In particular, Barto discusses various neural network solutions for the engineering problem of balancing a vertical pole on top of a moving cart, starting with his own previous work (Barto et al., 1983; Barto & Anandan, 1985). Network solutions to this and similar engineering control problems have developed concurrently with models by the same research group of Pavlovian conditioning, culminating in the *temporal difference* (TD) theory of prediction and control due to Sutton (1988). The TD theory is closely related to conditioning models involving time-derivative or time-difference, sometimes called differential Hebbian, learning rules (*e.g.*, Klopf, 1988; Sutton & Barto, 1991). Similar control problems have also been approached using back propagation techniques (*e.g.*, Werbos, 1988b; Nguyen & Widrow, 1990).

7.2. DECISION MAKING AND KNOWLEDGE REPRESENTATION

What, If Anything, Do Biological Organisms Optimize?

As the utility of neural networks in optimization tasks has been demonstrated, it is natural to ask how large a part optimization plays in the behavior of biological organisms. Klopf (1979, 1982) built a theory of brain function on the assumption that essentially *all* behavior can be explained by the optimization of a single variable. His variable was net stimulation of a subcortical brain area consisting of the reticular formation and part of the thalamus (*cf.* Figure 9.9 from Appendix 1). Klopf's outlook is reminiscent of ideas in the social sciences, such as the popular economic theory in which producers and consumers maximize "utility functions" (*e.g.*, Weintraub, 1979).

Yet other theorists, both in psychology and economics, have challenged the notion that all behavior is optimizing. Some animal learning data that argue against this notion are summarized in Levine (1983a) and Section 8 of Levine (1983b). One example is the partial reinforcement acquisition effect (PRAE), found by Gray and Smith (1969) and discussed in Chapter 3, whereby intermittent reinforcement is preferred to continuous reinforcement, presumably because of the element of surprise. The PRAE owes its existence to the choice of short-term over long-term satisfaction. The neural networks of Grossberg (1972a, b) and Grossberg and Gutowski (1987) include parameters that determine the relative strength of short-term dipole relief or frustration compared with longer-term pleasure or pain in energizing behavior. The relative influence of short- versus long-term reinforcement as motivators is also considered in Sutton (1988), Werbos (1988b), and Sutton and Barto (1991).

In humans, non-optimal behavior emerged in studies of decision making under risk by Tversky and Kahneman (1974, 1981, 1982) and Kahneman and Tversky (1982). These researchers found that in many choices relating to gain and loss estimation, preferences not only run counter to rational optimization but lack self-consistency over different linguistic framings of the choice. For example, when subjects are asked to consider two programs to combat a disease expected to kill 600 people, they tend to prefer the certain saving of 200 people to a 1/3 probability of saving all 600 with 2/3 probability of saving none. However, subjects also tend to prefer a 1/3 probability of nobody dying with a 2/3 probability of 600 dying to the certainty of 400 dying (Figure 7.14). The choices are identical in actual effect, but are perceived differently because of differences in frame of reference (comparing hypothetical states in one case with the state of all being alive, in the other case with the state of all dying). Tversky and Kahneman (1982), in explaining their own data, say that "choices involving gains are often risk averse while choices involving losses are often risk taking."

A similar challenge to optimization theory was mounted in economics by Heiner (1983, 1985). Heiner argued that economic behavior can be influenced by such non-optimizing factors as habit. Specifically, if the environment or the information it contains becomes too complex to understand well, and decisions must be made quickly, economic actors frequently resort to stereotyped actions regardless of their appropriateness.

In addition to affect and habit, organisms' reactions to events can be influenced by whether or not they are novel. Berlyne (1969), for example, showed that a rat can learn to press a lever if the lever-pressing is reinforced by a change in the ambient light intensity of its cage, regardless of the direction of the change.

Affect, Habit, and Novelty in Neural Network Theories

Does the influence of nonrational factors such as affect, habit, and novelty pose a difficulty for quantitative modeling of behavior? The answer to this question is of interest to the study of artificial as well as natural intelligence, because such factors may play a role in effective computation (*e.g.*, Dreyfuss, 1972; Dreyfuss & Dreyfuss, 1986; Johnson-Laird, 1988, Chapter 20). Only in the last few years have such phenomena been studied extensively in neural networks, but there are enough results to suggest that nonrational variables can be studied quantitatively as readily as rational ones. Hence, their study is likely to be a growth area in future neural network research.

Grossberg and Gutowski (1987) applied gated dipoles to explaining the Tversky-Kahneman data, discussed in the last subsection, on decision making under risk. Previously, Tversky and Kahneman themselves had proposed a variant of utility theory called prospect theory in which preferences are a nonlinear function of both gain (or loss) and its probability of occurrence. But prospect theory, while including some essential nonlinearities, excludes the context of statements and the past experience of decision makers. Grossberg and Gutowski's "affective balance" theory considers such dynamic variables.

Recall from Chapters 3 and 5 that gated dipoles (see Figure 3.7) provide a means of comparing current values of motivational or sensory variables with expected values of those same variables. Such expectation could be based either on recent past events or on verbally induced anticipation. It is the latter possibility that explains Tversky and Kahneman's data on the effects of linguistic framing on decisions (see Figure 7.14; mathematical derivations appear in Section 7.3).

Grossberg and Gutowski's explanation of Tversky and Kahneman's choice data is a significant advance but incomplete. Their network still optimizes a single variable, even if its optimization is not analogous to rational calculation in humans. In this case, the variable is net activity of the positive channel in a gated dipole, the dipole interpreted as relating to affect or motivation. More realistic explanations of decision processes are likely to depend on multiple decision criteria, including affect, habit, and novelty. *Which* criterion is used depends *both* on the cognitive task involved and on the current state of the organism (or network). There are extensive data on the influence of motivational state on learning (*e.g.*, Bower, 1981; Tikhomirov, 1983; Nottebohm, 1989).

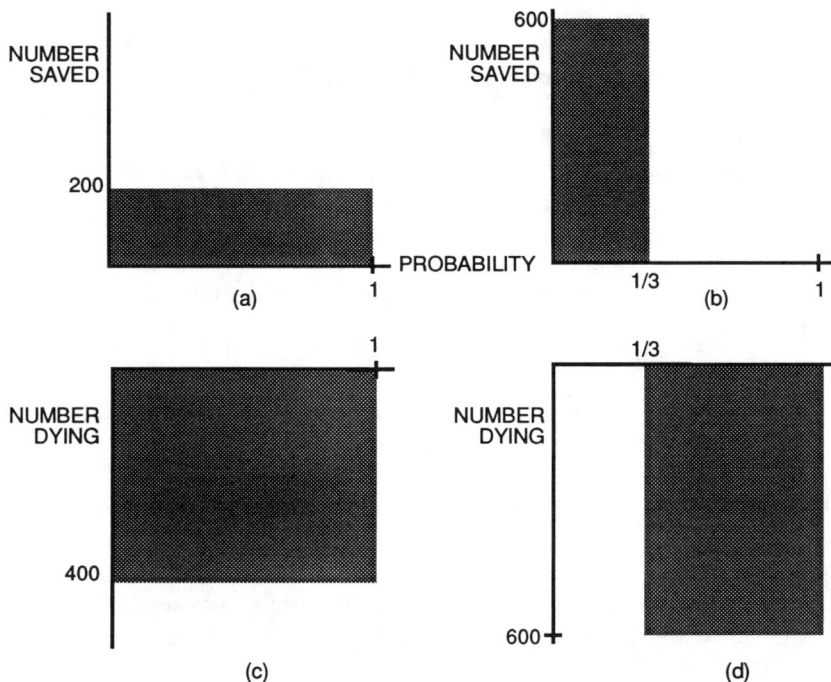

Figure 7.14. Effect of wording on preference as shown by Tversky and Kahneman. The symbols (a), (b), (c), and (d) refer to alternative ways to combat a disease expected to kill 600 people: (a) 200 will be saved for sure; (b) 600 will be saved with probability 1/3, and 0 will be saved with probability 2/3; (c) 400 will die for sure; (d) 600 will die with probability 2/3, and 0 will die with probability 1/3. Most subjects prefer (a) to (b) and (d) to (c), even though (a) and (c) are identical in actual effect, as are (b) and (d).

Neural network answers to these questions about decision processes are beginning to emerge from physiological and biochemical data on complex circuits including such brain regions as the association (particularly prefrontal) cortex, limbic system, basal ganglia, and parts of the midbrain. (All these regions are several synapses away from primary sensory or primary motor areas; that is, in the terminology of Rumelhart *et al.*, 1986, they consist entirely of hidden units!) As the networks modeling these regions become more realistic over the next several years, the resulting models are likely to have implications for treatment of neurological and psychiatric disorders. These models are also likely to have implications in artificial systems for solution of problems such as vexing context sensitivity and "intuition" (*cf.* Hewitt, 1986, or Winograd & Flores, 1987).

Returning to the question of the last section, whether humans and animals optimize some system variable, the answer is probably not a simple yes or no. Reinforcement learning theory predicts a wide range of observed behaviors, but not all of them. In fact, the same neural network principles, such as associative synaptic modification, competition, and opponent processing, can be used both in networks exhibiting reinforcement learning and in networks deviating from reinforcement learning (*cf.* Levine, 1989b).

Neural Control Circuits, Neurochemical Modulation, and Mental Illness

Grossberg (1984) discusses a variety of processes in a gated dipole network that bear some resemblance to particular mental or neurological disorders. These disorders include Parkinson's disease, some forms of schizophrenia, some forms of depression, and juvenile hyperactivity.

To understand Grossberg's approach to modeling mental illness, let us refer back to some properties of the gated dipole, as shown in Figures 3.7 and 3.8. The crucial variables are the activity x_5 of the channel that has been activated by a significant input, and the rebound activity x_6 of the opposite channel that occurs after the input is removed. The absolute and relative strength of these activities are regulated by the intensity and duration of the significant input, J, and by the intensity of the continuously active (*tonic*) nonspecific arousal input, I.

Grossberg (1972b, 1984) showed mathematically that if the nonspecific arousal I is within a certain range, the gated dipole network behaves in a fashion that is usually considered normal. Above or below that range, the network exhibits some pathologies that suggest symptoms of certain common mental disorders. These pathologies are next discussed qualitatively; a quantitative exposition appears in the 1984 article.

When the gated dipole network of Grossberg (1984) is underaroused, its threshold of response to limited-duration (*phasic*) inputs J is raised. That is, if I is too low, a larger J than usual is needed to obtain a positive on-channel output x_5. Paradoxically, once this threshold is exceeded, the on-reaction is hypersensitive to increments in J. Giving the network a "drug" that increases nonspecific arousal (analogous to an "upper") reduces these "symptoms" of hypersensitivity. But if too much of the "upper" is administered, the network can develop the opposite syndrome associated with overarousal, which will be discussed below.

Grossberg compared underarousal effects in his network to observed symptoms of both juvenile hyperactivity and Parkinsonism (see Table 7.1). Both illnesses are frequently treated by "uppers," which are typically drugs that enhance the efficacy of the neural transmitter *dopamine* — amphetamine for hyperactive children, and L-dopa for Parkinson patients. (For a list of neurotransmitters, see Table 9.1 of Appendix 1.) The side effects of too large doses of those drugs often include schizophrenic-like symptoms. Conversely, other drugs that suppress dopamine, drugs used to treat schizophrenics, have Parkinson-like side effects.

Based on a combination of these analogies and some mathematical properties of his dipole network, Grossberg (1984) made two testable predictions about sufferers from these two disorders, which had not been verified at the time. First, he suggested that hyperactive and Parkinson patients should exhibit a weak affective rebound. For example, they should have an abnormally small reaction to the cutting of a reward or punishment in half, and an abnormally small aftereffect to halving the brightness of a visual cue. Second, he suggested that the same sudden increments in nonspecific arousal that would cause an off-rebound in normals would cause increased on-channel activity in hyperactive and Parkinson patients. This could lead to dishabituation, thence distractibility, by irrelevant yet unexpected events. Grossberg proposed electroencephalographic tests of this hypothesis.

The effects of overarousal in the gated dipole of Grossberg (1984) are opposites of some underarousal effects. The threshold for response to a phasic input is reduced. But once the threshold is achieved, the network is abnormally *insensitive* to increments in input intensity. If the dipole is involved in motivation, such insensitivity is known psychiatrically as *flatness of affect*, and is characteristic of some kinds of schizophrenia. Overarousal can have effects in networks other than the gated dipole. For example, in a recurrent associative network that processes verbal lists (Grossberg & Pepe, 1970), a high level of nonspecific arousal can lead to fuzzy response categories and irrelevant associations (*e.g.*, punning), which are also common schizophrenic symptoms.

The article of Grossberg (1984) includes neurochemical analogs for some of his network variables. For example, overarousal is compared to excessive activity in the diffuse synapses from the substantia nigra, an area of the midbrain, to the cortex, limbic system, and corpus striatum (see Figure 9.8 for locations of these areas). (The input to the striatum, which is a motor control region, plays an important role in Parkinson's disease.) The synapses from the substantia nigra use the neurotransmitter dopamine. Grossberg mentions two other neurotransmitters, *norepinephrine* and *serotonin*, which are also broadcast by midbrain areas to other parts of the brain. Dopamine (DA), norepinephrine (NE), and serotonin (5HT) are collectively called *monoamines*.

Symptom of network underarousal	Symptom of Parkinsonism	Symptom of juvenile hyperactivity
High threshold to phasic cues	Difficulty in initiating movements	High thresholds in an EEG audiometry test (reduced by medication)
Hyperactivity above the threshold	Difficulty in terminating movements once begun	Hyperactivity above the threshold
Treatable by an "upper" drug	Treated by L-dopa, which enhances effects of dopamine	Treated by amphetamine, which enhances effects of catecholamines
Too much "upper" causes an overaroused syndrome	Too much L-dopa can elicit schizophrenic symptoms	Too much amphetamine can elicit psychosis
Hyposensitive to halving of reward or punishment intensity	Unknown	Unknown
Paradoxical dishabituation by unexpected events	Parkinson bracing to an unexpected push	Distracted by irrelevant sensory cues

Table 7.1. Comparison of underarousal effects in a gated dipole network with symptoms of Parkinsonism and of juvenile hyperactivity. (Modified from Grossberg, 1984, with permission of Elsevier Science Publishers.)

Hestenes (1991) carries this neurochemical theorizing further, by suggesting cognitive roles for all three monoamine transmitters. His theory is based on physiology of several brain areas, some of which are shown in the control circuit of Figure 7.15. Hestenes' idea of the role of DA is similar to the nonspecific arousal function posited by Grossberg (1984). To the two other transmitters, he posits more subtle roles in pattern processing. Serotonin had previously been thought to be an inhibitory transmitter that counteracts the excitatory effects of dopamine. But the hallucinations arising from the drug LSD, which inhibits serotonin binding at its usual postsynaptic receptor sites, suggested to Hestenes that 5HT is involved in pattern matching. In the framework of adaptive resonance theory (Carpenter & Grossberg, 1987a, b; see Section 6.3), 5HT modulation could influence the level of the vigilance that determines when two activity patterns from different brain areas are matched. As for norepinephrine, Hestenes proposes that it regulates selective attention, via connections to the *locus coeruleus* (main midbrain source of NE) from the limbic system and hypothalamus (areas in the reinforcement-drive circuit).

Hestenes also discusses possible biological bases for manic-depressive disorders. Following Swerdlow and Koob (1987), he proposes that manic-depressive illness results from malfunction of the *nucleus accumbens* (NAC), a gateway from limbic motivational areas to motor control areas in the basal ganglia (see Figure 7.16). Hence, NAC malfunction disrupts plan execution by reducing the influence of motivational inputs on motor outputs. The NAC in turn receives *both* dopamine and serotonin inputs from the midbrain, and its proper function seems to depend fairly precisely on the relative gains of those two transmitter inputs. This is the basis of his recommendation that manic-depressive disorders be treated with dietary supplements of tryptophan, an amino acid needed for the body's manufacture of 5HT.

Leven (1991), in the same collection as Hestenes' article, gives some hypotheses on the neural network dynamics involved in learned helplessness, a symptom of certain types of depression. His article fits data on the cognitive, affective, and visceral components of depression into a preliminary model of chemical transmitter activity based on previous work of Carpenter and Grossberg (1990).

Another mental disorder, or interrelated set of disorders, that has been modeled by neural networks is the syndrome resulting from damage to the prefrontal cortex (the association cortex of the frontal lobes). This syndrome includes perseveration in formerly, but no longer, rewarding behavior. Paradoxically, it also includes excessive attraction to novel objects. Both effects have been explained (Nauta, 1971) by weakened influence of positive or negative reinforcement on behavior; this occurs because the prefrontal cortex is the area of cortex with the strongest reciprocal connections to areas representing internal drive levels (the limbic system and hypothalamus).

An example of perseveration occurs in the *Wisconsin Card Sorting Test*. In this test, the subject is given a sequence of 128 cards, each displaying a number, a color, and a shape, and asked to match each card to one of four template cards (one red triangle, two green stars, three yellow crosses, four blue

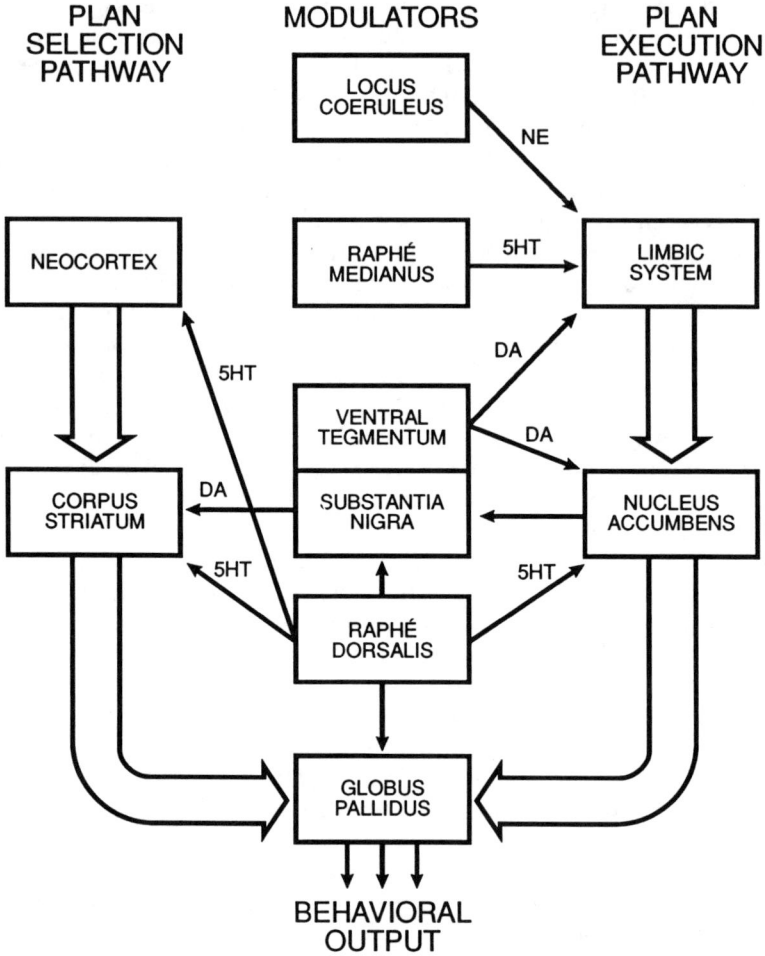

Figure 7.15. Behavioral control circuit, with some of its postulated brain regions and neural transmitters. The regions in the "modulator" column, other than the globus pallidus, are located in the midbrain. (Adapted from Hestenes, 1991, with permission of Lawrence Erlbaum Associates.)

MODULATORS

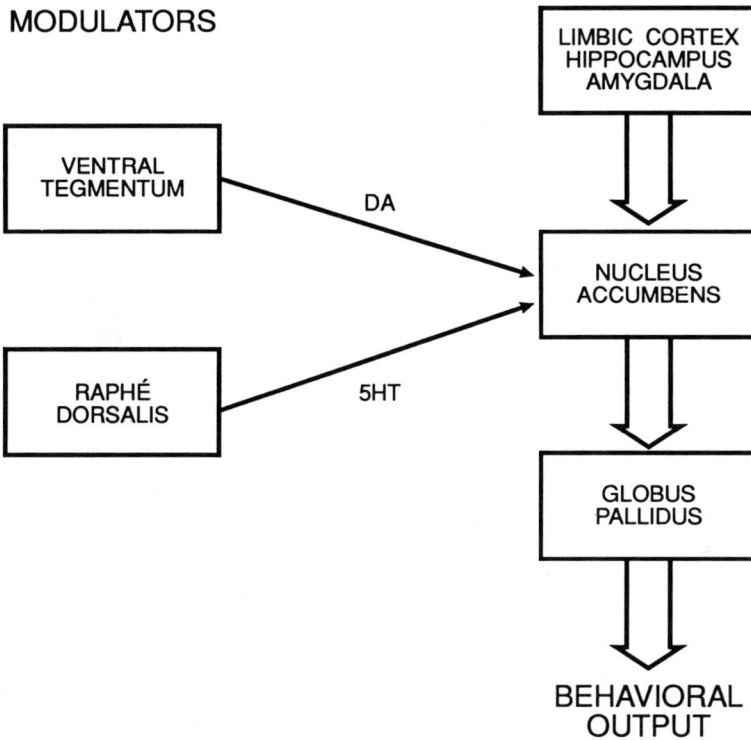

Figure 7.16. Closeup of part of the control circuit shown in Figure 7.15. This part details the role of the nucleus accumbens as a gateway from limbic motivational areas to pallidal motor control areas. (Adapted from Hestenes, 1991, with permission of Lawrence Erlbaum Associates.)

circles; see Figure 7.17). The experimenter then says whether the match is right or wrong, without saying why. After ten correct color matches, the experimenter switches the criterion to shape, without warning. After ten correct shape matches, the criterion is switched to number, then back to color, and so on. Milner (1963, 1964) showed that most patients with damage to a certain region of frontal cortex can learn the color criterion as rapidly as normals, but then cannot switch to shape.

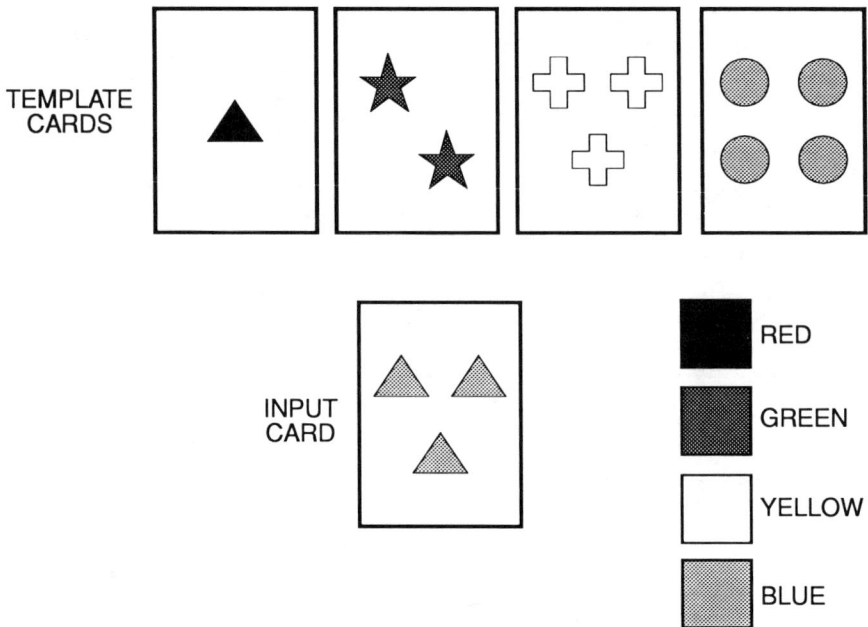

Figure 7.17. Cards used in the Wisconsin Card Sorting Test; the input card is matched to one of the four template cards. (Reprinted from Levine & Prueitt, 1989, with permission of Pergamon Press.)

Leven and Levine (1987) simulated the card sorting data using the network of Figure 7.18. In this network, based on ART (*cf.* Figure 6.20), the nodes in the field F_1 code features (numbers, colors, and shapes) whereas the nodes in F_2 code template cards. F_1 divides naturally into three subfields (number, color, and shape); corresponding to each subfield is a "habit node" and a "bias node." The habit nodes code how often classifications have been made, rightly or wrongly, on the basis of each feature. The bias nodes additively combine habit node activities with reinforcement signals (the experimenter's "Right" or "Wrong"), then gate the excitatory signals from F_1 to F_2. Table 7.2 shows results of simulating this network. A network parameter measuring the gain of reinforcement signals to bias nodes was varied. The network with high gain acted like Milner's normal subjects, whereas the network with low gain acted like Milner's frontal patients.

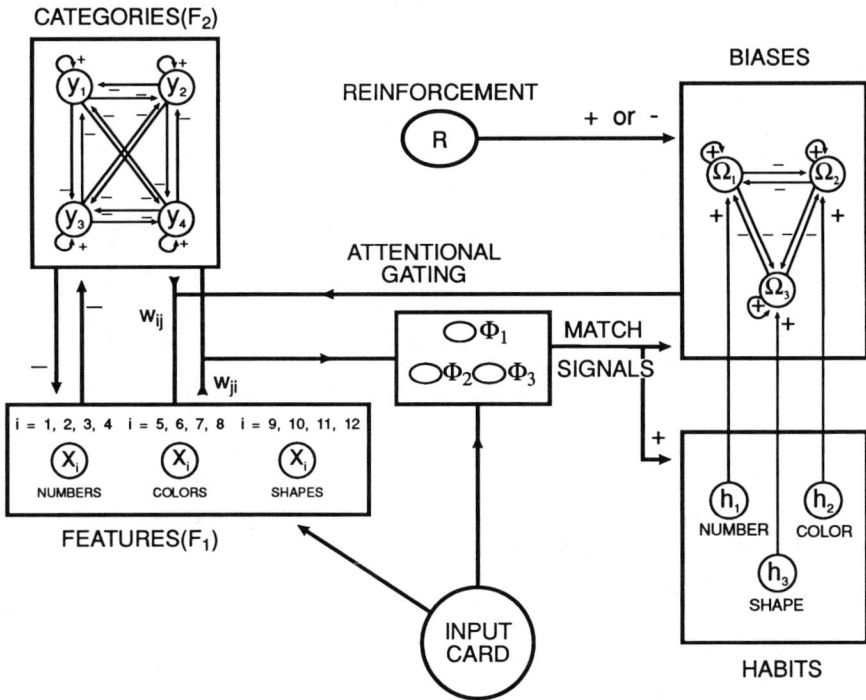

Figure 7.18. Network used to simulate card sorting data. Frontal damage is modeled by reduced gain of signals from the reinforcement node to bias nodes Ω_i (i=1 for number, 2 for color, 3 for shape). Bias nodes gate signals from feature to category nodes. Each bias node is influenced by the corresponding habit node, which encodes past decisions that used the given matching criterion. The match signal generators Φi send positive signals to habit nodes, and signals (of the same sign as the reinforcement) to bias nodes. (Adapted from Leven & Levine, 1987, copyright © 1987 IEEE; reprinted by permission of the publishers.)

	Criterion	Trial
$\alpha = 4$ ("Normal")	Color	13
	Shape	40
	Number	82
	Color (again)	96
	Shape (again)	115
$\alpha = 1.5$ ("Frontally damaged")	Color	13
	Thereafter, classified by color for all remaining trials	

Table 7.2. Results of simulations on the network of Figure 7.18. The parameter α measures the gain of signals from the reinforcement node to the bias nodes. The trial number listed is the first one on which the network achieved ten correct matches in a row based on the given criterion. (Reprinted from Levine & Prueitt, 1989, with permission of Pergamon Press.)

But perseveration due to frontal damage can be overridden by attraction to novelty, as in the monkey data of Pribram (1961). Pribram placed a peanut under a junk object several times, unobserved by a monkey. Each time this was done, he added a new object to the scene and waited for the monkey to choose which object to lift for food. On the first trial with a novel object present, normal monkeys tended to choose another object that had previously been rewarded, whereas monkeys with lesions of the ventral frontal cortex chose the novel object immediately.

Levine and Prueitt (1989) simulated the novelty data using the network of Figure 7.19. This network is an example of a *dipole field* (Grossberg, 1980). In a dipole field, each of several sensory stimuli has an "on" and an "off" channel structured like the two competing channels of a gated dipole (*cf.* Figure 3.7). In Figure 7.19, two such dipoles are shown, one corresponding to a previously rewarded object and the other to a novel object. The nodes $x_{1,5}$ and $x_{2,5}$, representing outputs of the "on" channels corresponding to each object, compete with each other and with nodes such as $x_{3,5}$ for other objects. The largest $x_{i,5}$ determines which object is approached.

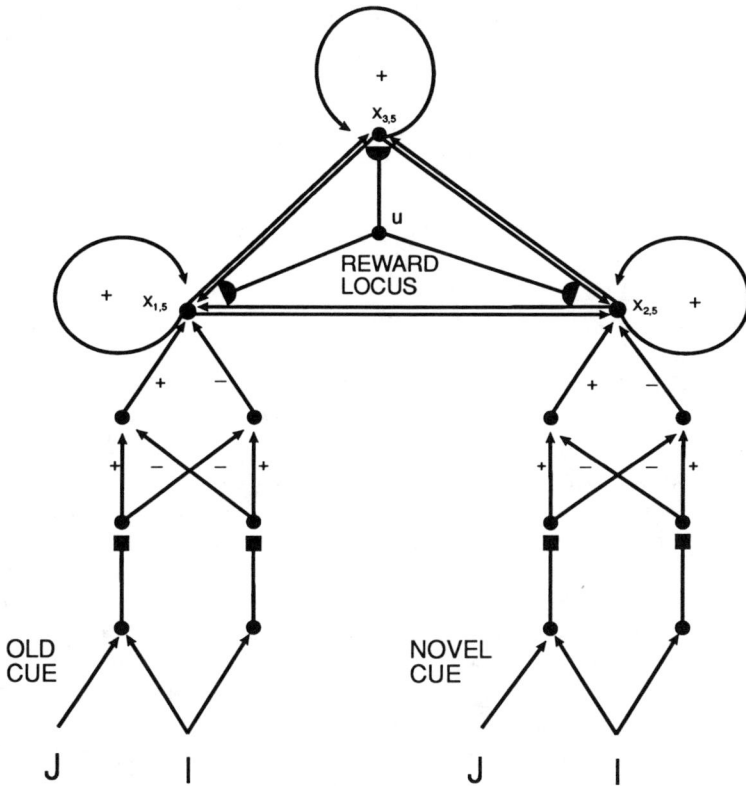

Figure 7.19. Dipole field used to simulate novelty data. Two gated dipoles are shown, corresponding to an old cue and a novel cue; their outputs $x_{i,5}$ represent tendencies to approach each cue. $x_{3,5}$ is the output of a third dipole, the rest of which is not shown. Competition among $x_{i,5}$ nodes is biased both in favor of previously rewarded cues, by adaptive connections with u, and in favor of novel cues, by transmitter depletion at "square" synapses. Relative strength of these biases depends on gain of reward signals to $x_{i,5}$'s, which (as in Figure 7.18) is lowered by frontal damage. (Adapted from Levine & Prueitt, 1989, with permission of Pergamon Press.)

With weak reward signals, as in frontally lesioned animals, the gated dipole mechanism causes the "on" channel for the old object to be more depleted than the "on" channel for the new object, since the old cue channel has been active longer. Hence $x_{2,5}$ is larger than $x_{1,5}$, and the new object is approached. With strong reward signals, as in normal animals, modifiable synapses between $x_{i,5}$ nodes and the food reward node cause $x_{i,5}$'s that have been previously associated with food to be selectively enhanced. This biases the competition in favor of $x_{1,5}$ enough to counteract transmitter depletion, and the old object is approached.

Frontal damage effects suggest detachment between what MacLean (*e.g.*, 1970) called the three layers of the *triune brain*: "reptilian" (habitual or instinctive), "old mammalian" (affective or emotional), and "new mammalian" (rational or verbal). The "triune brain" scheme, while now regarded as oversimplified, is qualitatively supported by such recent results as those summarized in Mishkin, Malamut, and Bachevalier (1984) and Mishkin and Appenzeller (1987). These researchers showed that one neural system subserves memories of the reinforcement value of stimuli, and another system subserves motor habits regardless of reinforcement. The memory system includes the hippocampus and amygdala (part of MacLean's old mammalian brain), and the habit system includes the corpus striatum (part of MacLean's reptilian brain.) The prefrontal cortex has at least two, and possibly three, functional subunits. Clearly, its functions cannot all be simulated by a single connection between motivational and cognitive areas. This cortical region has also been implicated in linking sequences of sensory or motor events across time (Fuster, 1980, 1985; Ingvar, 1985). This enables the organism to store an action plan and carry it through once it is initiated. The time-sequencing and motivational functions are interrelated; keeping a plan reverberating in STM depends on inhibiting competing sensory and motor representations that are motivationally irrelevant.

Levine (1986) and Levine *et al.* (1991) discuss possible time-sequencing networks. These networks include a rule whereby representations of longer sequences of stimuli or actions tend to suppress those of shorter sequences. A bias toward longer sequences is incorporated into the masking field (Cohen & Grossberg, 1986, 1987), which has been used to model the bias toward longer words in speech parsing (Cohen *et al.*, 1987; *cf.* Figure 7.13).

But another mechanism is needed to generate sequence representations in the first place. Figure 7.20 shows one such mechanism, combining an ART classification network with a form of the avalanche network (Grossberg, 1978b) for motor sequence performance. The idea of combining ART with the avalanche to classify *space-time patterns* (time sequences of spatial patterns) was due to Dawes (1989), who developed a different architecture for this task. Dawes integrated notions derived from Grossberg's work with other notions from stochastic process theory (the "innovations method," which bears some similarity to the novelty filter of Kohonen & Oja, 1976) and partial differential equations (the propagation of solitary waves, through an avalanche-like mechanism). This resulted in a two-layer feedback neural network for spatiotemporal pattern recognition. The role of the frontal lobes in making decisions among motor sequences has also been simulated by Dehaene and Changeux (1989).

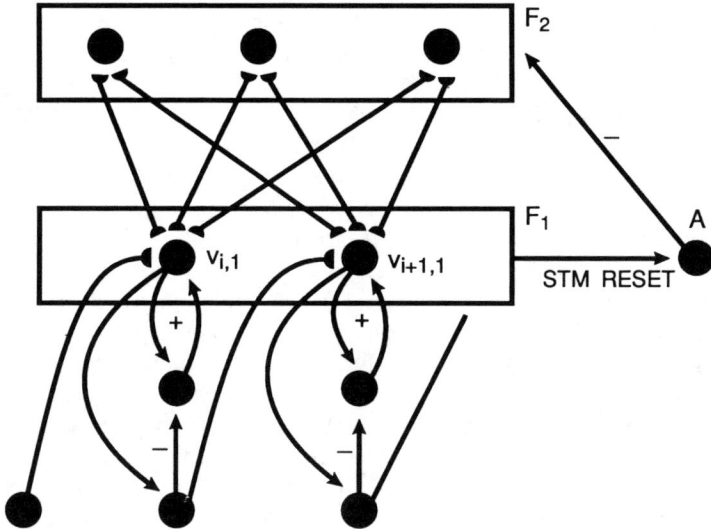

Figure 7.20. Possible network for learning space-time patterns. Layers F_1 and F_2 are joined to another subnetwork (due to Grossberg, 1978b) for learned sequential performance. For each $x_{i,1}$, there is an $x_{i,3}$ to keep it reverberating in short-term memory as long as needed, and an $x_{i,2}$ (influenced by an arousal source, not shown here) to shut off its reverberation. Each $x_{i,2}$ also influences the next stage $x_{i+1,1}$ of the sequence. (Reprinted from Levine *et al.*, 1991, with permission of Lawrence Erlbaum Associates.)

Some Comments on Models of Specific Brain Areas

The models discussed in the last subsection can be partially compared to biological processes in specific brain regions. At this stage of knowledge, the anatomical and physiological detail of these networks is still imprecise. In the near future, the increased coherence of available neurobiological data, due to recent technological advances in the laboratories (see Chapter 8 for a discussion), should permit more accurate anatomical labeling of network nodes.

In the late 1960's and early 1970's, many computational models were based on data about specific brain regions and tied these regions to specific functions. The builders of these models sought functional explanations for the characteristic anatomical structures of the regions they studied. These models were largely digital, based on linear threshold ideas, and some used elaborate counting arguments (see Levine, 1983, Section V for a review). Kilmer *et al.* (1969) related processing in the reticular formation to decisions among gross modes of behavior. Kilmer and Olinski (1974) related processing in hippocampus to decisions among actions within such behavioral modes. Marr (1969) related processing in the cerebellum to learning sequences of motor acts in appropriate stimulus contexts. Marr (1971) related processing in the hippocampus and limbic cortex to temporary memory storage. Finally, Marr (1970) related processing in the cerebral neocortex to pattern classification and goal direction (*cf.* Figure 9.9 for locations of all these regions).

Since the mid-1970's, models based on the structure of specific brain regions have largely been supplanted by models that start with cognitive functions and then seek appropriate structures to perform such functions. One exception has been models of the cerebellum, a region with a fairly well-defined set of functions (adaptive motor control) and a well-described set of interacting cell types. Pellionisz and Llinas (1982) and Pellionisz (1986) have modeled the cerebellum as a *tensor*, which is a mathematical transformation that maps some vectors into other vectors (*e.g.*, eye coordinates into head coordinates). Houk (1987) has modeled the cerebellum as an array of adjustable pattern generators. Recent physiological data on this brain region are summarized in Ito (1984) and Glickstein, Yeo, and Stein (1987).

With neural network theories now more firmly based on principles, efforts are emerging to integrate the function-based and structure-based types of model. Examples of such integration are found in the last subsection and in the introductory section of Chapter 5 (for example, the work of Zipser, 1986 and of Schmajuk & DiCarlo, 1991 on the role of the hippocampus in Pavlovian conditioning).

In the final subsection of this section, we return to higher-order functions, specifically, inference and semantic information processing. These studies bring neural network theories into contact with some classical problems of artificial intelligence. As yet, models of such processes have made relatively little contact with neuroscience, but ultimately they can provide intuition about functions of multisensory association areas in the neocortex.

Knowledge Representation: Letters and Words

The field of knowledge representation is a large area that lacks an exact definition. Broadly speaking, it refers to the representation of interrelationships among complex concepts (sometimes without a precise theory of how these concepts arise). The development of neural networks for knowledge representation is still in its early stages. One important subclass consists of those networks where nodes encode letters and words.

The oldest significant neural network model of letter and word recognition is the interactive activation model of McClelland and Rumelhart (1981) and Rumelhart and McClelland (1982). These researchers sought to explain the fact that recognition of letters is heavily dependent on context. For example, letters can be identified more accurately when they appear in words than when they appear in random sequences. This advantage of words partially extends to nonwords that are pronounceable by an English speaker (such as MAVE or REET). McClelland and Rumelhart conclude that modeling of these context effects has to involve both top-down and bottom-up processes, that is, feedback between "letter level" and "word level" nodes. A full sketch of the processing levels they posit for (both visual and auditory) word perception is shown in Figure 7.21a; the simplified (purely visual and three-level) model they actually use is shown in Figure 7.21b. (Recall the discussion in Section 6.2 of disambiguation by context at a different cognitive level, the semantic level, in the network of Anderson & Murphy, 1986.)

This same set of levels (feature, letter, word) has been the basis for more recent models by the Rumelhart-McClelland group (McClelland, 1985; Rumelhart & Zipser, 1985). In fact, these authors say in the preface of their popular book (Rumelhart & McClelland, 1986) that "The seeds of this book were sown in our joint work on the interactive activation model of word perception." McClelland's 1985 article, in particular, extends the earlier models to the simultaneous processing of multiple words. The network of that article reproduces some errors characteristic of humans attending to two words at once. Rumelhart and Zipser (1985) differ from the interactive activation model in that interlevel connections in their model are purely excitatory, whereas in McClelland and Rumelhart (1981), letter nodes can inhibit word nodes for words incompatible with the corresponding letter.

Grossberg and Stone (1986) state that this letter-word distinction poses a problem in the case of words of one letter: it has been shown (Wheeler, 1970) that those letters which are also words in English (A and I) are no easier to recall than other letters. If A and I were represented on both the letter and word levels, they would have a selective advantage over other letters, which is false.   Hence, the letter A and the word A need to have separate representations.

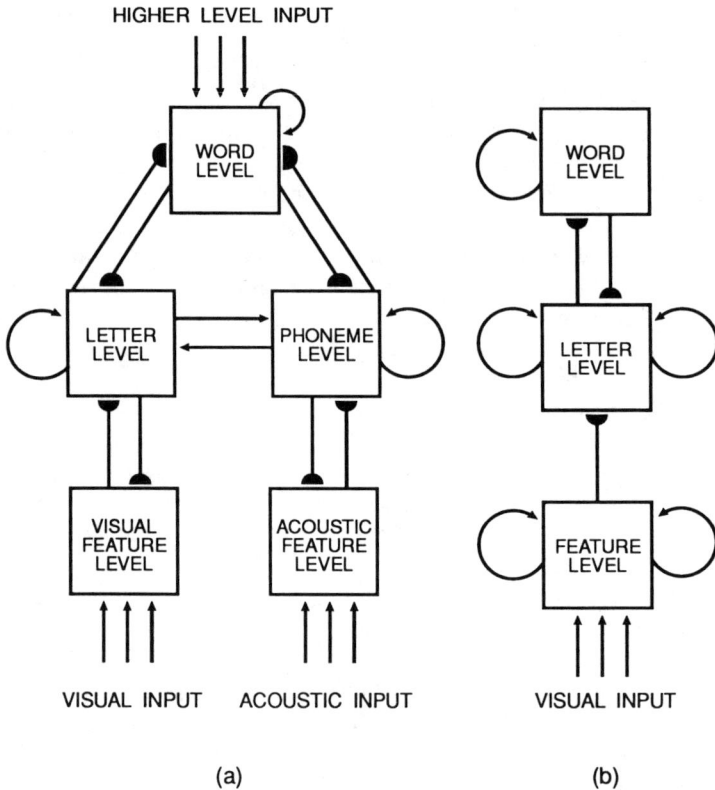

Figure 7.21. (a) Sketch of interconnections between processing levels involved in visual and auditory word perception.   (b) Simplified version of the network in part (a), involving only visual perception, actually used in simulations.   (From McClelland & Rumelhart, *Psychological Review* **88**, 375-407, 1981.   Copyright 1981 by the American Psychological Association.   Adapted by permission.)

These considerations led Grossberg and Stone to replace the concepts of letter and word nodes with the more abstract concepts of *item* and *list* nodes (*cf.* Figures 7.12 and 7.13).   An architecture for feedback between item and list levels is shown in Figure 7.22. The same architecture is used in speech recognition (Cohen & Grossberg, 1987; Cohen *et al.*, 1987). There the lists or "chunks" (auditory, in this case) need not be words, and there can also be competition between words when one is a "sublist" of the other (*e.g.*, "SELF" and "MYSELF").

The architecture of Figure 7.22 is called a "macrocircuit" by Grossberg and Stone (1986, p. 58). Each "box" in this "macrocircuit" contains a "microcircuit". The neural dynamics of each pair of adjacent, interconnected boxes are consistent with the adaptive resonance theory (ART) of Carpenter and Grossberg (1987a, b; *cf.* Figure 6.22). Feedback between A_3 and A_4, for example, is described by a combination of ART with the masking field of Cohen and Grossberg (1986, 1987; *cf.* Figure 7.13). This macrocircuit also incorporates an example of Piaget's circular reaction (see the motor control discussion in Section 7.1), here involving feedback between auditory and articulatory centers.

The Grossberg-Stone theory of word recognition also relies on factorization of pattern and energy (*cf.* the arm movement discussion in Section 7.1). In the word recognition domain, the distinction between pattern (what is processed) and energy (how strongly is it processed) takes the form of a distinction between *attentional priming* and *attentional gain control*. In the Grossberg-Stone theory, the priming stimulus ("pattern") is encoded at the F_2-to-F_1 synaptic weights, and gain control ("energy") from other nodes determines the relative amount of attention paid to that prime.

Using this macrocircuit, Grossberg and Stone explain a number of experimental results on word recognition. These results include reaction times to words versus nonwords after different primes (related words, unrelated words, or neutral stimuli like strings of X's); "word superiority" effects that can cause a tendency to misclassify nonwords that are like words except for one letter; and word frequency effects. The theory described by this network has begun to capture the context-sensitive nature of the psychological data.

Other researchers (Posner & Snyder, 1975a, b) have used a two-process attentional theory to explain effects of a previous priming stimulus on the accuracy and speed of word recognition. The Posner-Snyder theory distinguishes between spreading activation from the prime, which causes attention to be paid to stimuli similar to it, and limited capacity selective attention, which thereby prevents attention from being paid to other stimuli. Attentional priming is also similar to some effects studied with the CHARM (composite holographic associative recall model) of Eich (1982, 1985). Eich bases her model on two linear vector operations, convolution (see Section 4.5) for storage and correlation for recall. She uses this model to reproduce various experimental results showing that recall of words is enhanced by elaboration

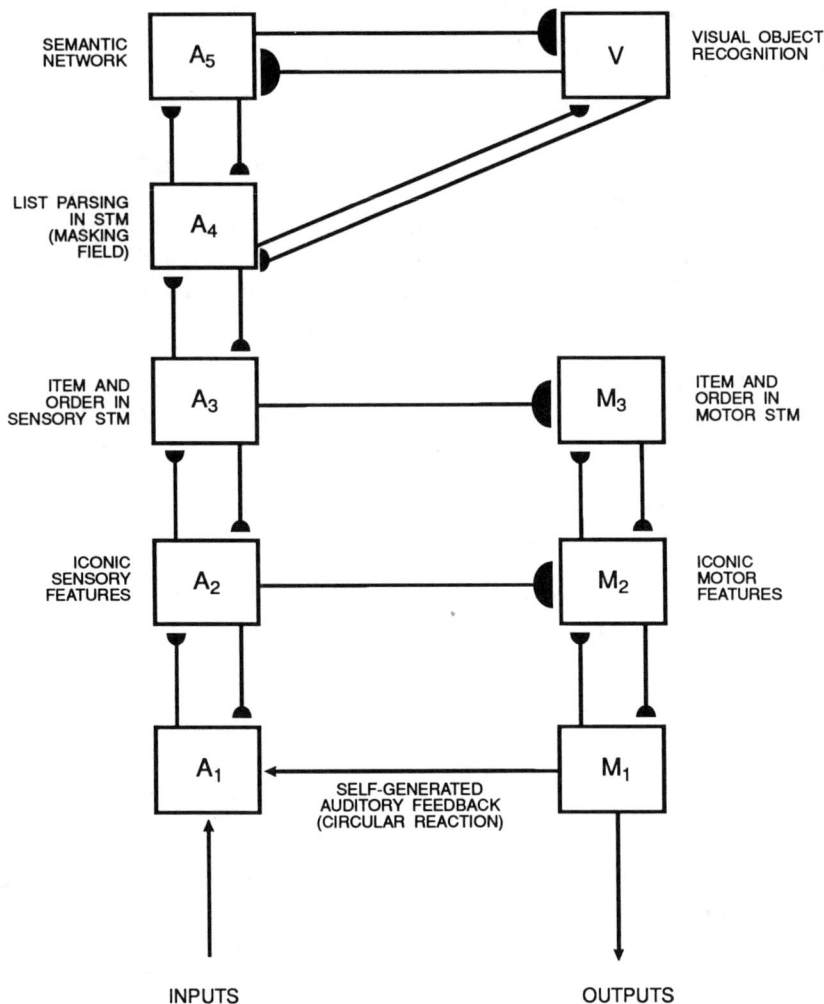

Figure 7.22. A large circuit governing recognition and recall, with interacting auditory (language processing), visual, and motor control subsystems. Parts of this circuit can be identified with parts of the circuit of Figure 7.12. (From Grossberg & Stone, *Psychological Review* **93**, 46-74, 1986. Copyright 1986 by the American Psychological Association. Adapted by permission.)

that involves words that are similar, either in meaning or in sound, and inhibited by elaboration that involves dissimilar words.

The work described in this subsection suggests that neural networks can be designed to perform such higher-order cognitive functions as forming concepts, building constitutive relations between concepts, and reasoning inferentially. This work also refutes a key argument of Fodor and Pylyshyn (1988), their statement that "Connectionists ... are committed to mental representations that don't have combinatorial structure" (p. 45).

Indeed, nodes in neural networks can learn to encode concepts of essentially arbitrary complexity if the networks contain sufficiently many levels, with nodes at each level learning spatial or spatiotemporal patterns of node activities at the previous level. This insight has emboldened many researchers to build connectionist networks described by differential or difference equations for interactions between concepts. In the near future, the combination of such efforts with the use of established design principles for lower-level category formation should yield networks that perform these same functions but have architectures that are more biologically plausible.

Knowledge Representation: Concepts and Inference

Smolensky (1986) built an abstract probabilistic network theory of inference, using the notion of *schemata*, a well-established concept in artificial intelligence and cognitive science. He defined schemata (Smolensky, 1986, p. 202) as "knowledge structures that embody our knowledge of objects, words, and other objects of comparable complexity" with the properties that "they have conceptual interpretations and that they *support inference*" (author's italics). He referred to the elementary units of which schemata are composed as *knowledge atoms*. These atoms (*e.g.*, words or consecutive two-letter combinations) are encoded at one of the two levels of nodes of his general network, which he called a *harmony network* for reasons to be developed below. The other level encodes the *representational features* (*e.g.*, letters) of which these atoms are composed.

Figure 7.23 shows a perceptual completion task and a harmony network that can perform that task. The task is to decide which letters are present in a word that has been partially obscured. The activated nodes in the harmony network are highly consistent: all the activations agree with the interpretation that the word is "MAKE." A primitive "harmony function" counts +1 for every agreement between an active atom and the value of a corresponding feature (*e.g.*, the atom "MA in the first two positions" and the feature "A in the second position") and -1 for every disagreement.

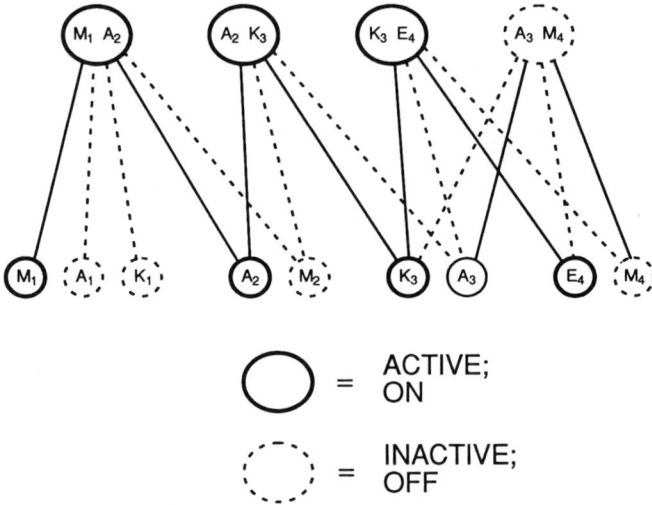

Figure 7.23. A harmony network for recognizing a partially occluded word as "MAKE." (Reprinted from Smolensky, 1986, with permission of MIT Press.)

Smolensky constructed more complex harmony networks that incorporate this harmony function in a probabilistic fashion. The probability of a system state is shown to follow the reverse of a Boltzmann distribution (see the discussion in Section 7.1 above); that is, the probability of a state is an exponentially *increasing* function of its harmony. Like the Boltzmann machine and other networks incorporating simulated annealing (again, see Section 7.1), the network has a "temperature" parameter that determines the growth rate of this probability distribution. The higher the temperature, the higher the probability of a low-harmony state. Hence, as in the Boltzmann machine, "heating" the network allows it to change from one steady state (in this case, an interpretation of the scene) to another, while "cooling" the network tends to settle it in the current state.

Cruz (1991) and Cruz, Hanson, and Tam (1987) have built a general class of artificial neural systems known as *knowledge representation networks* (KRN), an example of which is shown in Figure 7.24. Cruz defines the KRN in terms of hierarchies of *features*, which represent objects, conditions, or events; *relationships*, which act as connectives in defining complex features; and *operations*, which represent actions executable by the network. Cruz (1991) states that the representations of his concepts are *semantic*: that is, the connection strength between any two nodes is determined by their interrelationships. To give his example, the representation for "apple" has strong connections with the representations for "red" and "fruit." (Note that the "card-sorting" network of Figure 7.18 has similar properties.) The Cruz KRN has not been applied so far to any practical problems, but has been linked in a general way with goal direction and planning.

Kosko (1986a, 1987b) has designed what he calls *fuzzy cognitive maps*. The term "fuzzy" refers to the branch of mathematics known as fuzzy logic (Zadeh, 1965), where propositions are not necessarily true or false but have a truth value that can be any number between 0 and 1. In these maps, an example of which is shown in Figure 7.25, Kosko takes the relevant factors in a particular situation (*e.g.*, the AIDS epidemic or South African race relations) and represents each factor by a node in the network. The "fuzzy" connection weights between nodes represent the likelihood of a "presynaptic" event causing a "postsynaptic" event. Using these weights, the network can make dynamic choices about how to act in the situation.

Ryan, Winter, and Turner (1987) and Winter, Ryan, and Turner (1987) built an inference network based on adaptive resonance theory. Their adaptive resonance network is trained on a set of patterns that denote binary relationships of the form "R1 Rel R2," where "R1" and "R2" are two entities having the relationship "Rel"; in one of their examples, R1 is "elephant," Rel is "has color," and R2 is "gray." Hence, their feature layer is divided into three parts: nodes for entities R1, for relationships Rel, and for properties R2. In the network of Ryan *et al.* and Winter *et al.*, members of a class inherit properties of the class unless there is information to the contrary. For example, if elephants are gray and Clyde is an elephant, the network infers that Clyde is gray. But if Jumbo is an elephant and it is known that Jumbo is pink, this knowledge overrides the inference that would otherwise occur that Jumbo is gray.

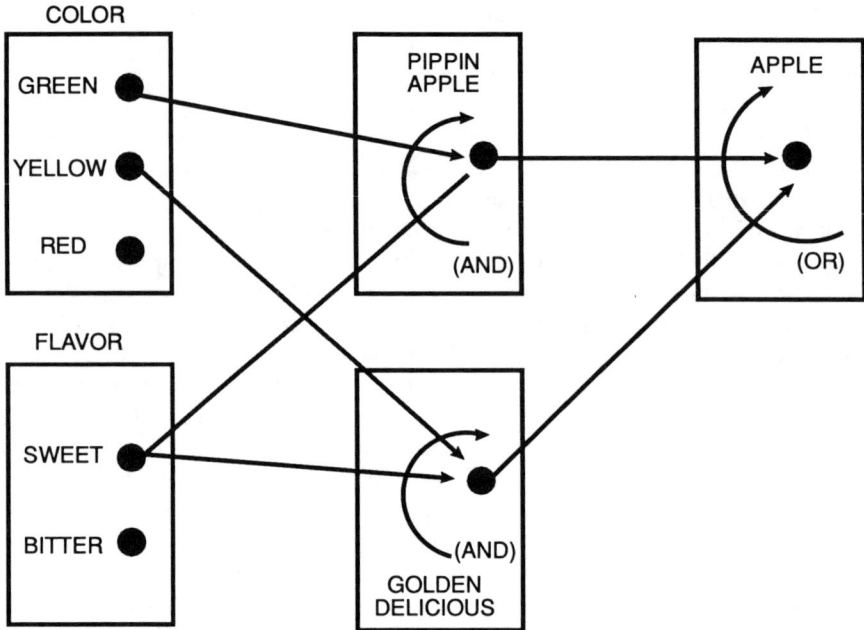

Figure 7.24. Example of a knowledge representation network (KRN). Arcs labeled (AND) and (OR) represent logical conjunction and disjunction, respectively. (Reprinted from Cruz, 1991, with permission of Lawrence Erlbaum Associates.)

Thagard (1989) has developed a connectionist inference scheme called ECHO, and applied ECHO in simulations to both scientific hypothesis testing and legal reasoning. The nodes in Thagard's networks represent either pieces of evidence or conclusions. Connections between nodes are excitatory if the concepts represented by the two nodes are logically consistent, and inhibitory if they are inconsistent. As evidence is evaluated, the activation of a particular theory or conclusion changes over time as a result of the network interactions. Thagard only suggests a mechanism for accepting or rejecting theories, not for modifying or synthesizing them.

Another principle frequently used in inference networks is the weighing of independent decisions by multiple subnetworks. This type of "voting" among modules has a long history in the neural network field (*e.g.*, von Neumann, 1951; Kilmer *et al.*, 1969). The best-known recent network using this principle is the one designed by Collins *et al.* (1988) to make mortgage insurance judgments.

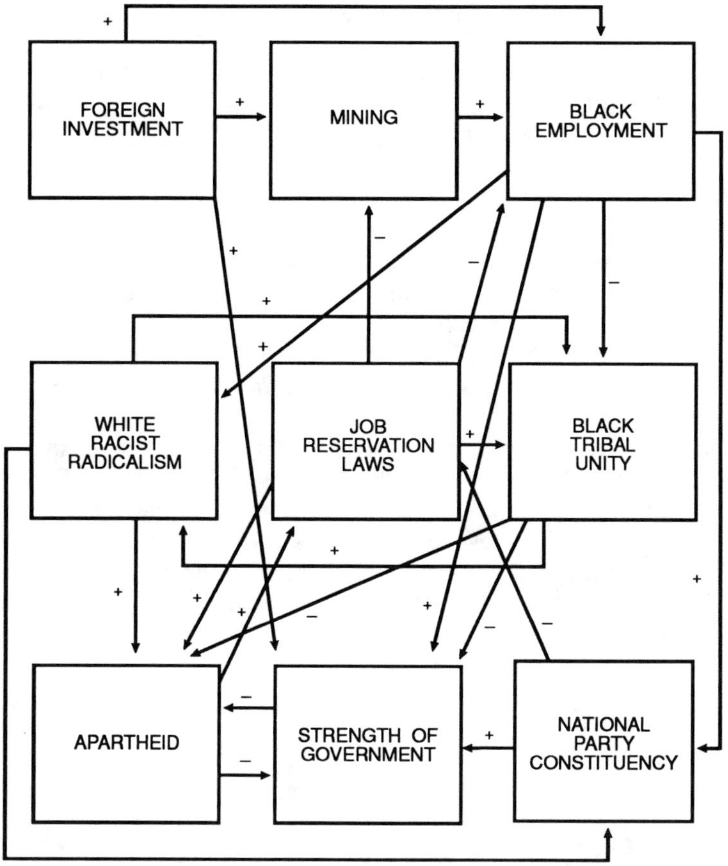

Figure 7.25. An example of a fuzzy cognitive map. The data base is South African politics and race relations. (Reprinted from Kosko, 1987b, copyright © IEEE, by permission of the publishers.)

Collins *et al.* based their network on the RCE (restricted coulomb energy) categorization model, an offshoot of the model of Reilly *et al.* (1982) discussed in Chapter 6.2. Reilly *et al.* (1987) previously constructed a general "multiple neural network" with several RCE modules. The specific form of this network that Collins *et al.* use is a 3-by-3 array of RCE modules, each having access to part of the database (information on the income, credit rating, and other relevant variables of mortgage insurance applicants). All modules are supervised by previously taught data on individuals who were unequivocally accepted or rejected for insurance, and then have to decide on actual cases which tend to be more ambiguous. Each module is hierarchically arranged in the manner of Figure 6.17 of this book. If the lower hierarchical levels, which make relatively coarse distinctions, agree in their judgments across modules, the higher levels are not engaged. In case of lower-level disagreements, the higher levels, which make finer distinctions, have to agree across modules.

A recent use of back propagation networks for knowledge-based inference has been developed by Fu (1989). Fu constructs conceptualization networks that are multilayer and involve logic conjunction. Rules are initially imposed in the form of network connections, but particular rules can be recognized as incorrect when the generalized back propagation rule is applied to a network constrained by correct knowledge. Such incorrect rules are then selectively removed.

The examples of networks discussed in this subsection are far from any biological theory of inference or of the formation of constructs (Kelly, 1955). But they illustrate that neural network methods are beginning to be used on high-level problems that have previously been dealt with, in a fragmentary manner, by symbolic processing programs from traditional artificial intelligence. The integration of this work with the biological control network theories discussed in preceding subsections should lead to some of the most exciting future advances in neural networks.

7.3. EQUATIONS FOR A FEW NEURAL NETWORKS PERFORMING COMPLEX TASKS

Hopfield and Tank's "Traveling Salesman" Network

The equations used in the solution of the Traveling Salesman Problem (TSP) by Hopfield and Tank (1984) are a modification of the continuous-time equations used by Hopfield (1984) and discussed in Section 4.5, namely

$$C_i \left(\frac{du_i}{dt} \right) = \sum_j w_{ij} x_j - \frac{u_i}{R_i} + I_i \qquad (4.15)$$

where $x_i = g_i(u_i)$ for some increasing, differentiable functions g_i. Since g_i is increasing, it has an inverse function g_i^{-1}; hence, one can write $u_i = g_i^{-1}(x_i)$. Recall that the system (4.15) has the Lyapunov or energy function

$$E = -\frac{1}{2} \sum_i \sum_j w_{ij} x_i x_j + \sum_i \frac{1}{R_i} \int_0^{x_i} g_i^{-1}(V) dV + \sum_i I_i x_i \qquad (4.16).$$

In the version of the network used in the TSP simulation, there are n rows and n columns, where n is the number of cities on the tour (*cf.* Figure 7.1). Hence the node activities x_i of (4.15) are replaced by doubly subscripted symbols x_{ij} where i represents a city name and j the position of that city in a tour. In order that the tour should favor shorter intercity distances, other things equal, the equations incorporate the distances d_{ik} between cities, which are in general unequal. The equations also incorporate inhibition within each row (*i.e.*, between representations of different cities at the same position) and within each column (*i.e.*, between representations of different positions for the same city).

All these considerations are incorporated in the connection strengths w_{ij} of Equation (4.15), which are labeled $w_{ij,kl}$ because of the double indices of city and position (using i and k for cities and j and l for positions). Using the standard mathematical terminology of $\delta_{ij} = 1$ if $i = j$ and 0 otherwise, Hopfield and Tank express these connection weights (unlearned) as follows:

$$w_{ij,kl} = -A\delta_{ik}(1-\delta_{jl}) - B\delta_{jl}(1-\delta_{ik})$$
$$-C - Dd_{ik}(\delta_{i,j+1} + \delta_{i,j-1}) \qquad (7.1).$$

The four terms in (7.1) denote, respectively, inhibitory connections within each row; inhibitory connections within each column; global inhibition; and distance dependence. The external input to the system is an excitatory bias designed to overcome the effects of the inhibitory connections, namely

$$I_{ij} = Cn \qquad (7.2).$$

Substituting (7.1) and (7.2) into (4.15), with all the R_i equal to the same value R and all the C_i equal to 1, yields the system of equations

$$\frac{du_{ij}}{dt} = -\frac{u_{ij}}{R} - A\sum_{l \neq j} x_{il} - B\sum_{k \neq i} x_{kj} - C[(\sum_k \sum_l x_{kl}) - n]^2$$
$$- D\sum_k d_{ik}(x_{i,k+1} + x_{i,k-1}) \qquad (7.3).$$

In equations (7.3), as in preceding articles, x_{ij} is equal to a sigmoid function of u_{ij}, in this case

$$g(u) = .5 \ (1 + \tanh(u/u_0)),$$

where "tanh" represents the hyperbolic tangent (that is, $(\exp(x)-\exp(-x))/(\exp(x)+\exp(-x)))$, and u_0 is a positive constant.

The energy or Lyapunov function for the TSP network defined by (7.3) is of course a form of the Lyapunov function in (4.16). Its exact expression is

$$E = -\frac{A}{2}\sum_i \sum_j \sum_{l \neq j} v_{ij}v_{il} - \frac{B}{2}\sum_j \sum_i \sum_{k \neq i} v_{ij}v_{kj}$$
$$- \frac{C}{2}[(\sum_i \sum_j v_{ij}) - n]^2 \qquad (7.4).$$

The Boltzmann Machine

The Boltzmann machine activates an updating rule very similar to the state updating rule of Hopfield (1982, 1984), discussed in Section 4.5. The modification of Hopfield's rule is, if the difference in energy between the 1 and 0 states of the kth node is ΔE_k, then regardless of its previous state, the state x_k of that node is 1 with probability

$$p_k = \cfrac{1}{1 + \exp\left(-\cfrac{\Delta E_k}{t} \right)} \tag{7.5}$$

and 0 otherwise, where T is the "temperature" parameter.

An example of the Boltzmann machine training procedure is described in Hinton and Sejnowski (1986, pp. 300-303); the derivation of the learning law is given in the Appendix of the same article. The training procedure alternates between two phases, a "+" and a "—" phase. During the "+" phase, the visible units are "clamped" into certain specified states. The hidden units are allowed to vary their states, according to a simulated annealing schedule. After annealing, the network is assumed to be close to an equilibrium, and it is run for several more iterations in order to estimate each of the values p_{ij}^+, the probability of both units i and j being in the "1" or "on" state. In the "—" phase of training, the visible and hidden units are both allowed to vary their states according to an annealing schedule. The probabilities of co-activation of pairs of units (whether visible or hidden) are estimated in the same way, yielding values p_{ij}^-.

After this whole sequence of annealings, the weights are changed, according to some equation of the form

$$\Delta w_{ij} = -a w_{ij} + b(p_{ij}^+ - p_{ij}^-) \tag{7.6}.$$

The weights decay at a slow rate and also change in proportion to the difference between probabilities of co-activation under "clamped" and "free ranging" conditions. The motivation for using this difference in probabilities is given in the Appendix of Hinton and Sejnowski (1986) and will now be sketched.

The probability difference term $p_{ij}^+ - p_{ij}^-$ in (7.6) is the derivative with respect to the weight w_{ij} of a certain objective function of the system (see the discussion of linear classification algorithms in Section 6.5). But the objective function used here is not the system energy function, as in the Hopfield models; rather, it is an information theoretic measure derived from the energy function. First let V_α be a fixed state vector for the visible units, and let H_β denote any possible state of the hidden units. If θ and μ denote all possible states of the visible and hidden units respectively, the Boltzmann probability distribution yields that the probability of the state V_α is

$$P^-(V_\alpha) = \sum_\beta P^-(V_\alpha \cap H_\beta) = \frac{\sum_\beta \exp\left(-\dfrac{E_{\alpha\beta}}{T}\right)}{\sum_{\theta,\mu} \exp\left(-\dfrac{E_{\theta\mu}}{T}\right)}$$

where T is the annealing temperature and $E_{\alpha\beta}$ is the energy of the system in state $V_\alpha \cap H_\beta$. That energy is in turn given by the familiar function used by Hopfield:

$$E_{\alpha\beta} = -\sum_{i<j} w_{ij} x_i^{\alpha\beta} x_j^{\alpha\beta}$$

where i and j could be indices for *either* visible or hidden units, and the superscript "αß" denotes the activity of node i or j in that state.

Now the state of the visible units in the clamped condition obeys a distribution $P^+(V_\alpha)$ which is independent of the weights w_{ij}. Using that fact, and the above equations, one can compute the gradient of the information measure

$$G = \sum_\alpha P^+(V_\alpha) \ln\left[\frac{P^+(V_\alpha)}{P^-(V_\alpha)}\right]$$

.

Heuristically, the function G is a measure of a gain in information from the free running condition to the clamped condition. Then it is shown that the partial derivative of G with respect to each weight w_{ij} is

$$-\frac{1}{T}[p_{ij}^+ - p_{ij}^-]$$

,

where

$$p_{ij}^+ = \sum_{\alpha,\beta} P^+(V_\alpha \cap H_\beta) x_i^{\alpha\beta} x_j^{\alpha\beta},$$

and p_{ij}^- is the corresponding sum of free running probabilities. Hence p_{ij}^+ is the sum of all probabilities of states of the visible and hidden units in which both nodes i and j are in the "on" state, in the clamped condition, and p_{ij} is the corresponding probability in the free running condition.

Grossberg and Kuperstein's Eye Movement Network

The saccade generator circuit described by Figure 7.5 obeys the following system of equations:

Long Lead Bursters:

$$\frac{dx_1}{dt} = -x_1 + I_1 - x_7 + x_7(0) \tag{7.7}$$

$$\frac{dx_2}{dt} = -x_2 + I_2 - x_8 + x_8(0) \tag{7.8}$$

Pausers:

$$\frac{dx_3}{dt} = -x_3 + x_4 - f(x_1) - f(x_2) \tag{7.9}$$

Arousal Cells:

$$x_4 = \text{constant}$$

Medium Lead Bursters:

$$\frac{dx_5}{dt} = -x_5 + x_1 + x_4 - g(x_2) - g(x_3) \tag{7.10}$$

$$\frac{dx_6}{dt} = -x_6 + x_2 + x_4 - g(x_1) - g(x_3) \tag{7.11}$$

Tonic Cells:

$$\frac{dx_7}{dt} = C(x_5 - x_6) \tag{7.12}$$

$$\frac{dx_8}{dt} = C(x_6 - x_5) \tag{7.13}$$

Motoneurons:

$$\frac{dx_9}{dt} = -x_9 + x_5 - x_6 + x_7 \qquad (7.14)$$

$$\frac{dx_{10}}{dt} = -x_{10} + x_6 - x_5 + x_8 \qquad (7.15)$$

The signal functions f and g were chosen to be functions of the form w/(a+w), which have occurred elsewhere in models of visual transduction. The terms $x_7(0)$ and $x_8(0)$ in Equations (7.7) and (7.8) are used to encode the inputs corresponding to the target position attained by the previous saccade.

VITE and Passive Update of Position (PUP) for Arm Movement Control

Bullock and Grossberg (1988, 1989) developed a theory for a vector of activities of nodes controlling interacting muscle groups. The actual form of their equations, however, uses the TPC, DV, and PPC nodes (see Figure 7.8) for a single muscle group, with the suggestion that the equations for antagonistic pairs and for synchronous groupings will be superimposed later.

For the VITE circuit with a single muscle group, let T, V, P, and G be the activities of the TPC, DV, PPC, and GO nodes respectively. Then the system of equations for these activities is

$$\frac{dV}{dt} = \alpha(-V + T - P) \qquad (7.16)$$

$$\frac{dP}{dt} = G \max(V,0) \qquad (7.17)$$

with α a positive constant. Bullock and Grossberg (1988, Appendix A) solve equations (7.16) and (7.17) under the simplifying assumption that G is a step function (0 before a certain time and 1 afterwards), and the results reproduce data on velocity profiles (*cf.* Figure 7.7) and speed-accuracy tradeoff. Later (Appendix 2 of the same article), they extend the results to arbitrary G functions, leading to explanations of some duration effects.

Bullock and Grossberg (1988, Appendix C) extend Equations (7.16) and (7.17) to allow for the possibility of passive update of position (*cf.* Figure 7.9). To Equation (7.17) for the PPC activity, they add an inhibitory signal from an "outflow-inflow interface" M, namely

$$\frac{dP}{dt} = G \max(P,0) + G_p \max(M,0) \qquad (7.18)$$

where G_p is a "passive gating" parameter that is large only when the muscle is in a passive (postural) state, measured by V being close to 0. The outflow-inflow interface, in turn, obeys an equation of the form

$$\frac{dM}{dt} = -\beta M + \gamma I - zP$$

with I an "inflow signal" denoting muscle displacement, β and γ positive constants, and z an adaptive gain control parameter which is governed by the equation

$$\frac{dz}{dt} = \delta G_p(-\epsilon z + \max(M,0)) \qquad .$$

Affective Balance and Decision Making Under Risk

Grossberg and Gutowski (1987) derived a quantitative measure of the affective value (perceived pleasantness or unpleasantness) of a given stimulus relative to a given context of preceding stimuli. Their derivation is based on the gated dipole equations discussed in Chapter 3:

$$\frac{dy_1}{dt} = -ay_1 + I + J \qquad\qquad \frac{dy_2}{dt} = -ay_2 + I$$

$$\frac{dw_1}{dt} = b(c-w_1) - ey_1w_1 \qquad \frac{dw_2}{dt} = b(c-w_2) - ey_2w_2$$

$$\frac{dx_1}{dt} = -fx_1 + gy_1w_1 \qquad\quad \frac{dx_2}{dt} = -fx_2 + gy_2w_2 \qquad (3.26).$$

$$\frac{dx_3}{dt} = -hx_3 + k(x_1-x_2) \qquad \frac{dx_4}{dt} = -hx_4 + k(x_2-x_1)$$

$$\frac{dx_5}{dt} = -mx_5 + (x_3-x_4)$$

In (3.26) as written, there is only a stimulus J to the "on" channel. Odd numbered activities and synaptic weights refer to the on channel and the even numbered activities and weights to the off channel (*cf.* Figure 3.7). Grossberg and Gutowski generalize (3.26) to include inputs to both channels, in this case, positively and negatively reinforcing inputs J^+ and J^- (*cf.* Grossberg & Schmajuk, 1987). They make algebraic calculations assuming that the variables in the network had reached a steady state, that is, they set derivatives equal to 0 in (3.26). If the decay rate is set to 1, a steady state assumption for y_1 yields $0 = dy_1/dt = -y_1 + (I + J^+)$, hence $y_1 = I + J^+$, the total input to the on channel. Likewise the total input to the off channel is $y_2 = I + J^-$.

Now consider the equations for w_1 and w_2, the depletable synapses within the on and off channels respectively, setting e = 1. Setting dw_1/dt to be 0 in (3.26) yields $0 = b(c-w_1)-y_1w_1$, and then solving for w_1 yields

$$w_1 = \frac{bc}{b+y_1} = \frac{bc}{b+I+J^+} \qquad (7.19).$$

Likewise, the off-channel transmitter equals

$$w_2 = \frac{bc}{b+I+J^-} \qquad (7.20).$$

Finally, let us calculate the actual on-response, or steady-state value of x_3. If g = h = 1, this is x_1-x_2. Now setting the derivatives of x_1 and x_2 to be 0, and the constants e and f to 1, causes x_1 to equilibrate to y_1w_1, and x_2 to y_2w_2. Hence the net on-response is

$$r^+ = (I+J^+)w_1 - (I+J^-)w_2 \qquad (7.21).$$

In a choice between two alternatives, the values of r^+ are computed, from (7.19), (7.20), and (7.21), separately for each alternative and then compared. We will sketch the part of Grossberg and Gutowski's derivation which uses these equations to explain the general phenomenon that "choices involving gains are often risk averse while choices involving losses are often risk taking" (Tversky & Kahneman, 1982). The linguistic framing effect shown in Figure 7.14 is a special case of this phenomenon, since subjects preferred the riskier choice when their context involved deaths and the safer choice when their context involved lives saved.

Let J_1^+ and J_1^- be the positive and negative inputs for the first alternative, and J_2^+ and J_2^- for the second alternative. (Probabilities are figured into input intensities.)   Assume that the system transmitters equilibrate to the first alternative. Then, by the same methods used to derive (7.21), the on-response to the first alternative is

$$r_1 = (I+J_1^+)w_1 - (I+J_1^-)w_2 \qquad (7.22),$$

whereas the on-response to the second alternative (where only inputs and not transmitters are altered) is

$$r_2 = (I+J_2^+)w_1 - (I+J_2^-)w_2 \qquad (7.23).$$

The choice depends on the sign of the quantity $\delta = r_1 - r_2$.

Now assume that there are two alternatives a_1 and a_2 of equal net value, that is, $J_1^+ - J_1^- = J_2^+ - J_2^- = D$. From (7.19), (7.20) (with the subscript "1" attached to the "J's"), and (7.22), it can be shown that

$$r_1 = bK(J_1^+ - J_1^-) = bKD \qquad (7.24)$$

where

$$K = \frac{bcD}{[(b+I+J_1^+)(b+I+J_1^-)]} \qquad (7.25).$$

From (7.19), (7.20), and (7.23), it can likewise be shown that

$$r_2 = K[(b+I)(J_2^+ - J_2^-) - I(J_1^+ - J_1^-) + J_2^+ J_1^- - J_2^- J_1^+] \qquad (7.26).$$

Combining (7.24)-(7.26) with the "equal net value" assumption $J_1^+ - J_1^- = J_2^+ - J_2^- = D$, the difference $\delta = r_2 - r_1$ becomes

$$\begin{aligned} \delta &= K[J_2^+ J_1^- - J_2^- J_1^+] = K[(J_2^- + D)J_1^- - (J_1^- + D)J_2^-] \\ &= D(J_1^- - J_2^-) \end{aligned} \qquad (7.27).$$

Now suppose the second alternative is riskier than the first, that is, $J_2^- >$ J_1^-. Then by (7.27), $\delta < 0$ if and only if $D > 0$. Since δ measures the relative preference for the second (riskier) alternative, this means that the less risky alternative is chosen if the alternatives are both favorable ($D > 0$), and not if the alternatives are both unfavorable.

EXERCISES FOR CHAPTER 7

*1. Simulate the "traveling salesman" of Hopfield and Tank (1985), using their equations

$$\frac{du_{ij}}{dt} = -\frac{u_{ij}}{R} - A\sum_{l \neq j} x_{il} - B\sum_{k \neq i} x_{kj} - C[(\sum_k \sum_l x_{kl}) - n]^2$$
$$- D\sum_k d_{ik}(x_{i,k+1} + x_{i,k-1})$$

(7.3)

where x_{ij} is the activity of the node representing the ith city and the jth position. Their parameters are A=B=500, R=1, C=200, D=500, u_0=.02, n=15, but they used 10 cities and 10 positions.

Use the sigmoid function $g(u) = .5 (1+\tanh (u/u_0)$. Choose the locations of the cities to be uniformly distributed within a two-dimensional square of edge length 1. Choose the initial values of u_{xi} to be uniformly distributed across the interval [u_{00}-.002, u_{00}+.002], where u_{00} is chosen so that $g(u_{00}) = .1$. (See p. 147 of Hopfield and Tank, 1985 for a justification for this choice of u_{00}.)

**2. Simulated annealing (Kirkpatrick et al, 1983; Hinton and Sejnowski, 1986) has been suggested as a way to get out of an "undesirable" local minimum and move toward a more "desirable" global minimum. Develop, and simulate if time permits, another algorithm to accomplish the same objective, with some variation on the following steps:

(a) Start with one of the simple homogeneous networks discussed in Chapter 4, whose solution trajectories can tend to one of several equilibria. An example would be the shunting competitive network of Grossberg (1973) or the additive network of Hopfield (1982).

(b) Add an extra "supervising" node to the network of part (a) with "knowledge" of the system's Lyapunov function and the value of the global minimum energy. If the system energy is in fact much larger than the global minimum, create an error signal from the supervising node to the subnetwork of part (a). Note: for the Grossberg equations

$$\frac{dx_i}{dt} = -Ax_i + Bf(x_i) - x_i\sum_{k=1}^{n} f(x_k)$$

the Lyapunov function (a special case of Cohen and Grossberg's) is

$$\frac{1}{2}\sum_{j,k=1}^{n} f(x_j)f(x_k) - \sum_{i=1}^{n}\int_0^{x_i} f'(y)[\frac{B}{y}f(y)-A]dy$$

(c) Transform the error signal from the supervising node to nonspecific random noise.

○3. A key issue relating to motor control of limbs is the mapping from extrinsic (task) coordinates to intrinsic (muscle/joint) coordinates. Recent data on human movement by Soechting and Flanders (1989) suggest that the first stage of this biological transformation is in fact linear. Specifically, subjects who are grasping at the position of an unseen, remembered target perform arm movements that are linearly related to target position. The same subjects when allowed to see a target make visual corrections of their movements, causing these movements to be nonlinearly related to target position.

(a) Comment on the utility of the back propagation algorithm, which can equally well learn linear and nonlinear mappings, as a potential model for these biological data.

(b) Suggest a model for these data using either the VITE model of Bullock and Grossberg or the tensor network model of Pellionisz and Llinas.

○4. Suggest some ways to model the Kahneman-Tversky preference reversal data (see Sections 7.2 and 7.3 above) without using gated dipoles. In particular, investigate how the Klopf hedonistic-neuron model (see Chapter 3) or drive-reinforcement model (see Chapter 5) could be modified to account for data of this sort. (Klopf, 1982 suggested that there could be hedonistic "inverse neurons" which "seek" negative stimulation, in addition to the more usual type that "seek" positive stimulation. That might be one possible avenue for working on this problem.)

○5. Formulate another choice problem based on data from the articles by Kahneman and Tversky referenced in this chapter. Simulate it using the affective balance (gated dipole) model.

○6. Behavioral data on lesions of the prefrontal cortex in monkeys (Pinto-Hamuy & Linck, 1965) suggest that frontally damaged monkeys can learn and perform an invariant sequence of movements but not a variable sequence. For example, they can learn to press a green light, a red light, and a letter "O" in that order, but cannot learn to press all three of those objects regardless of order. Suggest neural architectures to reproduce these data, along one of the following lines:

   (a) adaptive resonance (ART) with additional nodes;
   (b) back propagation (BP) with additional nodes;
   (c) brain-state-in-a-box (BSB) with additional nodes;
   (d) your own first principles (if possible, inspired by some known neuroanatomical or neurophysiological data).

○7. Suggest a connectionist model for social stereotyping, along the lines of one of the classification models discussed in Chapter 6. Hint: how can a network learn what features are appropriate to discriminate for a particular task (*e.g.*, abilities relevant for a job) and yet build a habit of discriminating on the basis of an inappropriate feature (*e.g.*, skin color, age, gender). Some clues may be found in the card sorting model of Leven and Levine (1987) (*cf.* Section 7.2 above).

○8. Discuss a neural network interpretation for a symptom of some mental illness. Examples include obsessive movements; compulsive thoughts; irrelevant verbal associations; biochemical addictions; or visual hallucinations. Interpret this symptom as a "pathology" in one of the neural network architectures described in previous chapters.

○9. *Semantic networks* form a class of knowledge representation techniques widely used in artificial intelligence. In semantic networks, nodes represent objects, concepts, or situations, while links between nodes represent the relations between them. An example of a set of ideas and the semantic network representing them is as follows:

   Clyde is a robin. A robin is a bird. Clyde owns Nest 1. Nest 1
   is a nest. Owner and ownee are related by ownership. Ownership
   is a situation. (See Figure 7.26).

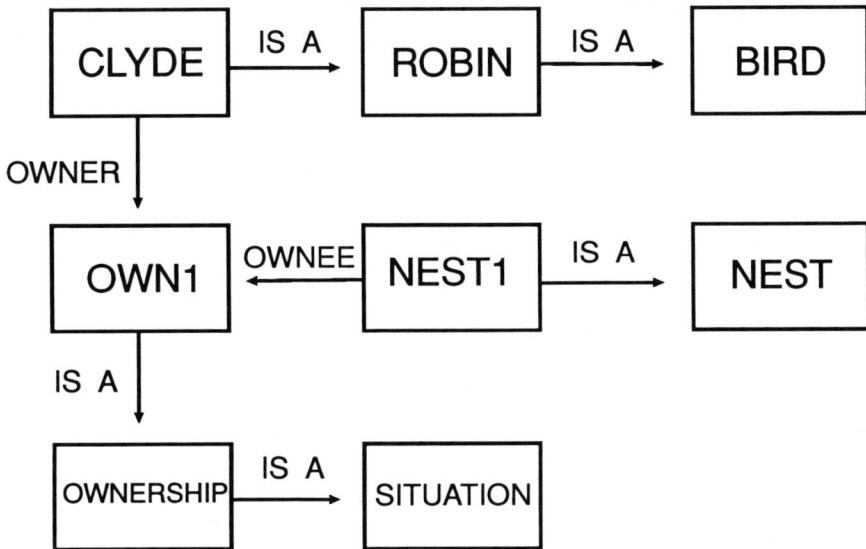

Figure 7.26. Example of a semantic net.

Meaning is assigned not, as in a neural network, by equations for node activities and connection strengths, but by procedures that manipulate the network. One example is *graph matching*. For example, suppose that we wish to answer the question *What does Clyde own?*. Then the fragment shown in Figure 7.27 is matched against the semantic network knowledge base. An "own node" that has an owner link to Clyde is sought. If a match is found, then the node that the "ownee" link points to answers the question.

Build a neural network to answer the same question, along the lines of either the interactive activation model of Rumelhart and McClelland, or the word recognition model of Grossberg and Stone.

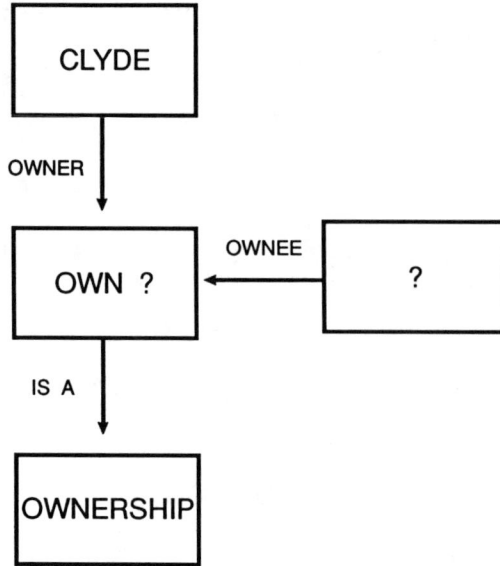

Figure 7.27. Example of a query to the net of Figure 7.26.

8

A Few Recent
Technical
Advances

Genius, in truth, means little more than the faculty of perceiving in an unhabitual way.

William James (*Principles of Psychology*)

We carry within us the wonders we seek without us: There is Africa and all her prodigies within us.

Sir Thomas Browne (*Religio Medici*)

The primary emphasis of this book has been on theory and general principles of organization. The book is intended as an introduction to neural and cognitive modeling not only in the sense of how modeling has been done by established researchers, but also in the sense of *how to* do modeling. To complete the picture, the reader needs some awareness of current progress in the field's two basic spheres of application: building neurally-inspired machines, and understanding actual brain function.

Hence, the main portion of the text closes with a look at some recent advances, first in the computing applications of neural networks and then in neurobiology and clinical neurology. In each field, I list a few of the most exciting advances. The choice of discoveries made in this chapter is an arbitrary sample rather than a "state of the art"; some readers will certainly think of other discoveries that are equally interesting but not included here.

8.1. SOME "TOY" AND REAL WORLD COMPUTING APPLICATIONS

Ever since the current growth phase of neural networks began (see Chapters 1 and 2 of this book for discussion), activity in all areas of the field, both theory and application, has been feverish. Since June, 1987, there have been six international neural network conferences. Attendance at each conference has been close to 2,000, with a slight majority of the attendees primarily interested in applications.

The article by Hecht-Nielsen (1986) provides a good picture of progress at that time in the computing applications of neural networks. In 1986, most applications of artificial neural systems (ANS) involved "toy problems" rather than real world problems. The one notable exception was the speech system known as NETtalk (see Section 7.1). Hecht-Nielsen also discussed the emergence, then recent, of some commonly used large-scale networks such as back propagation (see Section 6.2), brain-state-in-a-box (see Section 6.2), and adaptive resonance (see Section 6.3).

Hecht-Nielsen's article suggested that future progress in practical ANS design requires the actual construction of *neurocomputers*, that is, computers built along the lines (parallel and analog) of neural networks. This is because traditional digital computers take several hours, or even days, to run a complex neural network. But Hecht-Nielsen added that the cost of neurocomputers was staggeringly high. Four years later, while many companies in the United States, Europe, and Japan are working on various technologies for neurocomputers — including optical computing, VLSI, integrated circuits, and superconductivity among others — the cost has not been brought down appreciably.

Progress in ANS architecture itself, however, has been rapid since Hecht-Nielsen's article was published. The "toy" applications of ANS still outnumber the real world ones, and no research group or company has yet cornered the market on any specific real application. But significant new work, some of it marketable, has been done since 1986 in many areas of neural network application.

One of the more comprehensive reviews of neural net applications is Miller, Walker, and Ryan (1989), and a sense of the field can be obtained from that book's chapter titles. The areas discussed there are: sensor signal processing and data fusion; pattern classification, image processing, and machine vision; automated inspection; robotics and sensor-motor control; speech recognition and synthesis, and natural language; knowledge processing; financial applications; database retrieval; computer-based handwriting and character recognition; medical diagnosis, health care, and biomedical applications; manufacturing and process control; defense applications; other applications. A separate article, by Wenksay (1990), lists patents that have been based on neural network architectures. Space permits only cursory description of a few of these areas of application here.

The availability of supervised learning algorithms, notably back propagation and the RCE system (see Sections 6.2 and 7.3), has spurred the development of artificial neural networks devoted to fairly specific tasks. In other words, these algorithms have been the foundation for a variety of neural network "expert systems."

One area of application for such "expert systems" has been character recognition. Devices to recognize handwritten characters or numerals have most often employed back propagation networks, and have been closely studied by the United States Post Office (Kamangar & Cykana, 1988; Weidemann et al., 1989; Gong & Manry, 1989; Martin & Pittman, 1989). There have also been some preliminary successes at character recognition using the RCE and adaptive resonance theory networks.

A more dramatic application of neural network pattern recognition has been the design of an airport security system (Shea & Lin, 1989). A back propagation network was trained to detect the presence in luggage of nitrogen-based plastic explosives and tested at a major airport.

Closely related to pattern recognition is machine vision. Perhaps the most notable ANS for vision is the system based on the boundary and feature contour algorithms of Grossberg and Mingolla (1985a, b, 1987), discussed in Section 4.3. This system is not yet marketable but has been patented.

Like the earlier expert systems from artificial intelligence, neural network expert systems have also been employed for inference from known databases. The most striking example so far is the mortgage underwriter simulation of Collins et al. (1988) (see Section 7.2). In addition, neural network systems have been used to diagnose malfunctions in automobiles (Marko, James, Dosdall, & Murphy, 1989) and in jet engines (Dietz, Kiech, & Ali, 1989). There has also been some success in the use of ANS in medical diagnosis, such as the hybrid scheme involving back propagation combined with symbolic processing for diagnosing skin diseases (Yoon, Brobst, Bergstresser, & Peterson, 1989; see also Anderson, 1986 for related work).

Speech recognition and synthesis is another quite active area for application of neural network technology. The work of Rosenberg and Sejnowski (1986) and Sejnowski and Rosenberg (1986) on using back propagation networks for converting written into spoken English text was mentioned in Chapters 6 and 7. The reverse operation, converting spoken to written language, has also been done in neural networks. Kohonen (1988), for example, used some of his own previous networks to construct "phonetic typewriters," based on the spelling rules either of Finnish or of Japanese (the latter transliterated into Roman characters).

Some of the recent work in ANS for robotics and control was discussed in Section 7.1. In this area, particular success has been achieved using unsupervised algorithms. A notable example is the robot called INFANT, exhibiting eye-hand coordination, developed by Kuperstein (1988b) and Kuperstein and Wang (1990).

At this point in time, there is no limit to the imagination in the design of ANS applications. Students in my own neural networks classes have done final projects or written theses that have branched out into areas not covered above. This includes, for example, a traffic controller, using back propagation, that determines when a street light should change based on ongoing traffic information (Weathers, 1988). Also, there have been various efforts at composing music using neural networks. Elsberry (1989) combines a Hopfield net for choosing notes, a back propagation net for deciding if they conform to classical sequences, and an adaptive resonance net for detecting novelty. Kohonen (1989) applies to music his own earlier ideas on context-sensitive grammar; many of his compositions, in styles ranging from Bach to rock to romantic, were performed at the International Joint Conference on Neural Networks in January, 1990.

8.2. SOME NEUROBIOLOGICAL DISCOVERIES

Experimental work in neurobiology is progressing rapidly. The annual meetings of the interdisciplinary Society for Neuroscience draw close to 10,000 presenters annually. Publications such as *Science and Trends in NeuroSciences* give an indication of some of the exciting ongoing discoveries in this field.

Experimental data generation has been even more rapid in the last decade because of a few technical advances in the laboratory and clinic. These advances have included techniques for brain imaging: computerized tomography (CT or CAT), originally developed in the 1970's; and two other methods developed more recently, magnetic resonance imaging (MRI) and positron emission tomography (PET). Reviews of these techniques are found in many articles such as Andreasen (1988), which particularly stresses their use in psychiatric diagnosis, and Parks *et al.* (1989), which discusses their role in understanding neuropsychological test performance. CT and MRI are particularly useful for locating structural abnormalities, such as enlargement of ventricles (spaces within the brain) due to localized atrophy of brain tissue. PET uses positrons emitted from specific radioactive substances to locate brain substances, such as neurotransmitters, with which those substances interact. Hence, PET is expensive to operate (it requires connection with an active on-site cyclotron), but is by far the best technique for locating region-specific abnormalities in either transmitters or metabolism.

Another set of advances, which has yet to be used widely, is in techniques for electrical recording from neurons. Most studies of electrical activity in single neurons have been based on penetration with fine-tipped microelectrodes. This is the reason that the first major studies in cellular neurophysiology came from preparations that were relatively easy to penetrate in this manner. In particular, the squid giant axon was approachable because it reaches a diameter of up to 1 millimeter (the thickness of a pencil lead). By contrast, axons in human nervous systems are no thicker than a human hair.

Since the 1960's, there have been extensive single-cell studies in mammals, particularly cats and monkeys. Some of these studies have illuminated the functional organization of various sensory, motor, and motivational subsystems in the brain. A notable example is the series of studies by Nobel laureates Hubel and Wiesel (1962, 1963, 1965, 1968) of the selectivity of single cells in the cat and monkey visual cortex, with regard to position, orientation, and ocularity (see Table 4.1). More recently, there have been studies of single-cell responses in animals while they are involved in learning tasks. These include, for example, recordings from the auditory and motor cortices of cats learning to blink their eyes to a clicking sound (Woody, Vassilevsky, & Engel, 1970; Engel & Woody, 1972; Woody, Knispel, Crow, & Black-Cleworth, 1976; Sakai & Woody, 1980) and from the prefrontal cortices of monkeys learning to perform a delayed matching task (Rosenkilde et al., 1981; Fuster et al., 1982).

In the last few years, much of the interest of neurobiologists, as well as neural network modelers, has shifted from single neurons to populations of interconnected neurons. In fact, single-cell studies themselves point to the importance of anatomically localized functional groups. An example is the organization of the visual and somatosensory cortices into columns of cells with similar receptive fields (see Section 6.1 for a discussion of the visual case). The focus on populations and networks has spurred efforts to demarcate functional groups of neurons and measure average firing (action potential) frequencies in those groups. For this pupose, many investigators have developed multitipped electrodes that can record simultaneously from collections of 20 to 50 neurons. These investigators include Freeman (1975a, Chs. 4 and 7; 1978), Eichenbaum and Kuperstein (1985), Gross, Wen, and Lin (1985), and Sasaki, Bower, and Llinás (1989).

In the last few years, advances in both biochemical and electrophysiological techniques have contributed to a steady increase in the understanding of brain circuits and their associated chemical transmitters. This has enhanced the application of neurobiology to clinical problems, such as the treatment of drug addiction and the amelioration of mental illness.

Recent drug addiction studies have focused on the effects of substances such as cocaine, opiates, and amphetamine on neural transmitters along specific mammalian brain pathways. Much of the literature in this area is reviewed by Gawin and Ellinwood (1988), Koob and Bloom (1988), and Wise (1988).

In particular, there have been many studies on the interaction of addictive drugs with brain reward systems, particularly in rats. Results of Olds and Milner (1954), Olds (1955), and others led to mapping of a system of brain sites whose direct electrical stimulation is rewarding. This system follows pathways from the midbrain to the hypothalamus and then forward to the limbic system and cerebral cortex along fibers called the *medial forebrain bundle* (*cf.* Appendix 1). In recent years, experiments on intracranial self-stimulation have been extended to the actual use of cocaine or amphetamine injections as a reward. Results of such experiments tend to support the idea that the neurotransmitter dopamine (*cf.* Section 7.2 and Appendix 1) is necessary for normal operation of the reward system, and that several addictive drugs enhance cellular effects of dopamine in specific brain regions.

Recent studies of mental illness have likewise connected psychiatric disorders (as imprecisely as they are usually defined) with malfunctions of specific transmitter systems. Swerdlow and Koob (1987), and the associated commentaries, review much of the literature on mania and schizophrenia, usually associated with increases in extracellular dopamine at certain loci, and depression, usually associated with dopamine decreases. More of this literature, including effects of the neurotransmitter serotonin as well as dopamine, is reviewed in Hestenes (1991).

Some similarities between effects of schizophrenia and effects of frontal lobe damage are noted in Goldman-Rakic (1989). The frontal lobes are also implicated in obsessive-compulsive disorder (OCD) (Rapoport, 1989). OCD seems to involve metabolic abnormalities in pathways linking the frontal lobes to the caudate nucleus of the basal ganglia, a motor control area, and responds to treatment by a certain antidepressant which appears to potentiate the effects of serotonin.

The use of neural network theory as a tool by mainstream neurobiologists has yet to be the norm, but is growing rapidly. An increasing number of experimental laboratories include modeling as part of their ongoing research efforts. In addition, several conferences each year focus on specific topics in neurobiology, such as conditioning or vision, and bring together experimentalists and theorists. There are also many summer courses and workshops on computational neuroscience. Several new neural network journals have appeared since 1988; of these *International Journal of Neural Systems*, *Journal of Cognitive Neuroscience*, *Neural Computation*, and *Neural Networks* include strong biological components, as does the much older journal *Biological Cybernetics*.

All these developments seem to presage a much stronger partnership, probably within a decade, between theory and experiment in neuroscience and cognitive science. By this I mean a partnership such as exists in physics, where theories suggest experiments and experimental results refine theories.

APPENDIX 1: BASIC FACTS
OF NEUROBIOLOGY

The working neural or cognitive modeler, whether interested primarily in computing applications or in brain theory, should at least have a working knowledge of neurobiology on two levels. One is the level of neurons (brain or nerve cells), including their component parts — axons, dendrites, cell bodies or somata, and synapses — and the chemical transmitters at synapses. This includes knowledge of how conduction of nerve impulses is affected by the actions of the various ions in and around the cells. The other is the level of brain regions, their cognitive functions and the pathways between them. This should, at best, include some knowledge of how the nervous system has evolved from invertebrates to fish to other mammals to humans, both structurally and functionally.

This Appendix gives an extremely cursory summary of those biological facts which are, in my opinion, essential for the modeler to know. I refer the reader to other books whose coverage of these areas is far more detailed. Good general textbooks on all aspects of neural science include Shepherd (1983) and Kandel and Schwartz (1985); Shepherd's book is particularly strong on the cognitive aspects of neural structures. Thompson (1967) is a basic textbook on physiological psychology, with chapters on neuroanatomy and neurophysiology as well as chapters on such psychological topics as conditioning and memory. Carlson (1977) is a more recent source for the physiology relevant to behavioral and cognitive functions. Katz (1966) gives detailed, and still timely, descriptions of fundamental electrical and chemical processes at the neuronal level. A more recent and succinct treatment of these neuronal processes is found in Byrne and Schultz (1988). There are also many good textbooks on neuroanatomy of different brain regions, including Truex and Carpenter (1969) and Nauta and Feirtag (1986).

The Neuron

While the functional organization of the nervous system differs profoundly between squid, fish, rats, and humans, the organization of individual neurons differs much less. Hence, some classic studies on invertebrates have contributed greatly to our knowledge of general, including mammalian, neurophysiology. Particularly important is the work of Young (1936), Cole and

Hodgkin (1939) and others on the giant axon of the squid, which activates an escape reflex.

Figure 9.1 shows a schematic neuron. The main parts of it are the dendrites ("small branches"), which often receive electrical signals from other cells; the soma or cell body, which sums electrical potentials from many dendrites and also contains the cell's nucleus; and the axon, which conducts electrical signals and transmits them to other cells. The picture of Figure 9.1 is not universally accurate. Sometimes the axons are much shorter, relative to the other cell components, than the one shown there; short axons are particularly common in association areas of the human cerebral cortex. Also, the dendrites can sometimes be "senders" as well as "receivers." Still, this figure illustrates a "generic neuron" fairly well, and most of the exposition in this section assumes a neuron of this type.

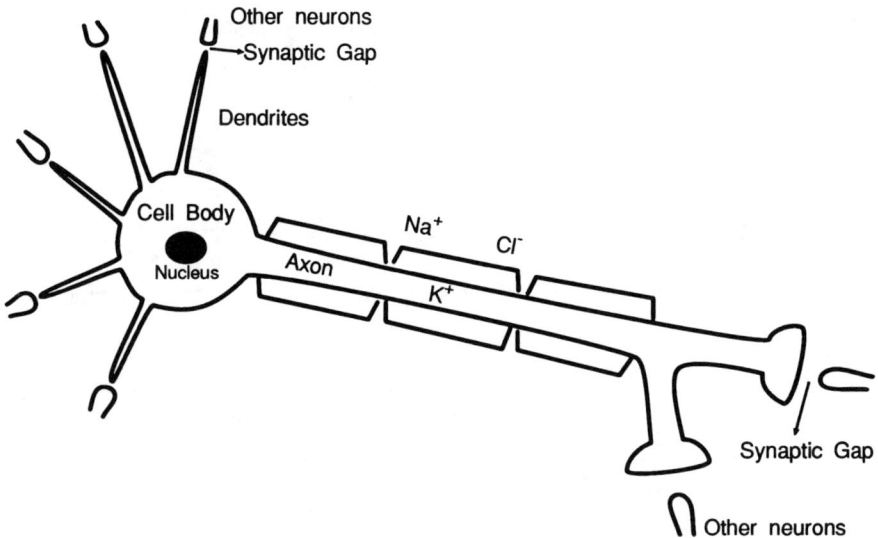

Figure 9.1. Schematic neuron. Main parts (axon, cell body or soma, dendrites, synapses) are labeled. Characteristic ions are shown where they are most prevalent, inside or outside the membrane. See text for details.

The significant variable for information transmission in a neuron is the electrical potential across the membrane of the cell's axon. This potential is determined by the intracellular and extracellular concentrations of three single-element ions, potassium (K^+), sodium (Na^+), and chloride (Cl^-) along with some compound ions. There are two distinct phases of transmembrane electrical activity, the *resting membrane potential* while the cell is not conducting an electrical impulse, and the *action potential* or actual "nerve impulse." The action potential, which is propagated down the axon, amounts to a reversal of electrical polarity from the resting phase; the inside is about 60 to 70 millivolts (mV) negative to the outside during the resting potential, but about 40 mV positive during the action potential. Figure 9.2 shows a typical action potential, also called a *spike* because of its shape.

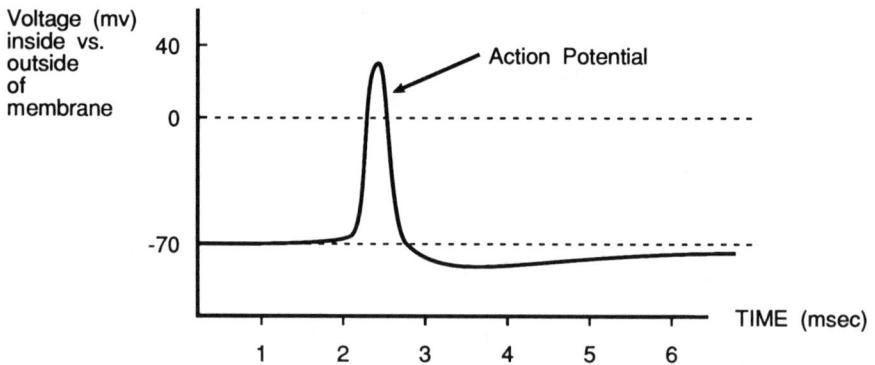

Figure 9.2. The action potential recorded across the membrane of a squid giant axon. (From Thompson, 1967, modified from Hodgkin & Huxley, 1939; reprinted with permission of Harper and Row Publishers.)

Figure 9.3 lists the intracellular and extracellular concentrations of various ions in the squid axon. Since the inside is electrically negative relative to the outside, it is somewhat surprising that the positive Na^+ (sodium) ion should be more concentrated on the outside, rather than rushing in to neutralize the polarity. This occurs because an active metabolic "pump," whose mechanism is not completely known, keeps most of the sodium outside in the resting state.

The processes that initiate the action potential temporarily shut off this pump, causing a reversal of the membrane potential.

The conduction of the nerve impulse, as Helmholtz and others in the last century discovered, is too slow to be merely electrical transmission as through a wire. Hence, this conduction must involve an active biochemical process. How the neuron changes from the resting to the excited state was essentially discovered in a series of experiments described, and quantitatively analyzed, by Hodgkin and Huxley (1952).

External

Internal

Na^+ 460

K^+ 10

Cl^- 540

Na^+	50
K^+	400
Cl^-	40 to 100
Isethionate$^-$	270
Aspartate$^-$	75

-60 mv inside

Figure 9.3. Ion concentrations (millimoles per liter) and potential difference across the membrane of the squid giant axon. (Reprinted from Bernard Katz, *Nerve, Muscle, and Synapse*, copyright 1966, with permission of McGraw-Hill Publishing Company.)

The cautionary note must be added, however, that impulses (action potentials) are *not* the only means of communication between neurons. The passive electrotonic spread of potentials from synapses has been found recently to play an important role, particularly in short-distance communication (see Shepherd, 1983, p. 102 for discussion). The potentials recorded by the electroencephalogram (EEG) result from this passive spread.

Early research on the squid giant axon showed that the action potential still occurs even if all the axoplasm (protoplasm on the inside of the axon) is squeezed out. It was concluded that the action potential is a *membrane* phenomenon. This research also showed that the action potential depends strongly on the presence of sodium ions in the extracellular medium. It was concluded that the potential change results from inward movement of sodium ions. The resting membrane is much less permeable to sodium ions than to potassium or chloride ions, but the action potential generation (excitation) process increases its permeability to sodium, allowing that inward movement of ions to take place.

The process of action potential generation is partially described by Shepherd (1983, p. 107): "a crucial property of the Na conductance ... is that it is involved in a positive feedback relation with the membrane depolarization. When the membrane begins to be depolarized, it causes the Na^+ conductance to begin to increase, which depolarizes the membrane further, which increases Na^+ conductance, and so on." If the cell membrane is depolarized from its resting state, either by an impulse from another neuron or by direct stimulating current, the cell will revert to its resting potential unless it reaches a *threshold* transmembrane voltage — typically in the neighborhood of -40 mv inside in the case of the squid axon. If the cell potential does reach that threshold, the aforementioned positive feedback will take place, ultimately leading to an action potential. This is the biological basis for all-or-none impulses (*cf.* Section 2.1).

An important consequence of the sodium-permeability mechanism is that the frequency of impulse generation is limited. For one or a few milliseconds after an impulse, no additional impulses can be generated; this is called the *absolute refractory period* (see the discussion of continuous models in Chapter 2). For several more milliseconds afterwards, there is a *relative refractory period* in which an action potential can be initiated, but the strength of current required to initiate an impulse is larger than normal.

The refractory periods result from the same ionic mechanism that terminates the impulse, namely, following of sodium conductance increase by an increase in the membrane's conductance of potassium (K^+). This leads to a movement of potassium ions outward, which reduces transmembrane potential back toward the resting level and thereby reduces sodium conductance. In addition, during the absolute refractory period, this outward flow of K^+ makes the membrane potential even more negative than normal, which further decreases the sodium conductance. The cell does not fully recover its ability to generate impulses until both ionic conductances are back to their resting levels.

Many, but not all, axons, particularly the longer ones, are covered by an electrically insulating layer known as the *myelin sheath*. This sheath is made of cells of a different type than neurons, known as *glial cells*. The action potential spreads like a wave down the axon, in a single direction (from dendrites toward outgoing synapses). In the case of myelinated axons, the conduction is along the outer membrane, "jumping" between holes in the myelin sheath known as the *nodes of Ranvier*, a process called *saltatory conduction*.

Thus far we have talked about movement of ions and conduction of electrical activity within a single neuron. As the impulse moves toward a synapse between two neurons, different processes take over.

Synapses, Transmitters, Messengers, and Modulators

The current view of nervous system organization did not become widely accepted until early in the twentieth century, with the work of Cajal and Sherrington (see Cajal, 1934, and Sherrington, 1947 for summaries). Before the work of those two pioneers, there were disputes between adherents of two doctrines, the "neuron doctrine," which held that the nervous system is composed of distinct cells, and the "reticular doctrine," which held that all the fibrous processes are continuous with each other. The "neuron doctrine" won with the discovery that many pairs of cells that are functionally interconnected are actually physically separated. This separation is known as the *synaptic gap*; its width is of the order of one to a few μ (1 μ = 10^{-6} meters).

The number of different types of junctions between cells is quite large (see, for example, Shepherd, 1983, pp. 73-75, and Figure 9.4). There are varying distances between cells, and there are both electrical synapses (where the action potential travels between cells by direct electrical conduction) and chemical synapses (where the conduction is mediated by a chemical transmitter that affects ionic conductances). The most characteristic junction type, particularly in mammalian brains, is the chemical synapse (see Figure 9.5).

The chemical synapse, unlike some of the other kinds of junctions, is unidirectional. The transmitting neuron is called the *presynaptic* cell, and the receiving neuron is called the *postsynaptic* cell. The difference between presynaptic and postsynaptic neurons is indicated by the greater thickness of the apposed presynaptic membrane, and the presence at the presynaptic side of swellings called *vesicles* which contain packets of chemical transmitter. A similar organization occurs at neuromuscular junctions, with an area of muscle playing the role of "postsynaptic neuron."

Even among chemical synapses, there is a dizzying variety. For example, there are two types, 1 and 2 (*cf.* Figure 9.5), with different vesicle shapes. It was once thought that most Type 1 synapses are excitatory and Type 2 synapses are inhibitory; this is still a useful heuristic, though it now admits a considerable number of exceptions. Also, synapses can be from the presynaptic axon to the postsynaptic dendrite (axodendritic); from axon to axon (axoaxonic), from axon to cell body (axosomatic); or, when the dendrite actually carries an action potential, from dendrite to dendrite (dendrodendritic). Of those cases, the axodendritic is the most common.

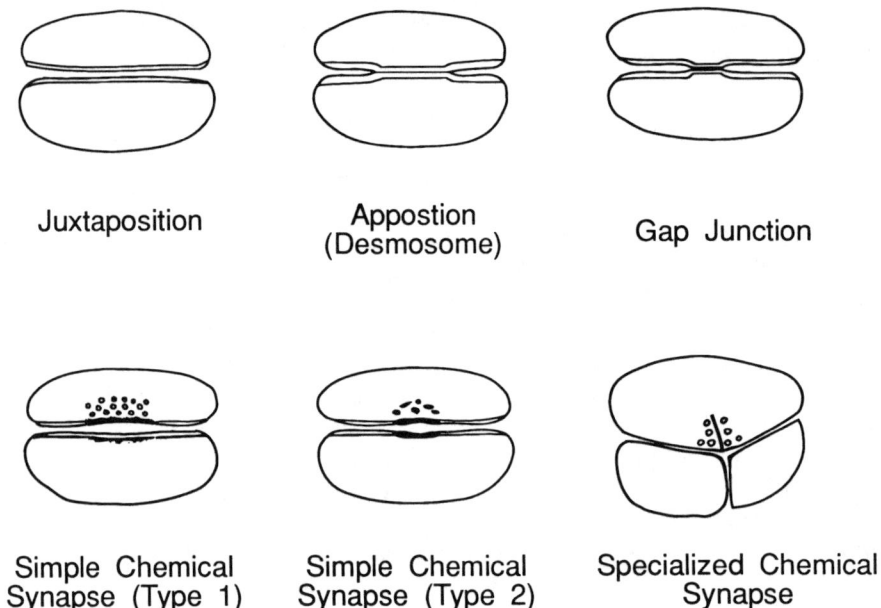

Juxtaposition   Appostion (Desmosome)   Gap Junction

Simple Chemical Synapse (Type 1)   Simple Chemical Synapse (Type 2)   Specialized Chemical Synapse

Figure 9.4. Types of junctions between nerve cells. (Reprinted from Shepherd, 1979, with permission of Oxford University Press.)

Typical chemical synapses, through whatever transmitter substance they release, cause a passive increase or decrease in the postsynaptic membrane potential, starting at the junction point (most commonly on a dendrite). In the case of excitatory synapses, the passive depolarization is termed the *excitatory postsynaptic potential* (EPSP). Similarly, inhibitory synapses cause a hyperpolarization (the opposite of depolarization) called the *inhibitory postsynaptic potential* (IPSP).

An EPSP occurs because transmitter causes a net inward movement of positive charge, by increasing membrane conductance to Na^+, K^+, and possibly other positive ions such as the calcium ion (Ca^{++}). The EPSP might or might not be large enough to depolarize the postsynaptic neuron to its firing threshold. In fact, the postsynaptic neuron typically has thousands of dendrites receiving synapses from different presynaptic cells. Hence its firing depends on the sum of EPSP's from these different dendrites minus the sum of IPSP's from other dendrites. IPSP's can occur as a result of increased conductance either for outward movement of positive charge (K^+) or for inward movement of negative charge (Cl^-). Some typical postsynaptic potentials are shown in Figure 9.6.

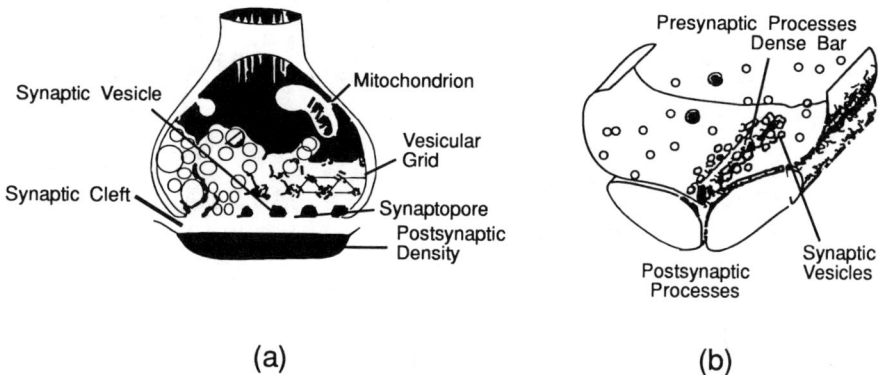

Figure 9.5. Structures of Type 1 (in (a)) and Type 2 (in (b)) chemical synapses. (From Shepherd, 1983, modified from Akert *et al.*, 1972, and Wood *et al.*, 1977; reprinted with permission of Oxford University Press.)

The first chemical neurotransmitter substance to be discovered was acetylcholine, which was identified in 1921 by Loewi as the substance used by the vagus nerve to decrease the heart rate. Subsequently, acetylcholine was found to be the transmitter substance used in other nerves connecting the brain to internal organs (autonomic nerves) and at the junctions between nerves and skeletal muscles. Later, it was also found to be one of the commonest transmitter substances at both excitatory and inhibitory synapses in the brain.

Other important neurotransmitters, many of them amino acids or derivatives of amino acids, are shown in Figure 9.7. The principal ones are acetylcholine (ACh); dopamine (DA); norepinephrine (NE, also known as noradrenaline or NA); serotonin (5HT); gamma-amino butyric acid (GABA); glutamate (GLU); and glycine (GLY). These transmitters can be excitatory or inhibitory, with the exception of GLY, GABA, and 5HT, which are almost always inhibitory, and GLU, which is always excitatory. (See the discussion in Section 7.2 of postulated cognitive roles for the *monoamine* transmitters — DA, NE, and 5HT — which are broadcast by certain midbrain regions out to vast areas of the neocortex and the limbic system.)

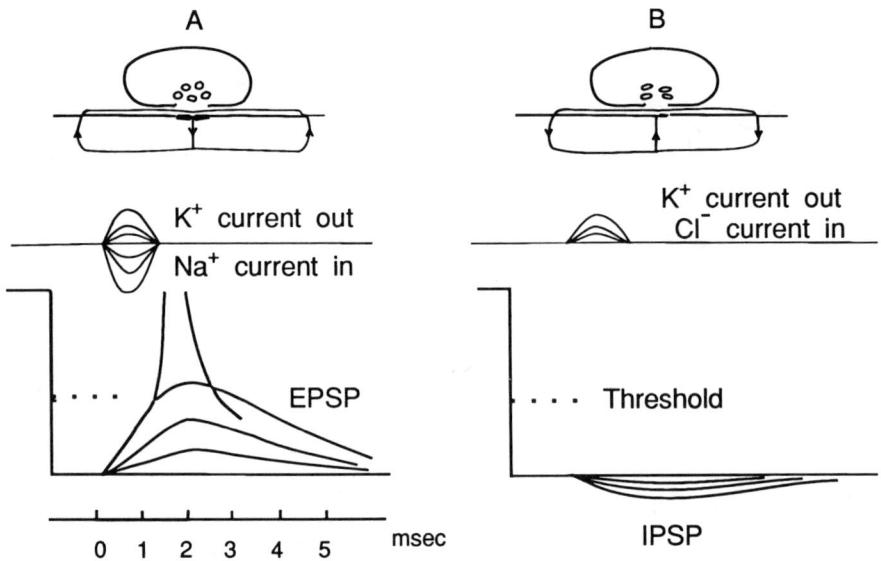

Figure 9.6. Typical postsynaptic potentials. Above: pre- and postsynaptic terminals, with net positive current flows shown by arrows for depolarizing (A) and hyperpolarizing (B) actions. Middle: time course of ionic current flows. Below: recordings of a typical EPSP (in A) and IPSP (in B). (Reprinted from Shepherd, 1983, with permission of Oxford University Press.)

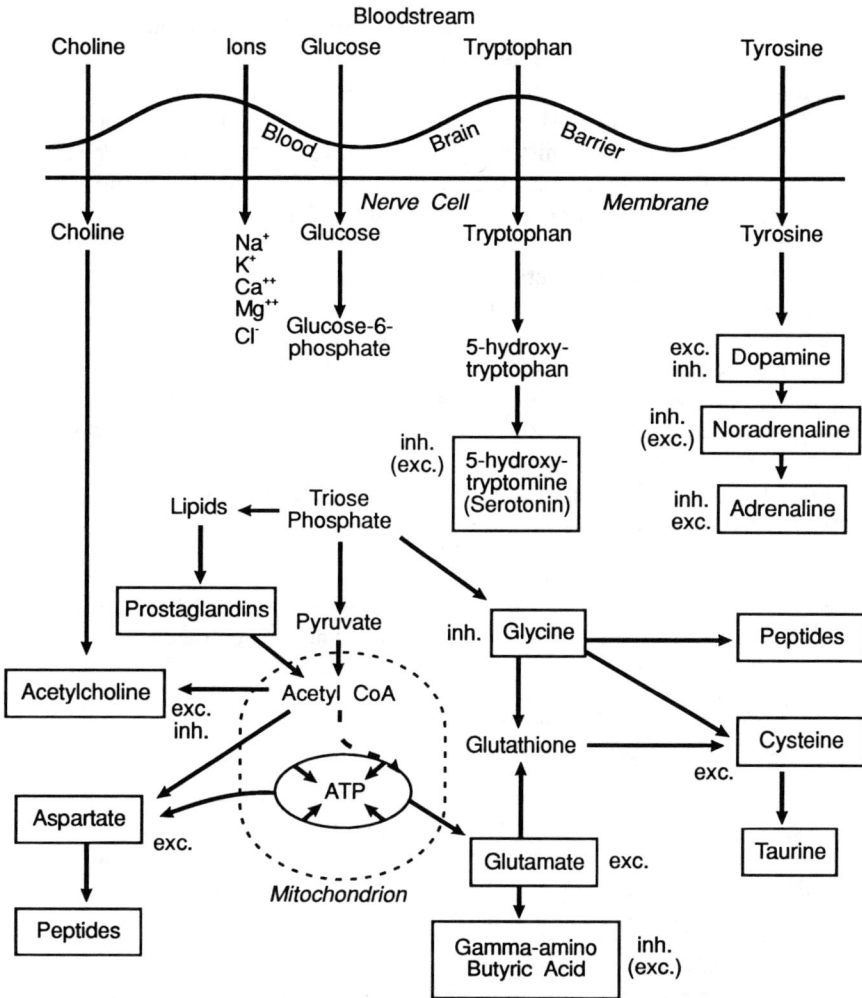

Figure 9.7. Summary of major neurotransmitters and related compounds, and some pathways involved in their transport from the bloodstream and their metabolism within the nerve cell. "Exc" amd "inh" denote excitatory or inhibitory nature of transmitters listed in boxes. (Adapted from Cooper, Bloom, & Roth, 1982, with permission of Oxford University Press.)

Space does not permit review of the complex chemical reactions involved in synaptic transmission. In general, the sequence of events is that presynaptic depolarization increases the movement of the calcium ion (Ca^{++}) near the synaptic gap, which in turn stimulates the release of transmitter from vesicles. This description is particularly good for *cholinergic* synapses, the ones using acetylcholine as their transmitter. Calcium plays a variety of other roles in neurochemistry. For example, Ca^{++} and the cyclic nucleotides (cAMP and cGMP) play crucial mediating roles in plastic changes at synapses (*cf.* the review by Byrne, 1987, discussed in Section 3.4, and the model of Gingrich & Byrne, 1985, discussed in Sections 5.2 and 5.3).

Not all chemical substances occuring in the brain are actual neurotransmitters. Table 9.1 summarizes the criteria that are generally agreed upon for a substance to be considered a transmitter. Many substances that are not, or not known to be, neurotransmitters play other important modulating roles in cellular reactions. Among these are the cyclic nucleotides discussed above, and the neuroactive peptides (see Gainer & Brownstein, 1981). The peptides include endorphins (morphine-like substances) which are associated with positive reinforcement.

1. *Anatomical*: presence of the substance in appropriate amounts in presynaptic processes.
2. *Biochemical*: presence and operation of enzymes that synthesize the substance in the presynaptic neuron and processes, and remove or inactivate the substance at the synapse.
3. *Physiological*: demonstration that physiological stimulation causes the presynaptic terminal to release the substance, and that iontophoretic application of the substance to the synapse in appropriate amounts mimics the natural response.
4. *Pharmacological*: drugs that affect the different enzymatic or biochemical steps have their expected effects on synthesis, storage, release, action, inactivation, and reuptake of the substance.

Table 9.1. Criteria for deciding whether a given substance is a neurotransmitter. (Reprinted from Shepherd, 1983 with permission of Oxford University Press.)

Invertebrate and Vertebrate Nervous Systems

Among invertebrates, nerve cells controlling movements in response to particular stimuli appear in most of the multicellular phyla, starting with the coelenterates (jellyfish and medusas). While learning has been studied in flatworms, the best developed invertebrate central nervous systems are in mollusks and arthropods. These nervous systems do not have brains in the sense that vertebrates do, but possess several *ganglia,* which are defined as concentrated areas involving several sensory and motor processing units.

Invertebrates are probably not capable of quite the same complexity of neural processing as are vertebrates. Yet invertebrate preparations have yielded basic multicellular studies of learning and conditioning (see Byrne, 1987 for a review) and of rhythmical firing patterns (*e.g.,* Selverston, 1976).

The vertebrate nervous system has a basic plan that has persisted in spite of major evolutionary changes. It is divided into the peripheral nervous system, consisting of nerves with connections to the rest of the body, and the central nervous system, consisting of the spinal cord and the brain. The peripheral nervous system has two main parts: skeletal and autonomic, the latter comprising nerve fibers that affect, and receive sensations from, internal organs.

Figures 9.8 and 9.9 show the brains of various vertebrates, going from lampreys (not proper vertebrates but chordates) up to humans. The characteristic divisions of forebrain, midbrain, and hindbrain are consistently maintained. In higher mammals, however, the forebrain balloons outward and then develops various folds (known as *convolutions* or *gyri*) to become the six-layered cerebral cortex. This cortex performs ever more sophisticated integrative functions, in feedback with the subcortical structures that change much less across species.

Functions of Vertebrate Subcortical Regions

Some of the more important large brain regions below the cerebral cortex are shown schematically in Figure 9.10. The following gross subdivisions are of functional importance:

> *Pons* and *medulla* — just above the spinal cord;
> *Midbrain* — just above the pons and medulla (the pons, medulla, and midbrain are usually considered to constitute the *brain stem*);
> *Thalamus* — deep inside the forebrain;
> *Hypothalamus* — below the thalamus;

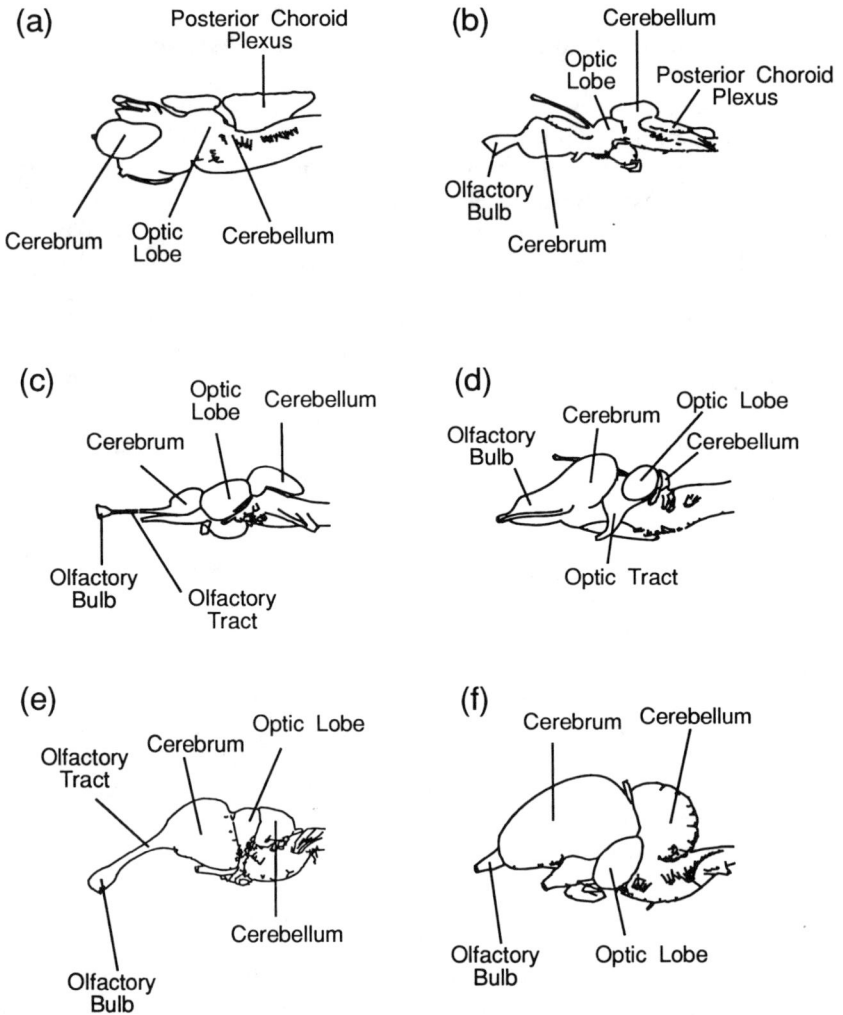

Figure 9.8. Diagram of vertebrate brain evolution: (a) lamprey; (b) shark; (c) codfish; (d) frog; (e) alligator; (f) goose. (Adapted from Romer & Parsons, 1977, with permission of W. B. Saunders Publishing Company.)

(g)

Neopallium

Cerebellum

Olfactory
Bulb

Olfactory   Paleopallium
Tract

(h)   Cerebral
Cortex
Olfactory   Cerebellum
Bulb

Olfactory
Tract

(i)

Cerebral
Cortex

Corpus
Callosum

Cerebellum

Fornix

Posterior
Choroid Plexus
Olfactory
Bulb

Pons

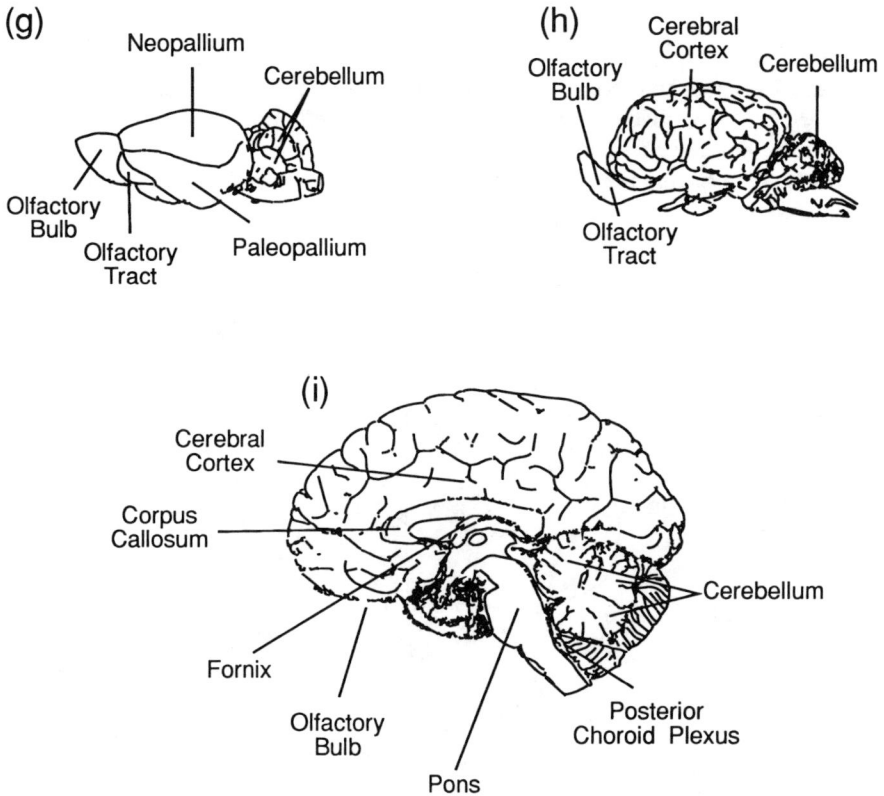

Figure 9.9. Continuation of Figure 9.8: (g) hedgehog; (h) horse; (i) human. "Neopallium" and "paleopallium" correspond to the cerebral cortex and limbic system. (Adapted from Romer & Parsons, 1977, with permission of W. B. Saunders Publishing Company.)

Limbic system — forming a border around much of the forebrain and midbrain;

Cerebellum — in back of the pons and underneath the rear (occipital) area of the cortex;

Basal ganglia — at the base of the forebrain.

We shall now briefly review some current knowledge of the cognitive functions of each of these large regions. This review should provide the reader with an intuitive "landscape," rather than revealed truths. The cautionary note must be added, of course, that complex behaviors involve circuits rather than isolated "centers." Also, just because stimulating a region promotes a behavior, or lesioning that region suppresses the behavior, it need not follow that the region's primary function is to perform that behavior (see Churchland, 1986 for a discussion).

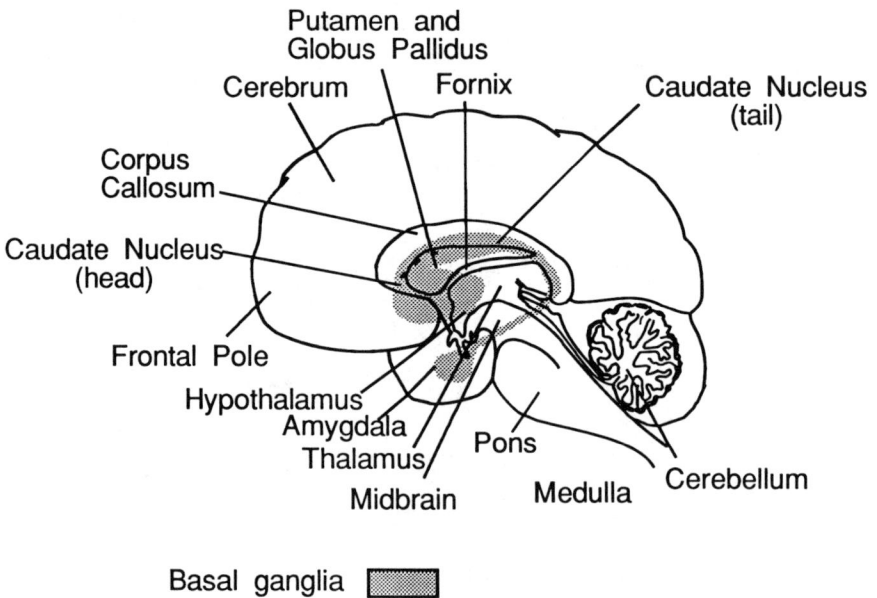

Figure 9.10. Medial view of the brain, showing locations of some of its major subdivisions. The amygdala is part of the limbic system (see Figure 9.12), and the fornix is a pathway linking the hypothalamus with parts of the limbic system. The corpus callosum is a pathway linking the two cerebral hemispheres. (Adapted from Thompson, 1967, with permission of Harper and Row Publishing Company.)

The pons and medulla include some fibers and cell nuclei from the autonomic nervous system. These areas, along with the midbrain, are also the locus of the *reticular activating system* (or *reticular formation*) which is involved in the regulation of sleep, waking, and arousal. (Recall the inclusion of "nonspecific arousal" in some neural networks discussed in earlier chapters, such as the one shown in Figure 3.8.) As Truex and Carpenter (1969, p. 316) state: "The term 'reticular formation' is a somewhat vague designation given a variety of special connotations; it originated in anatomy to describe portions of the brain stem core characterized structurally by a wealth of cells of various sizes and type ... enmeshed in a complicated fiber network." The reticular formation is usually considered to include some of the sources of modulating monoamine transmitters shown in Figure 7.15 (at least the raphé medianus and dorsalis and ventral tegmentum).

The thalamus is composed of different cell nuclei, most of which have one-to-one feedback connections with some part of the cortex. This includes areas of the cortex devoted to specific senses. For example, the lateral geniculate body of the thalamus is a way station for visual inputs from the optic nerve going to the visual cortex, while the medial geniculate body of the thalamus plays a similar function for the auditory cortex. It also includes multisensory association areas of the cortex; the mediodorsal nucleus of the thalamus, for example, has a one-to-one relationship with the prefrontal cortex. Many areas of the thalamus, including the mediodorsal nucleus, also have strong connections with the limbic system and hypothalamus, which are involved in emotional expression and processing of visceral information (see Figure 9.11).

The hypothalamus has extensive connections with the endocrine system. Hence, particular areas of the hypothalamus are involved in either feeding or mating behavior. The lateral hypothalamic area, for example, is part of a consummatory circuit for eating that also involves areas of the brain stem, some of which use the catecholamines (DA and NE) as neurotransmitters. The lateral hypothalamus is also a region whose direct electrical stimulation is rewarding (Olds, 1955). The ventromedial hypothalamus is opposite in effect to the lateral; stimulation of this area produces satiety and its lesion produces overeating. These two hypothalamic areas provided some of the inspiration for motivational dipoles in neural networks (Grossberg, 1972b; *cf.* Section 3.3).

The limbic system has been implicated in the emotional expression that accompanies such behavior as feeding, copulation, and aggression. This system includes several subregions going from the hippocampus and amygdala, just under the temporal lobes, through the cingulate gyrus, which is sometimes considered part of the cortex itself, to the septum, closer to the frontal area (see Figure 9.12). The full details of emotional expression involve a circuit linking the limbic system with parts of the hypothalamus, midbrain, and thalamus (see Figure 9.11). Precise conclusions about the role of each part of the circuit are lacking, in spite of a tremendous amount of data. Neural network models are likely to make contributions to sorting out all these findings (see the discussion in Section 7.2 of neurochemistry and mental illness).

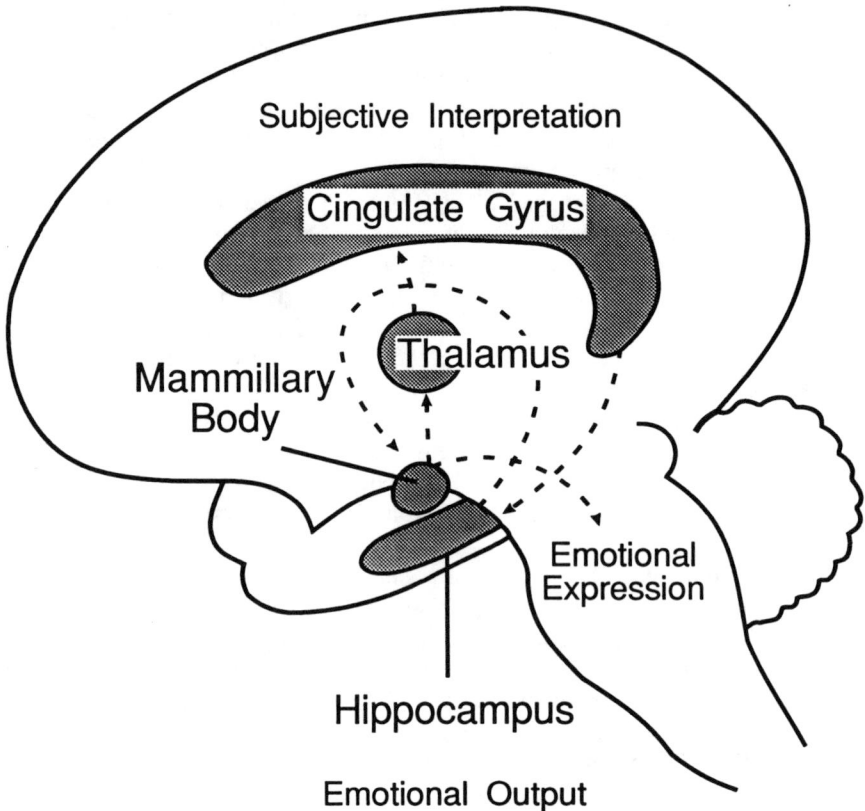

Figure 9.11. Schematic of the Papez circuit for emotional expression. (Reprinted from Shepherd, 1983, with permission of Oxford University Press.)

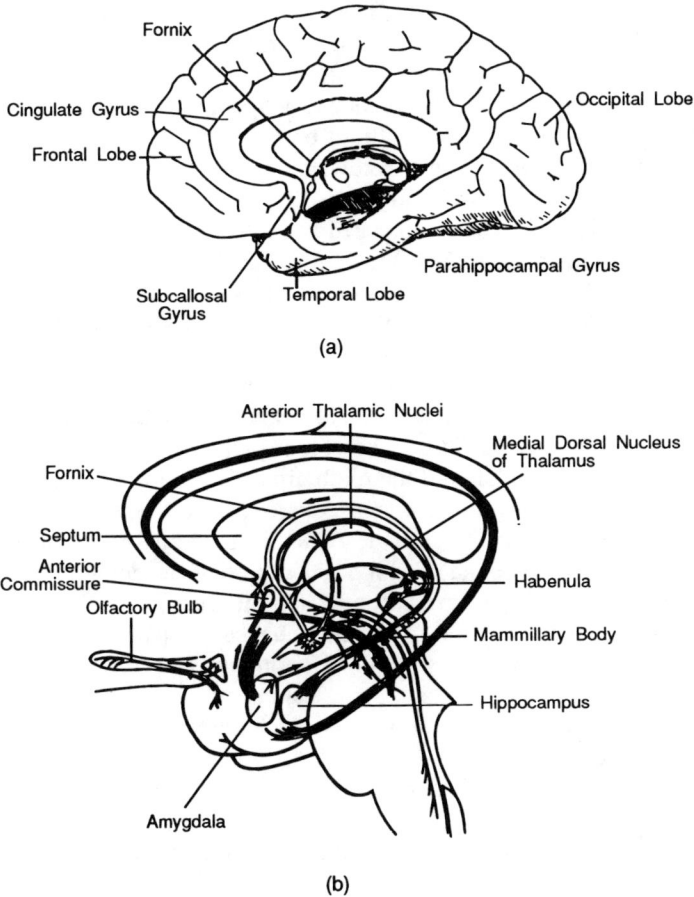

Figure 9.12. Medial view of the brain, highlighting the locations of major structures in the limbic system. (a) Location of the limbic lobe, the primitive area of cortex (cingulate and parahippocampal gyri) which encircles the brain stem. (b) Location of deeper limbic structures (*e.g.*, septum, hippocampus, and amygdala), most of which lie under the limbic lobe. (Adapted by permission of the publisher from Kandel & Schwartz, *Principles of Neural Science*. Copyright 1985 by Elsevier Science Publishing Co., Inc.).

The cerebellum and basal ganglia are both involved in different aspects of motor control (see the discussion of motor control network theories in Section 7.1). Reflex movements in vertebrates need involve only the spinal cord and parts of the brain stem. But for voluntary, adaptive movements, other centers are necessary, including the cerebellum, basal ganglia, and motor cortex. A partial schematic of motor control pathways in the brain and their connections with the spinal cord is shown in Figure 9.13.

The cerebellum has been implicated in contributing to the control of three large functions: muscle tone, balance, and sensory-motor coordination. It has been the subject of many neural network models (see the subsection on brain regions in Section 7.2) because its cell types and connections are easily identifiable and repeatable across species, and because its location makes it fairly accessible to neurophysiological study (see Eccles *et al.*, 1967).

The basal ganglia consist of the *striatum*, which includes the *caudate* and *putamen*, and the *globus pallidus*. The globus pallidus is the output end of this region, and projects to an area of the midbrain called the *substantia nigra*. The role of these areas in movement was discovered when it was noted that lesions in several parts of this region lead to characteristic motor disorders. Parkinson's disease is associated with degeneration of the dopamine (DA) input from the substantia nigra to the striatum; for this reason, the disease is often treated with the drug L-DOPA, which enhances dopamine activity. Cell degeneration in the striatum is found in Huntington's chorea, characterized by involuntary jerking movements. And lesions in the putamen and globus pallidus have been found with athetosis, characterized by slow writhing movements. The basal ganglia are extensively connected with the motor cortex and also, both directly and through the thalamus, with the prefrontal cortex.

Functions of the Mammalian Cerebral Cortex

The cerebral cortex, as noted in Figures 9.8 and 9.9, is the youngest brain region in the evolutionary sense. The well-developed six-layered cortex is present only in mammals, and one area, the prefrontal cortex, is six-layered only in primates. Also, as one moves up the scale of mammals, the cortex becomes ever more folded, with *sulci* or depressions, and *gyri* or areas of the surface between the sulci.

Figure 9.14 shows the major subdivisions of the primate cerebral cortex. The primary motor cortex (*cf.* Figure 9.13) is directly in front of the central sulcus in that figure. The somatosensory cortex, composed of subareas responding to touch or pressure at various parts of the body, is directly behind that same sulcus. The body is represented unequally in the somatosensory cortex, with face and hands having proportionately larger representation than

other areas. The primary visual cortex is in the occipital lobe and the primary
auditory cortex in the temporal lobe.

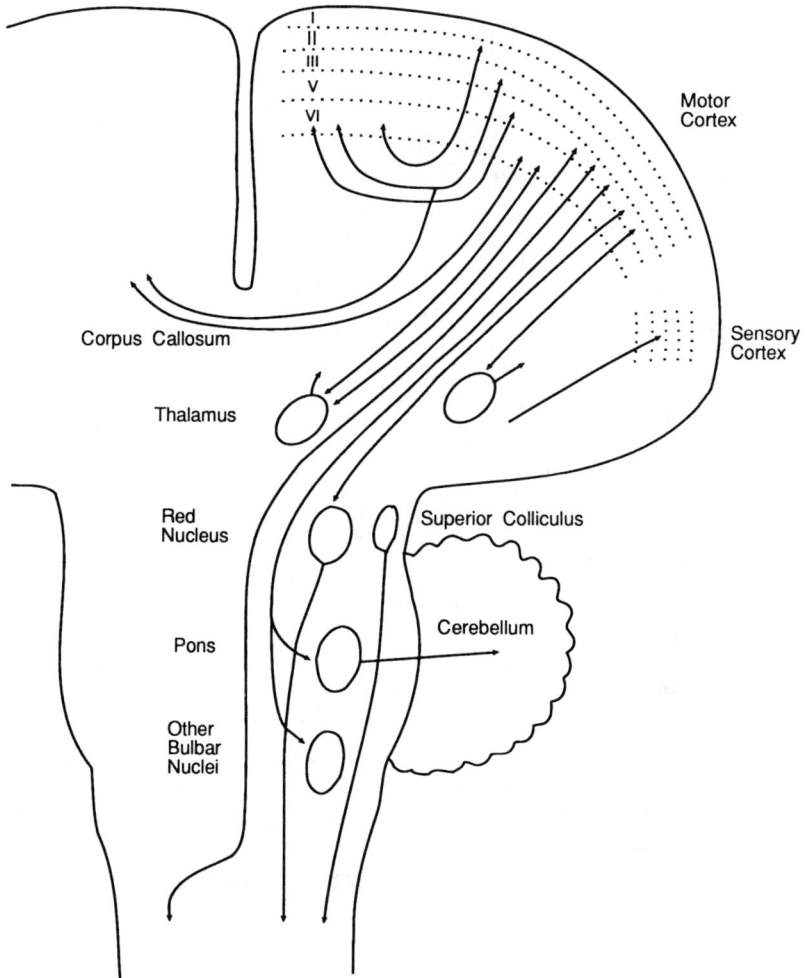

Figure 9.13. Summary of brain motor control pathways, some
descending toward the spinal cord. Roman numerals denote layers
of the cortex. (Reprinted from Shepherd, 1983, with permission
of Oxford University Press.)

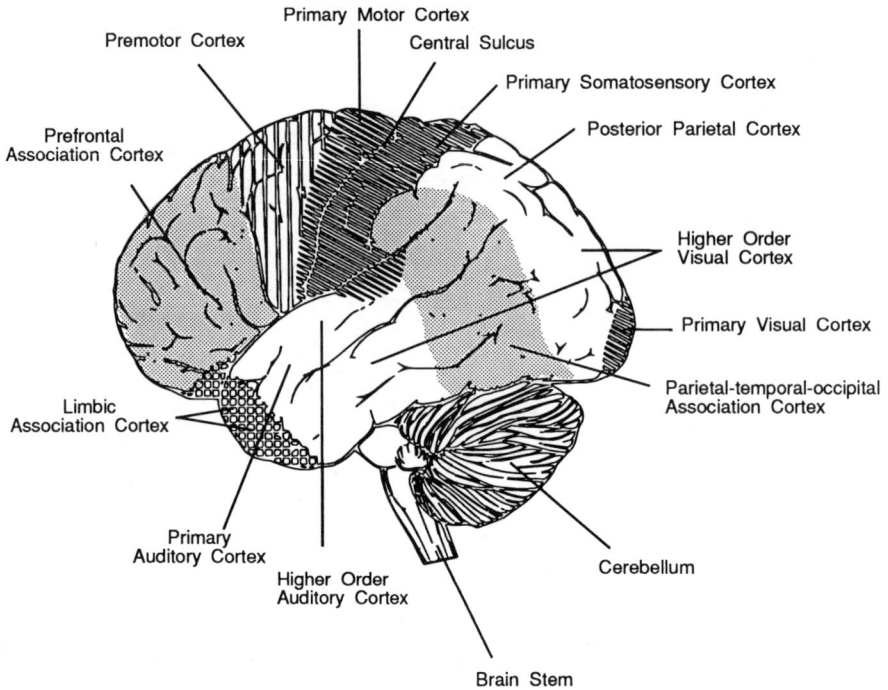

Figure 9.14. Schematic drawing of lateral and medial surfaces of the human brain, highlighting subdivisions of the cerebral cortex. "Primary" and "higher order," for sensory cortices, refer to processing stages, the primary being closest (synaptically) to the sensory input. "Primary motor" and "premotor" are motor control stages, the primary motor being closest to the motor output. (Adapted by permission of the publisher from Kandel & Schwartz, *Princ*iples of Neural Science. Copyright 1985 by Elsevier Science Publishing Co., Inc.).

The visual and auditory cortices are several synapses away from their corresponding sense organs (retina and cochlea). For illustration, Figure 9.15 shows the visual pathways, leading from the retina via the lateral geniculate to the primary visual cortex. Not shown in the figure are the three layers of retinal cells: receptors to bipolar to ganglion cells, influenced also by lateral connections from horizontal and amacrine cells. Not shown either are the three types of visual cortical cells — simple, complex, and hypercomplex — and the further relays from primary visual cortex, also known as Area 17, to secondary visual cortex or Area 18. The olfactory sense has a more primitive circuit: the receptors project directly into the olfactory bulb, which is in a part of the cerebral cortex having only two layers.

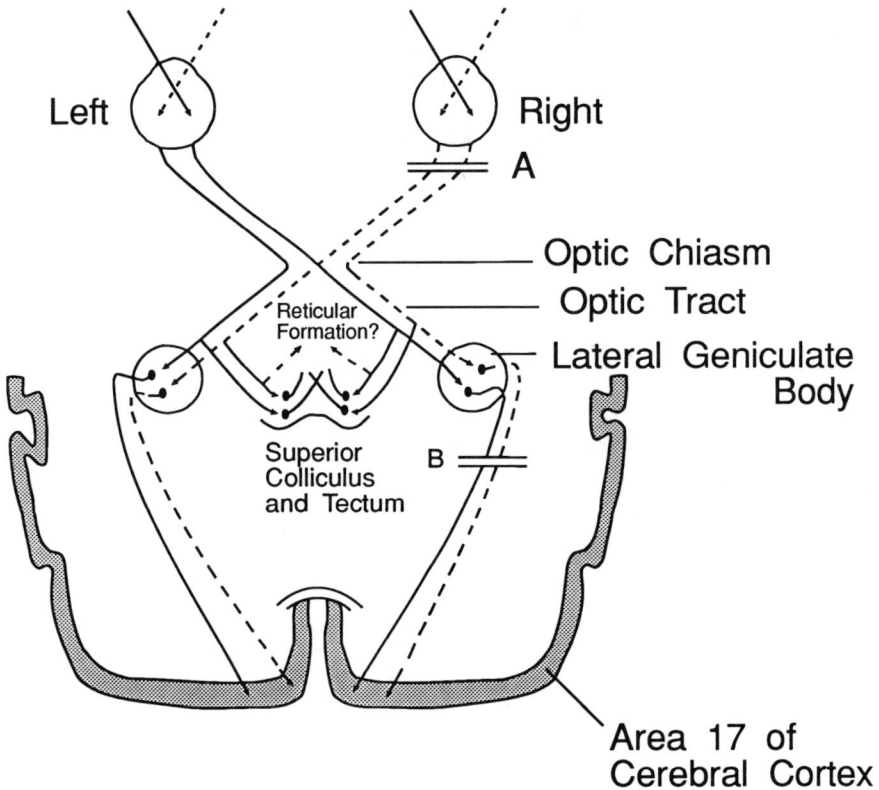

Figure 9.15. Schematic diagram of the mammalian visual system. Cutting the pathways at A eliminates input from the right eye. Cutting at B eliminates input from the right half of each eye. (Adapted from Gardner, 1958, with permission of W. B. Saunders Publishing Company.)

Finally, all of the sensory areas of the cortex send axons to association areas of the cortex — most of the temporal lobe and all of the parietal and frontal lobes. Many of these association areas have specific functions; for example, the regions known as Wernicke's area (in the temporal lobe) and Broca's area (in the frontal lobe near the boundary of the temporal) are important components of circuits for speech and language. Some specialized functions of the prefrontal cortex (furthest forward area of the frontal lobe), which is the only part of association cortex with extensive connections to the limbic system and hypothalamus, are discussed in Section 7.2.

APPENDIX 2:
DIFFERENCE AND DIFFERENTIAL
EQUATIONS IN NEURAL NETWORKS

The equations for neural networks involve changes over time in two types of variables — node activities and connection strengths. The simplest way to describe such changes is to assume they take place at discrete time intervals — every second, say, or every 250 milliseconds. In that case, time is measured in whole number intervals. Hence the equations derive the value of a particular variable at time t+1, with t being an integer (whole number), if the value of the same variable at time t is known. If the variable (activity or connection weight) is called x, we have the generic equation

$$x(t+1) = x(t) + \Delta x(t) \tag{10.1}$$

where Δ is a symbol that means "amount of change." Thus $\Delta x(t)$ represents the total of all changes in x within the given time period. Equation (10.1) is called a *difference equation* because the term $\Delta x(t)$ represents the difference between a variable at time t+1 and the same variable at time t.

In actual network models, the $\Delta x(t)$ term of (10.1) is replaced by some algebraic expression involving x(t) itself and other network variables (node activities or connection strengths). This expression reflects influences on the node or connection whose activity is x by excitation, inhibition, and modulation from the same node or connection, or from elsewhere in the network.

Example: The Sutton-Barto Difference Equations

One of the simpler examples of a neural network described by difference equations is the network of Sutton and Barto (1981). Figure 3.6 shows the major variables in that network: the conditioned stimulus traces x_i, the connection weights w_i, and the output y. The equations for that network, however, list some additional variables not shown in that figure: the eligibility traces $\bar{x}_i$ and the representation $\bar{y}$ of ongoing reinforcement node activity. Figure 10.1 expands the earlier figure to include these additional variables.

There is no difference equation for the stimulus traces $x_i(t)$ themselves, which simply reflect what is taking place in the sensory environment. But the

i^{th} eligibility trace obeys a difference equation reflecting the influence from the corresponding (i^{th}) stimulus trace. This is

$$\bar{x}_i(t+1) = \alpha\bar{x}(t) + x_i(t) \tag{3.23a}$$

where α is some number between 0 and 1.

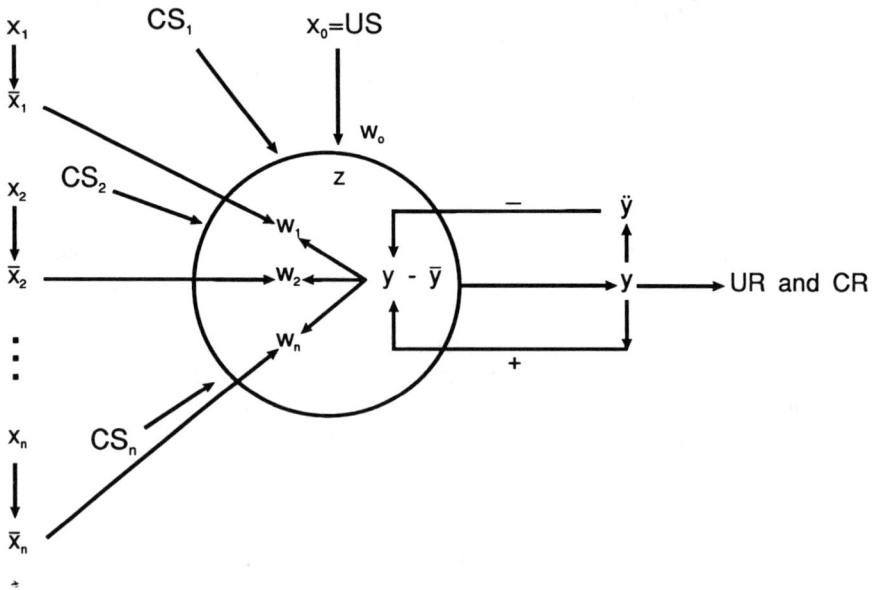

Figure 10.1. Extension of Figure 3.6 to include mutual influences among all the variables in Sutton and Barto's equations (3.23). Additional variables are the eligibility traces $\bar{x}_i$ and the ongoing reinforcement level $\bar{y}$; see text for details.

Let us analyze what (3.23a) says about the change in $\bar{x}_i$ over time. For definiteness, let us choose a specific value for α — say $\alpha = .8$. To find the difference $\Delta\bar{x}_i(t)$, we subtract $\bar{x}_i(t)$ from the expression for $\bar{x}_i(t+1)$. This yields

$$\begin{aligned}\bar{x}_i(t) &= \bar{x}_i(t+1) - \bar{x}_i(t) = \alpha\bar{x}_i(t) + x_i(t) - \bar{x}_i(t)\\ &= -(1-\alpha)\bar{x}_i(t) + x_i(t) = -.2\bar{x}_i(t) + x_i(t)\end{aligned} \tag{10.2}.$$

Equation (10.2) says that $\bar{x}_i$ is negatively influenced by its own decay back to a baseline, at a rate .2, and positively influenced by the actual stimulus $x_i(t)$. Some examples of runs with specific values, done using Lotus 123, are shown in Figure 10.2.

Now consider the equation for the ongoing reinforcement level $\bar{y}$. That equation is

$$\bar{y}(t+1) = \beta\bar{y}(t) + (1-\beta)y(t) \tag{3.23b}$$

where ß is another constant between 0 and 1.

Again, let us put a specific value for ß (say .6) into (3.23b) and analyze what that equation says about the change in $\bar{y}$ over time. Subtracting $\bar{y}(t)$ from the expression for $\bar{y}(t+1)$, we obtain

$$\begin{aligned}\Delta\bar{y}(t) &= (.6\bar{y}(t) + (1-.6)y(t)) - \bar{y}(t)\\ = (.6-1)\bar{y}(t) &+ (1-.6)y(t) = (1-.6)(y(t) - y(t)) =\\ .4&(y(t) - \bar{y}(t))\end{aligned} \tag{10.3}.$$

Equation (10.3) says that the crucial influence on the dynamics of $\bar{y}$ is the *difference* between actual and ongoing (or expected) amount of reinforcement. The factor .4 represents a learning rate, namely, the speed at which the expected value is updated.

How is the reinforcement level or output $y(t)$ calculated? Like the conditioned stimulus (CS) traces, this level is calculated instantaneously, based on the levels of all the CS's and of the unconditioned stimulus (US), and the connection weights from the CS representations to the output node.

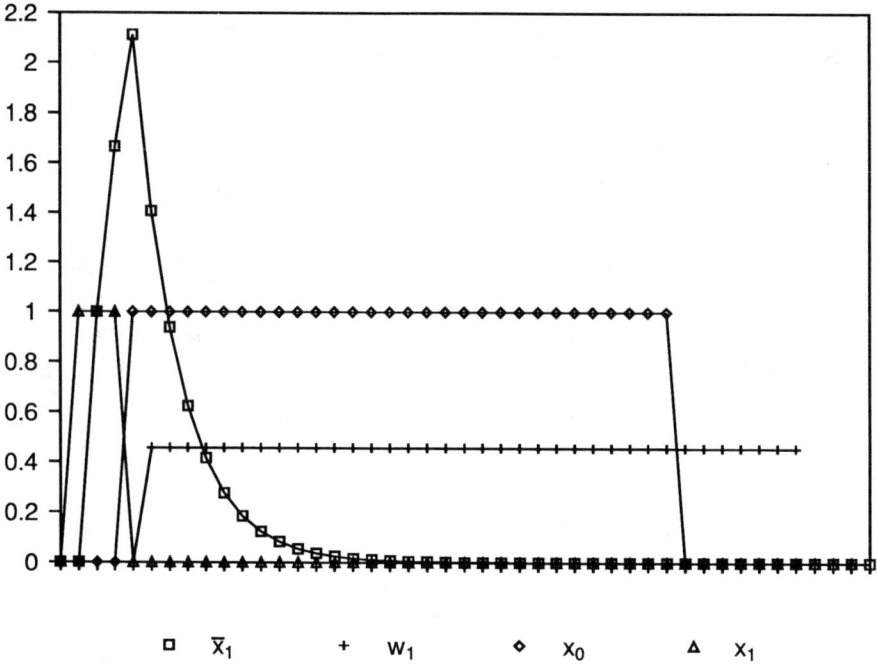

Figure 10.2. Graph of the variables in the Sutton-Barto equations (3.23) for the parameter values $\alpha=.67$, $\beta=0$, $c=.3$, and a particular sigmoid function f.

Let $z(t)$ represent the current level of the US. For the ith CS, the stimulus level is $x_i(t)$ and the connection weight $w_i(t)$. Hence, the combined signal from all the CS nodes is $z(t)$ plus the sum of all the products $x_i(t)\,w_i(t)$, written in the summation or *sigma* notation:

$$z(t) + \sum_i w_i(t)x_i(t),$$

where the sum is taken over all i between 1 and n. If, for example, n = 4, then

$$\sum_i w_i(t)x_i(t) = w_1(t)x_1(t) + w_2(t)x_2(t) + w_3(t)x_3(t) + w_4(t)x_4(t).$$

This combined CS signal is transformed by a sigmoid activation function f (see Figure 2.7b, which is reproduced in Figure 10.3). All these terms combine in the equation

$$y(t) = f(z(t) + \sum_i w_i(t)x_i(t)) \tag{10.4}.$$

OUTPUT

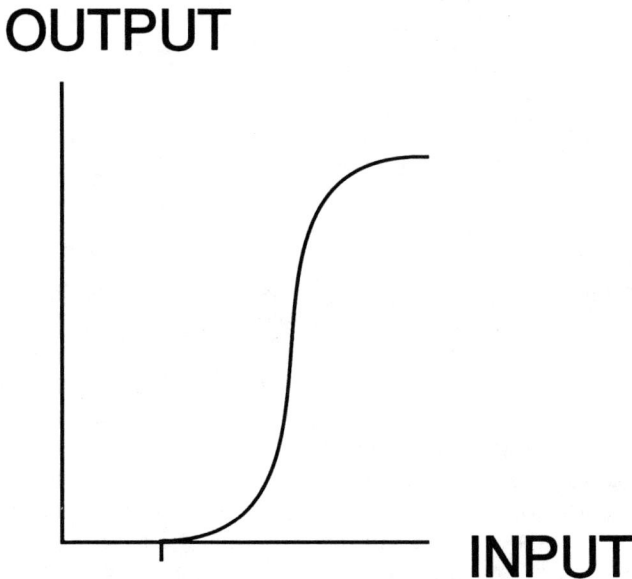

INPUT

Figure 10.3. Schematic sigmoid function (previously shown in Figure 2.7b).

Finally, we discuss the changes in the n synaptic connection strengths $w_i(t)$. The changes in these variables, denoted $\Delta w_i(t)$, are influenced by a learning rule. Recall from Section 3.3 that this is an associative rule whereby presynaptic (CS) activity is correlated not with absolute postsynaptic (US) activity, but rather with *change* in postsynaptic activity. Hence

$$w_i(t+1) = w_i(t) + c(y(t) - \bar{y}(t))\,\bar{x}_i(t) \tag{3.23c}$$

where c is a positive constant that denotes the rate of learning.

Differential Versus Difference Equations

The assumption made in the last section is that changes in the network take place at discrete time intervals (such as once every second). In biological neural systems, it is probably more realistic to assume that the interacting changes in the network take place continuously. Differential equations involve *derivatives*, or rates of change[1], of these variables, which are in turn approximations of the average Δf's for very small times, as will be explained below.

As an intuitive example of a derivative, one can look at what is actually measured by the speedometer of a car. As shown in Figure 10.4, speed is measured as distance covered divided by time elapsed. If f indicates the position on the road and t indicates the current time, then the speed of driving in any given time period is measured as change in f divided by change in t, or in the notation introduced above, as $\Delta f / \Delta t$. But over what length of time should speed be taken? At 12:00, you get different results if you measure the speed of travel since 11:50, or since 11:59, or since 11:59 and 50 seconds. It is for precise description of measurements such as these that Newton and Leibniz (independently) developed the idea of derivative, one of the two basic ideas of calculus, in the late eighteenth century.

Suppose the measured speed of the car (*cf.* Figure 10.4) traveling for 10 minutes up to noon is, say, 22 miles per hour, for 1 minute it is 21 miles per hour, for ten seconds it is 20.5 miles per hour, and for five seconds it is 20.2 miles per hour. One can say that as the time gets smaller and smaller, that is closer to an instant (zero length) of time, the speed during that time gets "closer

---

[1] Contrary to widespread popular opinion, mathematics textbooks — at least those published in the United States — ARE WRITTEN IN ENGLISH! Hence, such key definitions as "rate of change" (which is all too frequently filed and forgotten by calculus students) mean precisely what they intuitively sound like they mean.

and closer" to 20 miles per hour, which is called the limiting speed. ("Closer and closer" is an intuitive term, related to the more precise mathematical concept of *limit* which is discussed in Edwards and Penney, 1988; Swokowski, 1988; Thomas and Finney, 1988, or any other calculus textbook.)

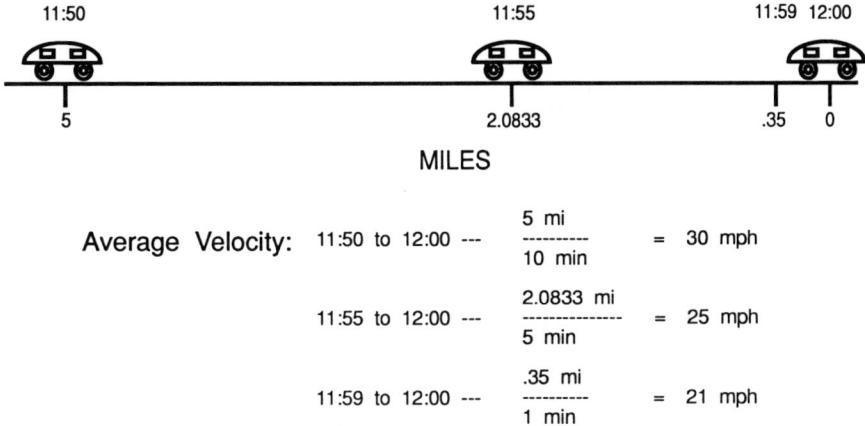

Figure 10.4. Schematic of positions at different times, just before 12:00 noon, of an automobile traveling eastward on a straight road, with a speedometer reading at noon of 20 miles per hour.

For any function that varies with time — such as a moving vehicle's position, or a node activity or a connection weight in a neural network — the *derivative* or *rate of change* of f is defined as the value that the quantity $\Delta f/\Delta t$ gets "closer and closer" to. For complex reasons based on the sociology of mathematics, there are *three* equivalent notations for the derivative of f with respect to time: df/dt, f′, and f dot. As shown in Figure 10.5, if the function is graphed with respect to time, and the curve is approximated near a given time by a straight line, then the derivative is indicated by the slope of that straight line, that is, how fast that line rises or falls as you move to the right.

Assuming sufficiently short time intervals, the same set of network interactions can be described by *either* a difference or a differential equation formulation. Take, for example, a single one of Sutton and Barto's equations, such as the equation for the eligibility trace $\bar{x}_i(t)$:

$$\Delta \bar{x}_i(t) = (.6\bar{x}_i(t) + (1-.6)x_i(t)) - \bar{x}_i(t) =$$
$$(.6-1)\bar{x}_i(t) + (1-.6)x_i(t) = (1-.6)(x_i(t) - \bar{x}_i(t)) = \qquad (10.2).$$
$$.4(x_i(t) - \bar{x}_i(t))$$

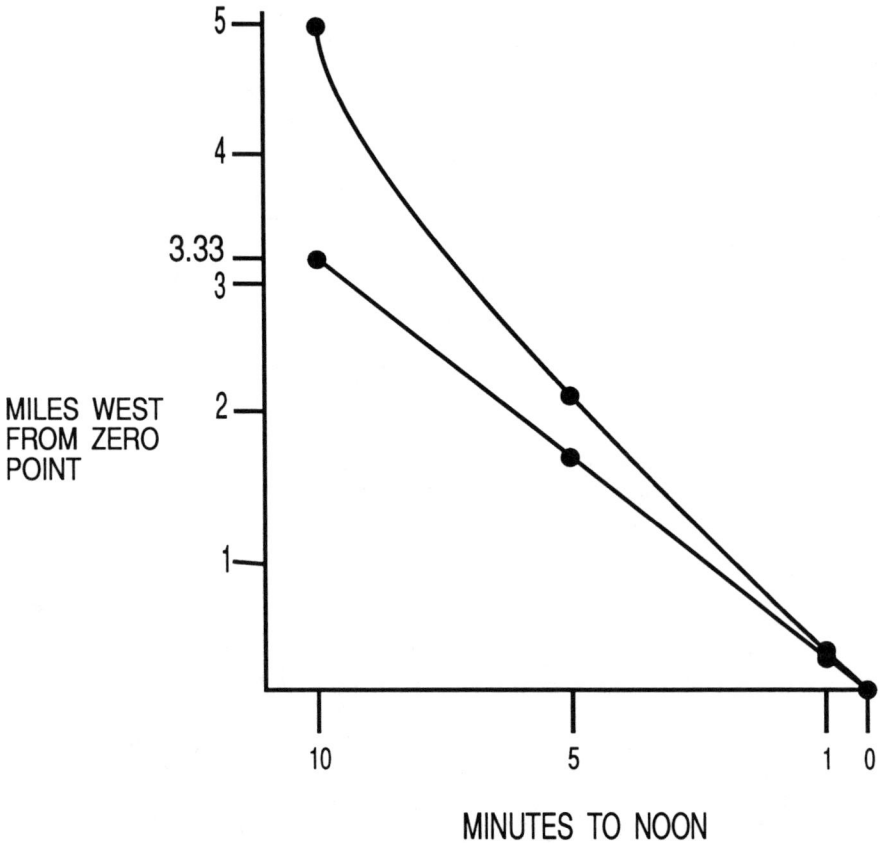

Figure 10.5. Curve in this graph represents the position of the car depicted in Figure 10.4, as a function of time. The line represents the linear approximation to that curve at the point (0,0). Its slope is (3.33 miles)/(10 minutes) = 1/3 miles/min = 20 miles/hour.

Since the time changes from t to t + 1, $\Delta t = 1$, so $\dfrac{\Delta \bar{x}_i}{\Delta t}$ is the same as $\Delta \bar{x}_i$.

As Δt gets small, $\dfrac{\Delta \bar{x}_i}{\Delta t}$ gets closer and closer to $\dfrac{d \bar{x}_i}{dt}$, the derivative of $\bar{x}_i$.
Hence, the differential equation form of (10.2) is

$$\frac{d\bar{x}_i}{dt} = -(1-\alpha)\bar{x}_i(t) + x_i(t)$$

Similarly, given a system of differential equations for the interacting variables in a neural network, each differential equation can be approximated by a difference equation, assuming the time steps are "small enough" for a good approximation. This is the basis for the *Euler method* of numerically solving differential equations, preferably on a computer.

There are other widely used methods of greater accuracy than the Euler, such as the *fourth-order Runge-Kutta method*, but the Euler method is serviceable for most network applications. Detailed descriptions of both these methods can be found in any introductory junior- or senior-level textbook on differential equations (*e.g.*, Rainville & Bedient, 1981; Braun, 1982; Boyce & DiPrima, 1986) or on numerical analysis (*e.g.*, Burden & Faires, 1985; Greenspan & Casulli, 1988). We shall give a capsule description of the Euler method in the next section, introducing it by example. The example we shall use is based on Grossberg's outstar equations, previously introduced in Section 3.2 above.

Outstar Equations: Network Interpretation and Numerical Implementation

Recall from Section 3.2 the network geometry called the *outstar* (Grossberg, 1968a), depicted in Figure 3.3. In an outstar, one node (or vertex, or cell population) v_1, called a *source*, projects to other nodes v_2, v_3, ..., v_n, called *sinks*. (The three dots after v_3 are a generally accepted notation for an indeterminate number of numbers or variables that fit into a general form.)

As discussed in Section 3.2, the source node activity x_1 is affected positively by the source node input I_i, and negatively by exponential decay back to a baseline rate (interpreted as 0). Recalling that the rate of change of x_1 as a function of time is described by its derivative, dx_1/dt, this leads to a differential equation of the form

$$\frac{dx_1}{dt} = -ax_1 + I_1 \qquad (3.13)$$

where a is a positive constant (the decay rate). The activities x_i or v_i, $i = 2, ...,$ n obey an equation similar to (3.13) with the addition of an effect of the source node activity. Hence

$$\frac{dx_i}{dt} = -ax_i(t) + bx_1(t-\tau)w_{1i}(t) + I_1(t),$$
$$i = 2, ..., n \qquad (3.14)$$

where b is another positive constant (*coupling coefficient*) and τ is a transmission time delay.

The synaptic weights, or long-term memory traces, w_{1i} at the source-to-sink synapses, in one version of the theory, have a passive decay which is counteracted by correlated activities of x_1 (with a time delay) and x_i, thus

$$\frac{dw_{1i}}{dt} = -cw_{1i} + ex_1(t-\tau)x_i \qquad (3.15).$$

But if x_1 is interpreted as encoding the sound A and x_i as encoding the sound B, Equation (3.15) implies that the association between A and B is weakened while the network is not actively hearing A. Hence, Grossberg modified this equation to make the association decay when A is presented without being followed by B, but remain constant when A is not presented at all. This change can be achieved by replacing (3.15) (with the time delay τ set to 0) by

$$\frac{dw_{1i}}{dt} = x_1(-cw_{1i} + ex_i) \qquad (3.16)$$

so that w_{1i} remains unchanged while $x_1 = 0$ but decreases while $x_1 > 0$ and $x_i = 0$.

We next go through a simple example of the outstar equations using the simple Euler method. In our example, there are only two sink nodes — x_2 and x_3, with corresponding synaptic weights (from the source node) w_2 and w_3. We also assume there is no decay of memory in the absence of source node stimulation, that is, we use Equations (3.13), (3.14) (for i = 2 and 3), and (3.16) (for i = 2 and 3) — five equations in all. As for the constants in those equations, set a = 5, b = 1, c = .1, d = 1, τ = 0. So the specific forms of the equations become

$$\frac{dx_1}{dt} = -5x_1(t) + I_1(t)$$

$$\frac{dx_2}{dt} = -5x_2(t) + x_1(t)w_{12}(t) + I_2(t)$$

$$\frac{dx_3}{dt} = -5x_3(t) + x_1(t)w_{13}(t) + I_3(t) \qquad (10.7).$$

$$\frac{dw_{12}}{dt} = x_1(t)(-.1w_{12}(t)+x_2(t))$$

$$\frac{dw_{13}}{dt} = x_1(t)(-.1w_{13}(t)+x_3(t))$$

Now to solve Equations (10.7) numerically, it only remains to set the inputs I_1, I_2, and I_3, and the starting values of x_1, x_2, x_3, w_{12}, and w_{13}. We set up an example in which the source node input arrives at regular intervals, followed by inputs to the sink nodes which remain in a regular proportion. Hence, in the terminology of Section 3.2, these sink node inputs form a *spatial pattern*. In particular, let I_1 = 2 on every tenth time step, starting with the first, and 0 on all other time steps. Let pattern $I_i = \theta_i I$, where θ_2 = .7 and θ_3 = .3 on time steps directly *after* those times when I_1 = 2, and 0 on other time steps (*cf.* Figure 10.6). (Hence I_2 = 1.4 and I_3 = .6 when they are not zero.) Suppose every time step is of length .1 (which is the largest value typically used in numerical examples; an actual step size of 1 leads to too much inaccuracy).

Consider, for example, Equation (10.7a) for the source node activity. Suppose this node is inactive, that is, has activity 0, at time 0. Then the simple Euler method says that change in x_1 over the time .1 divided by change in time (which is .1) will be represented by the right-hand side of (10.7a), namely, it will equal $-5x_1 + I_1$, at the previous time. To calculate the new value of x_1, then, we obtain

$$x_1(.1) = x_1(0)+\Delta x_1(0) =$$
$$x_1(0) + .1(-5x_1(0)+I_1(0)) \qquad (10.8).$$

(Equation (10.8) is an approximation to the original differential equation (10.3), but the computer program treats it as an exact statement). Substituting 0 for every occurrence of $x_1(0)$ in (10.8), and 2 for I_1 (since the input is on for that time step), we derive

$$x_1(.1) = 0 + .1(-5(0) + 2) = .2.$$

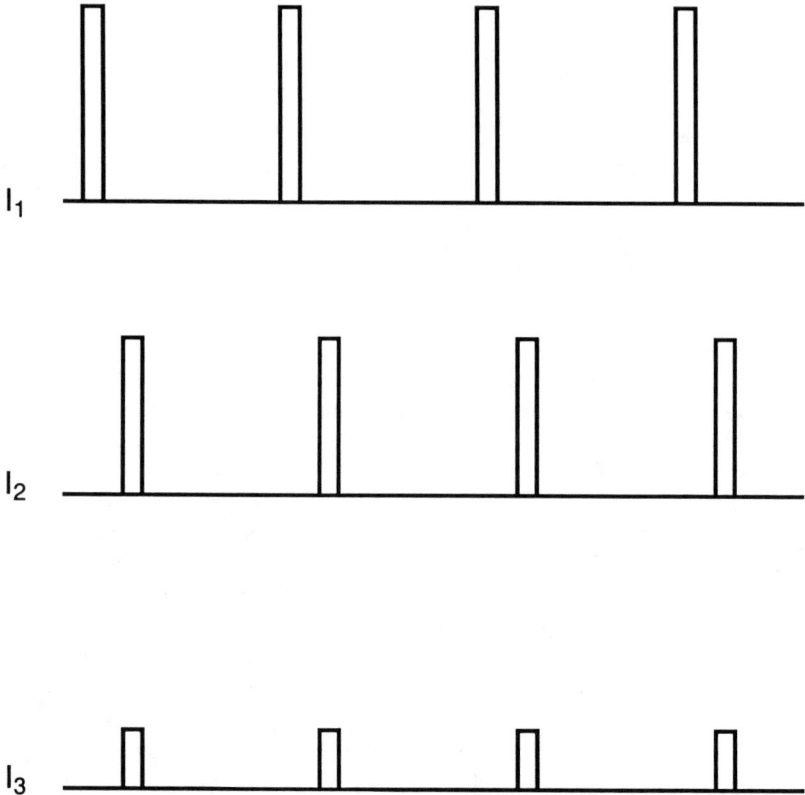

Figure 10.6. Examples of inputs to an outstar network with one source node x_1 and two sink nodes x_2 and x_3. The sink node inputs form a spatial pattern (*i.e.*, remain proportional) and uniformly lag behind the source node inputs, both occurring at regular time intervals. This leads to learning of the association between x_1 activation and the given spatial pattern.

The simple Euler method applies this same process repeatedly at later time steps. That is, (10.8) generalizes to

$$x_1(t+1) = x_1(t) + .1(-5x_1(t) + I_1(t))$$ (10.9)

for any time t. At the next 9 time steps, the input I_1 is 0. So we obtain

$$x_1(.2) = x_1(.1) + .1(-5x_1(.1)) = .2 + .1(-5)(.2) = .1;$$
$$x_1(.3) = x_1(.2) + .1(-5x_1(.2)) = .1 + .1(-5)(.1) = .05,$$
$$etc.$$

More generally, the rule for the Euler method is for any variable, call it y, and a time step of size Δt,

$$(y \text{ } at \text{ } time \text{ } t + \Delta t) = (y \text{ } at \text{ } time \text{ } t) + (\Delta t) \times (expression \text{ } on \text{ } the \text{ } right \text{ } hand \text{ } side \text{ } of \text{ } the \text{ } differential \text{ } equation \text{ } for \text{ } y)$$ (10.10).

The last parenthetical expression in (10.10) is some function of all the network variables, with the values of those variables at time t substituted in. Figure 10.7 shows the values of the five variables in our outstar system over 30 time steps.

To illustrate how differential equations are transformed into difference equations, let us apply (10.10) to the equations (10.7). We list both C and FORTRAN versions of a program segment written to solve these equations numerically. The program must start with initial values for the variables, which can either be obtained from a random number generator over some range, or read in arbitrarily. (Since the interesting phenomenon in outstars is the convergence of relative x's and relative w's to θ's, the initial values of x_2 and x_3 should *not* be set proportional to θ_2 and θ_3, and the same for the initial values of w_{12} and w_{13}.) For simplicity, we shall set them arbitrarily here. The program calculates these variables over 5000 time steps with step size .1.

C VERSION

```
*include (stdio.h)

main ()
```

```
{
    FILE *fp;
    char *arg;
    /* declaration above needed to open a file */
    float x1=0.0, x2=0.2, x3=0.5, w12=0.2, w13=0.5;
    float i1, i2, i3;
    float x1old, x2old, x3old, w12old, w13old;
    int k, i;
    /* open file*/
    arg = "top";
    fp = fopen(arg, "w");
    /*          */
    fprint (fp, "                  Outstar dataset\n");
        k = 1;
        i = 1;
        while (i < 50)
        {
            fprintf(fp, " %8.6f %8.6f ", w12, w13);
            fprintf(fp, '  %8.6f %8.6f %8.6f \n", x1, x2, x3);
            x1old = x1;
            x2old = x2;
            x3old = x3;
            w12old = w12;
            w13old = w13;
            if (k == 1)
                i1=2.0;
                else
                i1=0.0;
            if (k == 2)
                {
                i2=1.4;
                i3=0.6;
                }
                else
                i2=i3=0.0;
            x1=x1+0.1*(-5.0*x1old+i1);
            x2=x2+0.1*(-5.0*x2old+x1old*w12old+i2);
            x3=x3+0.1*(-5.0*x3old+x1old*w13old+i3);
            w12=w12+0.1*x1old*(-.1*w12old+x2old);
            w13=w13+0.1*x1old*(-.1*w13old+x3old);
            if (k == 10)
                k=1;
```

```
    else
    k++;
 i++;
    }
}
```

FORTRAN VERSION

```
X1=0.
X2=.2
X3=.5
W12=.2
W13=.5
```

C K IS A COUNTER THAT MOVES FROM 1 TO 10 AND THEN GOES
C BACK TO 1. ON TRIALS WHEN K=1, THE OUTSTAR SOURCE
C RECEIVES AN INPUT. ON TRIALS WHEN K=2, THE OUTSTAR
C SINK NODES RECEIVE INPUTS.

```
REAL I1,I2,I3
K=1
DO 1 I=1,5000
X1OLD=X1
X2OLD=X2
X3OLD=X3
W12OLD=W12
W13OLD=W13
IF (K .EQ. 1) THEN
  I1=2.
  ELSE I1=0.
  ENDIF
IF (K .EQ. 2) THEN
  I2=1.4
  I3=.6
  ELSE I2=I3=0.
  ENDIF
X1=X1+.1*(-5.*X1OLD+I1)
X2=X2+.1*(-5.*X2OLD+X1OLD*W12OLD+I2)
X3=X3+.1*(-5.*X3OLD+X1OLD*W13OLD+I3)
W12=W12+.1*X1OLD*(-.1*W12OLD+X2OLD)
```

```
W13=W13+.1*X1OLD*(-.1*W13OLD+X3OLD)
IF (K .EQ. 10) THEN
   K=1
   ELSE K=K+1
   ENDIF
STOP
END
```

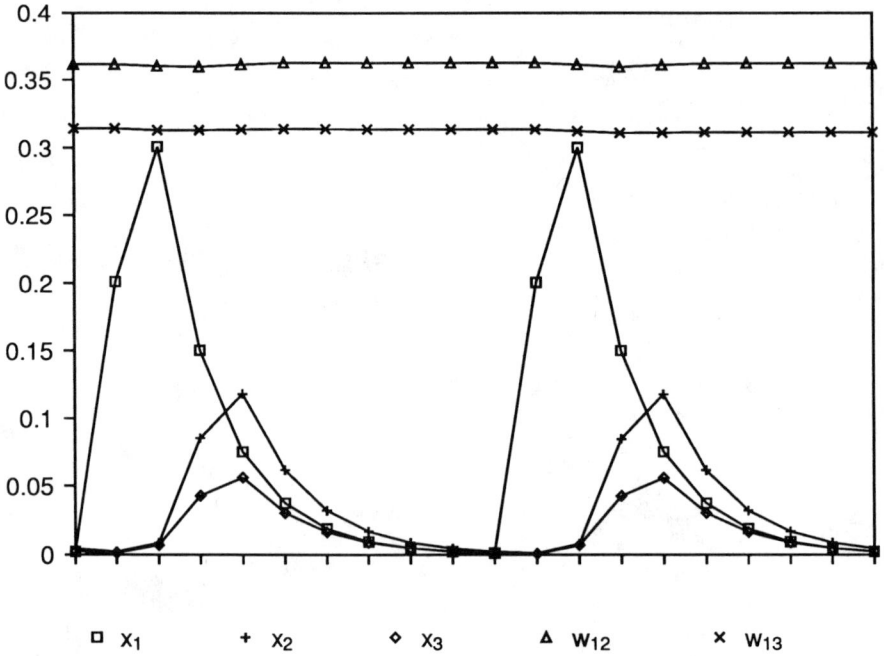

Figure 10.7. Graph of the outstar variables over time for the parameters shown in Equations (10.8) and representative initial values.

In the above programs, the "old" values (values at the previous time step) of all the variables are preserved so that the variables can all be updated in succession. Various differential equation solving packages are available that do all the updating simultaneously. More information about these packages can be obtained from several sources, including the book of Press, Flannery, Teukolsky, and Vetterling (1988) which has versions, and accompanying diskettes, in C, PASCAL, and FORTRAN.

The next two subsections of this Appendix are not necessary for the student to perform the simulation exercises in the book, but aid the student in following some of the mathematical discussions elsewhere in the text. The section on the chain rule for derivatives is provided as background for the derivation of the back propagation algorithm in Section 3.5. The section on dynamical systems is provided as background for the discussions of equilibrium states and energy functions at various points in Sections 4.2, 4.5, and 7.1.

The Chain Rule and Back Propagation

The chain rule determines the derivative of a composite function, that is, a function whose argument is itself a function of another variable. Some examples of composite functions in biological applications are given in Gentry (1978, pp. 250-253). In one of his examples, a nerve impulse is translated into a muscular movement. The muscle reaction is a function of the number of acetylcholine ions liberated at neuromuscular junctions by the nerve impulse, and the number of ions liberated is itself a function of the number of millivolts in the impulse.

If f is a function of some variable y, and y is in turn a function of another variable x, then the derivative of f *as a function of x*, written df/dx, is the product of the derivative of f as a function of y with the derivative of y as a function of x. That is

$$\frac{df}{dx} = \left(\frac{df}{dy}\right)\left(\frac{dy}{dx}\right)$$

(10.11).

If f, instead of being a function of a single variable y, is a function of several variables called y_1, y_2, ..., y_n, each a function of x, the rule (10.11) generalizes to one involving the *partial derivatives* of f. For each i, the partial derivative $\frac{\partial f}{\partial y_i}$ is defined as the rate of change of f as y_i is varied, with all the other variables kept constant. Then the derivative of the composite function f becomes the sum of contributions from the variables y_1, y_2, ...,y_n, thus:

$$\frac{df}{dx} = \sum_{i=1}^{n}\left(\frac{\partial f}{\partial y_i}\right)\left(\frac{dy_i}{dx}\right)$$

(10.12).

Both (10.11) and (10.12) are frequently used to obtain derivatives of complex expressions in neural network equations. For example, they are used in the derivation, seen in Section 3.5, of the changes of weights to hidden units in the three-layer back propagation network, given the changes of weights to output units. A detailed justification follows now for some of the steps in the earlier derivation.

First, recall that the jth output unit (on the pth pattern) receives a signal equal to the linear sum of the outputs y_{pi} from the hidden layer weighted by the connections w_{ij}. This signal is called

$$net_{pj} = \sum_i w_{ij} y_{pi} \qquad (3.9).$$

If f is the activation function of unit j, then the output of unit j is

$$y_{pj} = f(net_{pj}) = f(\sum_i w_{ij} y_{pi}) \qquad (3.24).$$

Recall, also, that the total error in the pth output pattern is measured in terms of deviation of the output pattern vector from a target pattern vector (t_{p1}, t_{p2}, ..., t_{pn}). Since deviation from the target could be either positive or negative, the differences are squared, leading to a total error signal

$$E_p = \frac{1}{2} \sum_j (t_{pj} - y_{pj})^2 \qquad (3.25).$$

Then the response change δ_{pj} should be based on how much the jth unit contributes to the incorrectness of the response. This is done by taking the negative derivative of the total error E_p as a function of y_{pj}.

The t_{pj} in (3.25) are constants, and for a given output unit j, the only part of the expression (3.24) that changes with the output signal y_{pj} is the part corresponding to that j, namely, $1/2 (t_{pj} - y_{pj})^2$. From standard formulas for derivatives of polynomial functions (e.g., Swokowski, 1988), we obtain that the derivative of E_p with respect to the output signal y_{pj} is

$$\frac{\partial E_p}{\partial y_{pj}} = - y_{pj}$$

But for calculating the changes in weights to hidden units, it is necessary to get the derivative of the error not as a function of y_{pj}, but of the signal net_{pj} from the hidden layer. Using the chain rule (10.16), that derivative is the product

$$\frac{\partial E_p}{\partial net_{pj}} = \frac{\partial E_p}{\partial y_{pj}} \frac{dy_{pj}}{dnet_{pj}} \quad .$$

By the last equation, this translates to

$$\frac{\partial E_p}{\partial net_{pj}} = - y_{pj} \left(\frac{dy_{pj}}{dnet_{pj}} \right) \quad .$$

But by (3.24), the output signal is the function f (usually sigmoid) applied to net_{pj}, so that $dy_{pj}/d\ net_{pj} = f'(net_{pj})$, where " ′ " denotes derivative. If this change in the error with respect to the net signal (which determines a weight change) is called δ_{pj}, then

$$\delta_{pj} = (t_{pj} - y_{pj}) f'(net_{pj}) \qquad\qquad (3.10a).$$

If the jth unit is instead a hidden unit, then again using the chain rule, we obtain from (3.24), (3.25), and (3.10a) (" ∂ " denoting partial derivative) that

$$\delta_{pj} = - \frac{\partial E_p}{\partial net_{pj}} = -f'(net_{pj}) \left[\frac{\partial E_p}{\partial y_{pj}} \right] \quad .$$

If k is the generic index of output units that receive projections from hidden unit j, we obtain, again by the chain rule and previous equations, a value for the above expression in brackets, namely

$$\frac{dE_p}{dy_{pj}} = \sum_k \left[\frac{dE_p}{dnet_{pk}} \right] \left[\frac{dnet_{pk}}{dy_{pj}} \right]$$

$$= \sum_k \left[\frac{dE_p}{dnet_{pk}} \right] w_{jk} = - \sum_k \delta_{pk} w_{jk}$$

Combining the above two expressions, we obtain finally that if unit j is a hidden unit,

$$\delta_{pj} = f'_j(net_{pj})\sum_k \delta_{pk} w_{jk} \qquad\qquad (3.10b).$$

Dynamical Systems: Steady States, Limit Cycles, and Chaos

A *dynamical system* is defined as the movement through time of solution trajectories for a system of differential or difference equations for interacting variables (see, for example, Hirsch & Smale, 1974 for more details). Each trajectory is described by a vector composed of the values of all the variables in the system at any given time. If n is the number of variables, these vectors can be treated as points in an n-dimensional space. Most of the discussion in this section is about dynamical systems based on differential equations; for difference equations, the mathematics is more difficult, and the system is more likely to exhibit chaotic behavior (see Frauenthal, 1980, Ch. 6, or Smital, 1988, Ch. 3 for one of the classic examples).

Of course, if n is larger than 3, an n-dimensional space is an abstract mathematical object that cannot be drawn. But in many cases (*e.g.*, Anderson *et al.*, 1977; Hopfield, 1982; Cohen & Grossberg, 1983), the network being studied is homogeneous in its structure, so that the number of nodes has little effect on qualitative behavior. For such systems, taking the number n of nodes to be 2 or 3 allows one to draw pictures of the dynamics of the network over time (*cf.* Figures 4.9 and 4.12). Such qualitative studies of time dynamics can also be useful for networks that are not homogeneous but have homogeneous subnetworks whose activities are described by a time-varying vector. One example is the vector of weights from any given category node to the n feature nodes in an adaptive resonance network (Carpenter & Grossberg, 1987a). Another example is the vector of input-to-hidden-unit weights in a back propagation network (Rumelhart *et al.*, 1986).

A system of differential equations defining a dynamical system usually cannot be solved in closed form, that is, with the solutions expressed as combinations of elementary functions like exponentials, logarithms, polynomials, and trigonometric functions. But, frequently, numerical simulations can be supplemented by theorems about the system's *asymptotic behavior*, that is, what the vector of system variables approaches as time gets large. For most neural networks (as for systems derived from other physical and biological applications), the activities and connection strengths have upper and lower bounds. Hence, the time-varying vector of system variables remains within some "box" or hyper-rectangle in n-dimensional space. The asymptotic behavior of such bounded systems usually falls into one of three categories:

1. EQUILIBRIUM. The system vector approaches a single point in n-dimensional space (*cf.* Figure 4.9a). Such a point is called an *equilibrium state* (or *steady state*, or *rest point*, or *critical point*) of the system. Many dynamical systems have only a finite number of possible equilibrium states. In neural networks, each steady state corresponds to a possible stable activity pattern of the network (*cf.* Sections 4.2 and 7.1).

2. LIMIT CYCLE. The system vector approaches a periodic orbit in n-dimensional space (*cf.* Figure 4.9b). In neural networks, periodic orbits are sometimes used to model cyclical processes in the nervous system. Examples include models of reverberating memory in the cortex and thalamus (Wilson & Cowan, 1973); of hallucinations in visual perception (Ermentrout & Cowan, 1980); of circadian rhythm generation in the hypothalamus (Carpenter & Grossberg, 1985); and of rhythmical movements in crustaceans (Selverston, 1976).

3. CHAOS. In bounded two-dimensional dynamical systems, the Poincare-Bendixson Theorem (Hirsch & Smale, 1974) shows that convergence to an equilibrium point or to a limit cycle are the only possibilities. The proof of that theorem relies on some facts of two-dimensional geometry (*e.g.*, a curve in the plane has a distinct "inside" and "outside," a result known as the Jordan Curve Theorem) and is no longer valid in three or more dimensions. In three or more dimensions, the system vector can asymptotically wander through n-dimensional space in a fashion that appears to be random but is actually deterministic (*e.g.*, Lorenz, 1963). This phenomenon is widely known as *chaos*, and has been suggested as a basis for behavioral variability in nervous systems (*e.g.*, Skarda & Freeman, 1987; Mpitsos, Burton, Creech, & Soinila, 1988).

Some information about qualitative behavior of a dynamical system can be obtained from studying the functions defining the equations. In general, if a system involves n interacting variables $x_1(t)$, $x_2(t)$, ... , $x_n(t)$, the rate of change of each of the variables $x_i(t)$ is some function of $x_i(t)$ itself and all the other variables. Most systems defining neural networks are *autonomous*, that is, the functions do not depend on time. Hence, if we call the function f_i, we have a system of differential equations of the form

$$\frac{dx_i}{dt} = f_i(x_1, x_2, ..., x_n) \qquad (10.13).$$

For a neural network, the function f_i in (10.13) denotes the combination of all the excitatory and inhibitory influences on x_i if x_i is a node activity, and of positive and negative influences on x_i if x_i is a connection weight. An equilibrium state is a state, or value of the vector $(x_1, x_2, ..., x_n)$, at which all these relative influences are "balanced," that is, $f_i = 0$ for all i, i =1, 2, ..., n.

Techniques for studying the qualitative behavior of a system of equations of the form (10.13) all involve consideration of the functions f_i and their derivatives. Whether f_i is positive or negative at a given point in n-dimensional space determines the direction of change of x_i if the system state reaches that point.

In particular, if $\vec{x} = (x_1, x_2, ..., x_n)$ is an equilibrium point, it is of interest whether solutions of the equations (*trajectories*) starting at points near $\vec{x}$ approach $\vec{x}$ or move away from $\vec{x}$ as time gets large. In the former case, $\vec{x}$ is called an *asymptotically stable* equilibrium; in the latter case, $\vec{x}$ is *unstable*. (There is also an intermediate case, an equilbrium that is *stable* but not asymptotically stable. In that case, nearby trajectories stay in the vicinity of $\vec{x}$ without actually approaching $\vec{x}$.) The stable equilibria are the ones that can actually be reached by the system, and are therefore the ones of interest for applications.

The criteria for stability are discussed in any advanced differential equations textbook (*e.g.*, Hirsch & Smale, 1974; Miller & Michel, 1982). One of these criteria involves the matrix of partial derivatives of the functions f_i, which is called the *Jacobian matrix* of the system of equations, at $\vec{x}$. Recall from Section 6.5 that the *eigenvectors* of a matrix A are n-dimensional vectors $\vec{y}_i$ such that $A\vec{y}_i$ is a constant multiple of $\vec{y}_i$. The constant which is multiplied is called an *eigenvalue* of the matrix A (Jordan, 1986a). The equilibrium point $\vec{x}$ is asymptotically stable if all the eigenvalues of the Jacobian matrix at that point (which may be real or complex) have negative real parts, and unstable if any of the eigenvalues have positive real parts. Hence, the eigenvalues indicate the direction of flow of solution trajectories close to an equilibrium.

If no eigenvalues have positive real parts, but some of them are 0 or purely imaginary, the direction of this flow is ambiguous. Hence, under those conditions, one must resort to other methods for determining stability. One of the most important of these is the method of *Lyapunov functions* (*cf.* Section 4.2). Recall that a Lyapunov function (sometimes spelled Liapunov, Liapounov, or Liapounoff) is defined as a function of the system variables that is decreasing along system trajectories. More precisely, let V $(x_1, x_2, ..., x_n)$ be any real-valued function of the state vector $\vec{x}$. Then if $x_1, x_2, ..., x_n$ satisfy the system of differential equations (10.13), the chain rule (see the last subsection) shows that the derivative of V along solutions of the system is

$$\frac{dV}{dt} = \sum_{i=1}^{n} \left(\frac{\partial V}{\partial x_i} \right) \left(\frac{dx_i}{dt} \right) = \sum_{i=1}^{n} \left(\frac{\partial V}{\partial x_i} \right) f_i \qquad (10.14).$$

The expression on the right-hand side of (10.14) is a function of the vector $\vec{x}$. If this expression is always nonpositive over the range of state vectors reachable by the system, this means that the function V is a Lyapunov function, always nonincreasing along trajectories. Under those conditions, a variety of theorems constrains the motion of trajectories to approach equilibria.

REFERENCES

(Chapters in parentheses denote chapters *of this book* in which references appear.)

Aarts, E. H. L., & Korst, J. H. M. (1987). Boltzmann machines and their applications. In J. W. deBakker, A. J. Nijman & P. C. Treleaven (Eds.), *Parallel Architectures and Languages Europe*. Lecture Notes in Computer Science (Vol. 1), 258. Berlin: Springer-Verlag. (Ch. 7)

Aarts, E. H. L., & Korst, J. H. M. (1989). *Simulated Annealing and Boltzmann Machines*. New York: John Wiley and Sons. (Ch. 7)

Ackley, D. H., Hinton, G. E., & Sejnowski, T. J. (1985). A learning algorithm for Boltzmann machines. *Cognitive Science* 9, 147-169. (Ch. 7)

Akert, K., Pfenninger, K. H., Sandri, C., & Moore, H. (1972). Freeze etching and cytochemistry of vesicles and membrane complexes in synapses of the central nervous system. In G. D. Pappas & D. B. Purpura (Eds.), *Structure and Function of Synapses* (pp. 67-86). New York: Raven. (Appendix 1)

Amari, S.-I. (1971). Characteristics of randomly connected threshold element networks and network systems. *Proceedings of the IEEE*, **59**, 35-47. (Ch. 2)

Amari, S.-I. (1972). Characteristics of random nets of analog neuron-like elements. *IEEE Transactions on Systems, Man, and Cybernetics* **2**, 643-657. (Ch. 2)

Amari, S.-I. (1974) A method of statistical neurodynamics. *Kybernetik* **14**, 201-215. (Ch. 2)

Amari, S.-I. (1977a). Dynamics of pattern formation in lateral-inhibition type neural fields. *Biological Cybernetics* **27**, 77-87. (Ch. 4)

Amari, S.-I. (1977b). A mathematical approach to neural systems. In J. Metzler (Ed.), *Systems Neuroscience* (pp. 67-117). New York: Academic. (Ch. 4, 6)

Amari, S.-I. (1977c). Neural theory of association and concept formation. *Biological Cybernetics* **26**, 175-185. (Ch. 6)

Amari, S.-I. (1980). Topographic organization of nerve fields. *Bulletin of Mathematical Biology* **42**, 339-364. (Ch. 6)

Amari, S.-I., & Arbib, M. A. (1977). Competition and cooperation in neural nets. In J. Metzler (Ed.), *Systems Neuroscience* (pp. 119-165). New York: Academic. (Ch. 4, 7)

Amari, S.-I., & Arbib, M. A., Eds. (1982). *Competition and Cooperation in Neural Nets. Lecture Notes in Biomathematics*, Vol. 45. New York: Springer-Verlag. (Ch. 4)

Amari, S.-I., & Takeuchi, M. (1978). Mathematical theory of category detecting nerve cells. *Biological Cybernetics* **29**, 127-136. (Ch. 4, 6)

Amit, D. J., Gutfreund, H., & Sompolinsky, H. (1985). Spin-glass models of neural networks. *Physical Review A* **32**, 1007-1018. (Ch. 2)

Andersen, P., & Eccles, J. C. (1962). Inhibitory phasing of neural discharge. *Nature* **196**, 645-647. (Ch. 4)

Andersen, P., Gross, G. N., Lomo, T., & Sveen, O. (1969). Participation of inhibitory and excitatory interneurones in the control of hippocampal cortical output. In M. Brazier (Ed.), *The Interneuron (pp. 415-465).* Los Angeles: University of California Press. (Ch. 4)

Anderson, B. J., Lee, S., Thompson, J., Steinmetz, J., Logan, C., Knowlton, B., Thompson, R. F., & Greenough, W. T. (1989). Increased branching of spiny dendrites of rabbit cerebellar Purkinje neurons following associative eyeblink conditioning. *Society for Neuroscience Abstracts*, **15**, 640. (Ch. 2, 3, 5)

Anderson, J. A. (1968). A memory storage model utilizing spatial correlation functions. *Kybernetik* **5**, 113-119. (Ch. 3)

Anderson, J. A. (1970). Two models for memory organization using interacting traces. *Mathematical Biosciences* **8**, 137-160. (Ch. 3)

Anderson, J. A. (1972). A simple neural network generating an interactive memory. *Mathematical Biosciences* **14**, 197-220. (Ch. 3)

Anderson, J. A. (1973). A theory for the recognition of items from short memorized lists. *Psychological Review* **80**, 417-438. (Ch. 3, 6)

Anderson, J. A. (1983). Cognitive and psychological computation with neural models. *IEEE Transactions on Systems, Man, and Cybernetics* **SMC-13**, 799-815. (Ch. 6)

Anderson, J. A. (1986). What neural networks are and what neural networks can do. In: D. Z. Anderson (Ed.), *Neural Networks and Neuromorphic Systems — A Workshop Sponsored by the National Science Foundation: Viewgraph Reproductions.* (Ch. 8).

Anderson, J. A., & Mozer, M. (1981). Categorization and selective neurons. In G. Hinton & J. A. Anderson (Eds.), *Parallel Models of Associative Memory (pp. 213-236).* Hillsdale, NJ: Lawrence Erlbaum Associates. (Ch. 6)

Anderson, J. A., & Murphy, G. L. (1986). Psychological concepts in a parallel system. *Physica D* **22**, 318-336. (Ch. 2, 4, 6, 7)

Anderson, J. A., Penz, P. A., Gately, M. T., & Collins, D. (1988). Radar signal categorization using a neural network. *Neural Networks* 1, Suppl. 1, 422. (Ch. 6)

Anderson, J. A., & Silverstein, J. W. (1978). Reply to Grossberg. *Psychological Review* **85**, 597-603. (Ch. 6)

Anderson, J. A., Silverstein, J. W., Ritz, S. A., & Jones, R. S. (1977). Distinctive features, categorical perception, and probability learning: some applications of a neural model. *Psychological Review* **84**, 413-451. (Ch. 4, 6, Appendix 2)

Andreasen, N. (1988). Brain imaging: applications in psychiatry. *Science* **239**, 1381-1388. (Ch. 8)

Angeniol, B., DeLaCroix Vaubois, G., & LeTexier, J.-Y. (1988). Self-organizing feature maps and the traveling salesman problem. *Neural Networks* **1**, 289-293. (Ch. 7)

Anninos, P. A. (1972a). Mathematical models of memory traces and forgetfulness. *Kybernetik* **10**, 165-167. (Ch. 2)

Anninos, P. A. (1972b). Cyclic modes in artificial neural nets. *Kybernetik* **11**, 5-14. (Ch. 2)

Anninos, P. A., Beek, B., Csermely, T. J., Harth, E. M., & Pertile, G. (1970). Dynamics of neural structures. *Journal of Theoretical Biology* **26**, 121-148. (Ch. 2, 4).

Annis, R. C., & Frost, B. (1973). Human visual ecology and orientation anisotropies in acuity. *Science* **182**, 729-731. (Ch. 4)

Aparicio, M., IV, & Strong, P. (1991). Pavlovian conditioning as simulated annealing. In D. S. Levine & S. J. Leven (Eds.), *Motivation, Emotion, and Goal Direction in Neural Networks (pp. 1-37)*. Hillsdale, NJ: Lawrence Erlbaum Associates, in press. (Ch. 5)

Ashby, W. R., Foerster, H. von, & Walker, C. C. (1962). Instability of pulse activity in a net with threshold. *Nature* **196**, 561-562. (Ch. 2)

Athale, R., Friedlander, C. B., & Kushner, B. G. (1985). Attentive associative architectures and their implementation to optical computing. *Proceedings of the SPIE* **625**, 179-185. (Ch. 5)

Atkeson, C. G., & Hollerbach, J. M. (1985). Kinematic features of unrestrained vertical arm movements. *Journal of Neuroscience* **5**, 2318-2330. (Ch. 7)

Barlow, H. B., Blakemore, C., & Pettigrew, J. D. (1967). The neural mechanism of binocular depth discrimination. *Journal of Physiology (London)* **193**, 327-342. (Ch. 4)

Barto, A. G. (1990). Connectionist learning for control: an overview. In T. Miller, R. S. Sutton & P. J. Werbos (Eds.), *Neural Networks for Control*. Cambridge, MA: MIT Press, in press. (Ch. 7)

Barto, A. G., & Anandan, P. (1985). Pattern recognizing stochastic learning automata. *IEEE Transactions on Systems, Man, and Cybernetics* **15**, 360-375. (Ch. 3, 6, 7)

Barto, A. G., & Sutton, R. S. (1982). Simulation of anticipatory responses in classical conditioning by a neuron-like adaptive element. *Behavioural Brain Research* **4**, 221-235. (Ch. 5)

Barto, A. G., Sutton, R. S., & Anderson, C. W. (1983). Neuron-like elements that can solve difficult learning control problems. *IEEE Transactions on Systems, Man, and Cybernetics* **13**, 835-846. (Ch. 5, 6, 7)

Bear, M. F., Cooper, L. N., & Ebner, F. F. (1987). A physiological basis for a theory of synapse modification. *Science* **237**, 42-48. (Ch. 2, 6).

Berlyne, D. E. (1969). The reward-value of indifferent stimulation. In J. T. Tapp (Ed.), *Reinforcment and Behavior* (pp. 179-214). New York: Academic Press. (Ch. 7)

Beurle, R. L. (1956). Properties of a mass of cells capable of regenerating pulses. *Philosophical Transactions of the Royal Society of London, Series B* **250**, 55-84. (Ch. 2)

Bienenstock, E. L., Cooper, L. N., & Munro, P. W. (1982). Theory for the development of neuron selectivity: orientation specificity and binocular interaction in visual cortex. *Journal of Neuroscience* **2**, 32-48. (Ch. 4, 6).

Blakemore, C., Carpenter, R. H. S., & Georgeson, M. A. (1970). Lateral inhibition between orientation detectors in the human visual system. *Nature* **228**, 37-39. (Ch. 4)

Blakemore, C., & Cooper, G. F. (1970). Development of the brain depends on the visual environment. *Nature* **228**, 477-478. (Ch. 4, 6)

Blazis, D. E. J., Desmond, J. E., Moore, J. W. & Berthier, N. E. (1986). Simulation of the classically conditioned nictitating membrane response by a neuron-like adaptive element: a real-time variant of the Sutton-Barto model. *Proceedings of the Eighth Annual Conference of the Cognitive Science Society* (pp. 176-186). Hillsdale, NJ: Lawrence Erlbaum Associates. (Ch. 5)

Bliss, T. V. P. & Lomo, T. (1973). Long-lasting potentiation of synaptic transmission in the dentate area of the anaesthetized rabbit following stimulation of the perforant path. *Journal of Physiology* (London) **232**, 331-356. (Ch. 1, 3)

Blomfield, S. (1974). Arithmetical operations performed by nerve cells. *Brain Research* **69**, 115-124. (Ch. 4)

Bower, G. H. (1981). Mood and memory. *American Psychologist* **36**, 129-148. (Ch. 5, 7)

Bower, G. H., Gilligan, S. G., & Monteiro, K. P. (1981). Selectivity of learning caused by adaptive states. *Journal of Experimental Psychology (General)* **110**, 451-473. (Ch. 5)

Bower, J. M., & Llinás, R., Simultaneous sampling of the responses of multiple, closely adjacent, Purkinje cells responding to climbing fiber activation. *Society for Neuroscience Abstracts* **9**, 607. (Ch. 8)

Box, G. E. P., & Muller, M. E. (1958). A note on the generation of random normal deviates. *Annals of Mathematical Statistics* **29**, 610-611. (Ch. 6)

Boyce, W. E., & Di Prima, R. C. (1986). *Elementary Differential Equations*. New York: John Wiley and Sons. (Appendix 2)

Braun, M. (1982). *Differential Equations and Their Applications*. New York: Springer-Verlag. (Appendix 2)

Brindley, G. S. (1967). The classification of modifiable synapses and their use in models of conditioning. *Proceedings of the Royal Society of London, Series B* **168**, 361-376. (Ch. 5)

Brindley, G. S. (1969). Nerve net models of plausible size that perform many simple learning tasks. *Proceedings of the Royal Society of London, Series B* **174**, 173-191. (Ch. 5)

Brown, T. H., Chapman, P. F., Kairiss, E. W., & Keenan, C. L. (1988). Long-term synaptic potentiation. *Science* **242**, 724-728. (Ch. 3)

Bullock, D., & Grossberg, S. (1988). Neural dynamics of planned arm movements: emergent invariants and speed-accuracy properties during trajectory formation. *Psychological Review*, **95**, 49-90. (Ch. 3, 7)

Bullock, D, & Grossberg, S. (1989). VITE and FLETE: neural modules for trajectory formation and postural control. In W. A. Hershberger (Ed.), *Volitional Action (pp. 253-298)*. Amsterdam: North-Holland/Elsevier. (Ch. 7)

Buonomano, D. V., Baxter, D. A., & Byrne, J. H. (1990). Small networks of empirically derived adaptive elements simulate some higher-order features of classical conditioning. *Neural Networks*, **3**, 507-523. (Ch. 5)

Burden, R. L., & Faires, J. D. (1985). *Numerical Analysis*. Boston: Prindle, Weber, and Schmidt. (Appendix 1)

Byrne, J. H. (1987). Cellular analysis of associative learning. *Physiological Reviews* **67**, 329-439. (Ch. 3, 5, Appendix 1)

Byrne, J. H., & Schultz, S. G. (1988). *An Introduction to Membrane Transport and Bioelectricity*. New York: Raven. (Appendix 1)

Cajal, S. Ramon y (1934). Les preuves objectives de l'unité anatomique des cellules nerveuses. *Trob. Lab. Inest. Biol. Univ. Madrid* **29**, 1-37. (Translation: Purkiss, M. V. & Fox, C. A., Madrid: Instituto "Ramon y Cajal," 1954. (Appendix 1)

Carlson, N. R. (1977). *Physiology of Behavior*. Boston: Allyn and Bacon, 1977. (Appendix 1)

Carpenter, G. A., & Grossberg, S. (1985). A neural theory of circadian rhythms: split rhythms, after-effects, and motivational interactions. *Journal of Theoretical Biology* **113**, 163-223. (Appendix 2)

Carpenter, G. A., & Grossberg, S. (1987a). A massively parallel architecture for a self-organizing neural pattern recognition machine. *Computer Vision, Graphics, and Image Processing* **37**, 54-115. (Ch. 4, 5, 6, 7, Appendix 2)

Carpenter, G. A., & Grossberg, S. (1987b). ART 2: self-organization of stable category recognition codes for analog input patterns. *Applied Optics* **26**, 4919-4930. (Ch. 4, 6, 7)

Carpenter, G. A., & Grossberg, S. (1989). Search mechanisms for adaptive resonance theory (ART) architectures. *International Joint Conference on Neural Networks*, Washington, DC, June, 18-22, 1989 (Vol. I, pp. 201-205). Piscataway, NJ: IEEE. (Ch. 6)

Carpenter, G. A., & Grossberg, S. (1990). ART 3: Hierarchical search using chemical transmitters in self-organizing pattern recognition architecture. *Neural Networks* **3**, 129-152. (Ch. 6, 7)

Chandrasekharan, B., Goel, A., & Allemang, D. (1988). Connectionism and information processing abstractions. *AI Magazine*, July, 1988 (pp. 24-34). (Ch. 7)

Chowdhury, D. (1986). *Spin Glasses and Other Frustrated Systems*. Princeton, NJ: Princeton University Press. (Ch. 2)

Churchland, P. S. (1986). *Neurophilosophy*. Cambridge, MA: MIT Press. (Appendix 1)

Cognitive Science, Volume 9 (Special Issue on Connectionism). (Ch. 1)

Cohen, M. A. (1988). Sustained oscillations in a symmetric cooperative-competitive neural network: disproof of a conjecture about content-addressable memory. *Neural Networks* **1**, 217-221. (Ch. 4)

Cohen, M. A., & Grossberg, S. (1983). Absolute stability of global pattern formation and parallel memory storage by competitive neural networks. *IEEE Transactions on Systems, Man, and Cybernetics* **13**, 815-826. (Ch. 3, 4, Appendix 2)

Cohen, M. A., & Grossberg, S. (1984). Some global properties of binocular resonances: disparity matching, filling-in, and figure-ground synthesis. In P. Dodwell & T. Caelli (Eds.), *Figural Synthesis* (pp. 117-152). Hillsdale, NJ: Lawrence Erlbaum Associates. (Ch. 4)

Cohen, M. A., & Grossberg, S. (1986). Neural dynamics of speech and language coding: Developmental programs, perceptual grouping and competition for short term memory. *Human Neurobiology* **5**, 1-22. (Ch. 7)

Cohen, M. A., & Grossberg, S. (1987). Masking fields: a massively parallel neural architecture for learning, recognizing, and predicting multiple groupings of patterned data. *Applied Optics*, **26**, 1866-1891, 1987. (Ch. 4, 7)

Cohen, M. A., Grossberg, S., & Stork, D. (1987). Recent developments in a neural model of real-time speech analysis and synthesis. *IEEE First International Conference on Neural Networks (Vol. IV, pp. 443-453)*. San Diego: IEEE/ICNN. (Ch. 7)

Cole, K. S., & Hodgkin, A. L. (1939). Membrane and protoplasm resistance in the squid giant axon. *Journal of General Physiology* **22**, 671-687. (Appendix 1)

Collins, E., Ghosh, S., & Scofield, C. (1988). An application of a multiple neural network learning system to emulation of mortgage underwriting judgments. *IEEE International Conference on Neural Networks, 1988* (Vol. II, pp. 459-466). (Ch. 6, 7, 8)

Cooper, J. R., Bloom, F. E., & Roth, R. H. (1982). *The Biochemical Basis of Neuropharmacology.* New York: Oxford University Press. (Appendix 1)

Cowan, J. D. (1970). A statistical mechanics of nervous activity. In M. Gerstenhaber (Ed.), *Lectures on Mathematics in the Life Sciences* (Vol. 2, pp. 1-57). Providence, RI: American Mathematical Society. (Ch. 2)

Cruz, C. A. (1991). Knowledge-representation networks: goal-direction in intelligent neural systems. In D. S. Levine & S. J. Leven (Eds.), *Motivation, Emotion, and Goal Direction in Neural Networks* (pp. 369-409). Hillsdale, NJ: Lawrence Erlbaum Associates, in press. (Ch. 7)

Cruz, C. A., Hanson, W. A., & Tam, J. Y. (1987). Computational network environment. *IEEE First International Conference on Neural Networks* (Vol. III, pp. 531-538). San Diego: IEEE/ICNN. (Ch. 7)

Dalenoort, G. J. (1983). Grossberg's "cells" considered as cell assemblies. *The Behavioral and Brain Sciences* **6**, 662-663. (Ch. 4)

DARPA Neural Network Study (1988). Alexandria, VA: AFCEA International Press. (Preface, Ch. 1)

Dawes, R. (1989). The parametric avalanche: continuous estimation and control with a neural network architecture. *International Joint Conference on Neural Networks, Washington, DC, June 18-22, 1989* (Vol. II, p. 579). Piscataway, NJ: IEEE. (Ch. 7)

Dawes, R. (1991). Perfect memory. In D. S. Levine & S. J. Leven (Eds.), *Motivation, Emotion, and Goal Direction in Neural Networks* (pp. 411-425). Hillsdale, NJ: Lawrence Erlbaum Associates, in press. (Ch. 6)

Dehaene, S., & Changeux, J.-P. (1989). A simple model of prefrontal cortex function in delayed-response tasks. *Journal of Cognitive Neuroscience* **1**, 244-261. (Ch. 7)

Denker, J. S., Ed. (1986). *Neural Networks for Computing.* AIP Conference Proceedings. New York: American Institute of Physics, Vol. 151. (Ch. 7)

Deregowski, J. B. (1973). Illusion and culture. In R. L. Gregory & G. H. Gombrich (Eds.), *Illusions in Nature and Art* (pp. 161-192). New York: Scribner, 1973. (Ch. 4)

Dev, P. (1975). Perception of depth surfaces in random-dot stereograms: a neural model. *International Journal of Man-Machine Studies* **7**, 511-528. (Ch. 4)

Dietz, W. E., Kiech, E. L., & Ali, M. (1989). Jet and rocket engine fault diagnosis in real time. *Journal of Neural Network Computing* **1**, 5-18. (Ch. 8)

Dowling, J. E. (1987). *The Retina: an Approachable Part of the Brain.* Cambridge, MA: Harvard University Press. (Ch. 4)

Dowling, J. E., & Boycott, B. B. (1966). Organization of primate retina: electron microscopy. *Proceedings of the Royal Society of London, Series B* **166**, 80-111. (Appendix 1)

Dreyfuss, H. L. (1972). *What Computers Can't Do: A Critique of Artificial Intelligence.* New York: Harper and Row. (Ch. 7)

Dreyfuss, H. L., & Dreyfuss, S. (1986). *Mind Over Machine.* New York: Free Press. (Ch. 7)

Duda, R. O., & Hart, P. E. (1973). *Pattern Classification and Scene Analysis.* New York: John Wiley and Sons. (Ch. 3, 6)

Durbin, R., & Willshaw, D. (1987). An analogue approach to the traveling salesman problem using an elastic net method. *Nature* **326**, 689-691. (Ch. 7)

Eccles, J. C., Ito, M., & Szentagothai, J. (1967). *The Cerebellum as a Neuronal Machine.* New York: Springer. (Ch. 4, Appendix 1)

Edelman, G. M. (1987). *Neural Darwinism.* New York: Basic Books. (Ch. 2, 6)

Edwards, C. H., Jr., & Penney, D. E. (1988). *Calculus and Analytic Geometry* (2nd edition). Englewood Cliffs, NJ: Prentice-Hall. (Appendix 2)

Eich, J. M. (1982). A composite holographic recall model. *Psychological Review* **89**, 627-661. (Ch. 7)

Eich, J. M. (1985). Levels of processing, encoding specificity, elaboration, and CHARM. *Psychological Review* **92**, 1-38. (Ch. 7)

Eichenbaum, H., & Kuperstein, M. (1985). Unit-activity, evoked-potentials, and slow waves in the rat hippocampus and olfactory bulb recorded with a 24-channel microelectrode. *Journal of Neuroscience Methods* **15**, 703-712. (Ch. 8)

Ellias, S. A., & Grossberg, S. (1975). Pattern formation, contrast control, and oscillations in the short-term memory of shunting on-center off-surround networks. *Biological Cybernetics* **20**, 69-98. (Ch. 4, 6)

Elsberry, W. R. (1989). Integration and hybridization in neural network modeling. Unpublished M.S. thesis, University of Texas at Arlington. (Ch. 1, 8)

Engel, J., Jr., & Woody, C. D. (1972). Effects of character and significance of stimulus on unit activity at coronal-pericruciate cortex of cat during performance of conditioned motor response. *Journal of Neurophysiology* **35**, 220-229. (Ch. 8)

Ermentrout, G. B., & Cowan, J. D. (1980). Large scale spatially organized activity in neural nets. *SIAM Journal on Applied Mathematics* **38**, 1-21. (Ch. 4, Appendix 2)

Feldman, J. A. & Ballard, D. H. (1982). Connectionist models and their properties. *Cognitive Science* **6**, 205-254. (Ch. 1)

Fender, D. H., & Julesz, B. (1967). Extension of Panum's fusional area in binocularly stabilized vision. *Journal of the Optical Society of America* **57**, 819-830. (Ch. 4)

Ferster, D. & Koch, C. (1987). Neuronal connections underlying orientation selectivity in cat visual cortex. *Trends in NeuroSciences* **10**, 487-492. (Ch. 6)

Finkel, L. H. & Edelman, G. M. (1985). Interaction of synaptic modification rules within populations of neurons. *Proceedings of the National Academy of Sciences* **82**, 1291-1295. (Ch. 6)

Flash, T. & Hogan, N. (1985). The coordination of arm movements: an experimentally confirmed mathematical model. *Journal of Neuroscience* **5**, 1688-1703. (Ch. 7)

Fodor, J. A. & Pylyshyn, Z. W. (1988). Connectionism and cognitive architecture: a critical analysis. In S. Pinker & J. Mehler (Eds.), *Connections and Symbols (pp. 3-71)*. Cambridge, MA: MIT Press. (Ch. 4, 7)

Frauenthal, J. C. (1980). *Introduction to Population Modeling*. Boston: Birkhauser. (Appendix 2).

Freeman, W. J. (1972a). Waves, pulses, and the theory of neural masses. *Progress in Theoretical Biology* **2**, 87-165. (Ch. 2)

Freeman, W. J. (1972b). Linear analysis of the dynamics of neural masses. *Annual Review of Biophysics and Bioengineering* **1**, 225-256. (Ch. 2)

Freeman, W. J. (1975a). *Mass Action in the Nervous System*. New York: Academic Press. (Ch. 2, 8)

Freeman, W. J. (1975b). Parallel processing of signals in neural sets as manifested in the EEG. *International Journal of Man-Machine Studies* **7**, 347-369. (Ch. 2)

Freeman, W. J. (1978). Spatial patterns of an EEG event in the olfactory bulb and cortex. *Electroencephalography and Clinical Neurology* **44**, 586-605. (Ch. 8)

Freeman, W. J. (1983). Experimental demonstration of "shunting networks," the "sigmoid function," and "adaptive resonance" in the olfactory system. *The Behavioral and Brain Sciences* **6**, 665-666. (Ch. 4)

Fu, L.-M. (1989). Integration of neural heuristics into knowledge-based inference. *Connection Science* **1**, 327-342. (Ch. 7)

Fukushima, K. (1975). Cognitron: a self-organizing multilayered neural network. *Biological Cybernetics* **20**, 121-136. (Ch. 6)

Fukushima, K. (1980). Neocognitron: A self-organizing neural network model for a mechanism of pattern recognition unaffected by shift in position. *Biological Cybernetics* **36**, 193-204. (Ch. 2, 6)

Fukushima, K., & Miyake, S. (1982). Neocognitron: A new algorithm for pattern recognition tolerant of deformation and shifts in position. *Pattern Recognition* **15**, 455-469. (Ch. 6)

Fuster, J. M. (1980). *The Prefrontal Cortex.* New York: Raven. Reprinted in 1989. (Ch. 7)

Fuster, J. M. (1985). The prefrontal cortex: mediator of cross-temporal contingencies. *Human Neurobiology* **4**, 169-175. (Ch. 7)

Fuster, J. M., Bauer, R. H., & Jervey, J. P. (1982). Cellular discharge in the dorsolateral prefrontal cortex of the monkey during cognitive tasks. *Experimental Neurology* **77**, 679-694. (Ch. 4, 8)

Gainer, H. & Brownstein, M. J. (1981). Neuropeptides. In G. J. Siegel, R. W. Albers, B. W. Agranoff, & R. Katzman, *Basic Neurochemistry.* Boston: Little Brown. (Appendix 1)

Gardner, E. (1958). *Fundamentals of Neurology.* Philadelphia: W. E. Saunders. (Appendix 1)

Garey, M. R., & Johnson, D. S. (1979). *Computers and Intractibility.* New York: W. H. Freeman. (Ch. 7)

Gawin, F. H., & Ellinwood, E., Jr. (1988). Cocaine and other stimulants: actions, abuse, and treatment. *New England Journal of Medicine* **318**, 1173-1182. (Ch. 8)

Gelperin, A., Hopfield, J. J., & Tank, D. W. (1985). The logic of *Limax* learning. In A. Selverston (Ed.), *Model Neural Networks and Behavior.* (pp. 237-261). New York: Plenum. (Ch. 5)

Geman, S. (1979). Some averaging and stability results for random differential equations. *SIAM Journal on Applied Mathematics* **36**, 86-105. (Ch. 2)

Geman, S. (1980). The law of large numbers in neural modeling. In S. Grossberg (Ed.), *Mathematical Psychology and Psychophysiology* (pp. 91-105). Providence, RI: American Mathematical Society. (Ch. 2)

Geman, S. & Geman, D. (1984). Stochastic relaxation, Gibbs distribution, and the Bayesian restoration of images. *IEEE Transactions on Pattern Analysis and Machine Intelligence* **6**, 721-741. (Ch. 4, 7)

Gentry, R. D. (1978). *Introduction to Calculus for the Biological and Health Sciences.* Reading, MA: Addison-Wesley. (Appendix 2)

Georgopoulos, A. P., Kalaska, J. F., Caminiti, R., & Massey, J. T. (1984). The representation of movement direction in the motor cortex: Single cell and population studies. In G. M. Edelman, W. E. Gall & W. M. Cowan (Eds.), *Dynamic Aspects of Neocortical Function* (pp. 501-524). New York: John Wiley and Sons. (Ch. 7)

Gibson, J. H., & Radner, M. (1937). Adaptation, after-effect, and contrast in the perception of tilted lines. I. Quantitative studies. *Journal of Experimental Psychology* **20**, 453-467. (Ch. 4)

Gingrich, K. J., & Byrne, J. H. (1985). Simulation of synaptic depression, posttetanic potentiation, and presynaptic facilitation of synaptic potentials from sensory neurons mediating gill-withdrawal reflex in *Aplysia*. *Journal of Neurophysiology* **53**, 652-669. (Ch. 5, Appendix 1)

Gingrich, K. J. & Byrne, J. H. (1987). Single-cell model for associative learning. *Journal of Neurophysiology* **57**, 1705-1715. (Ch. 5, Appendix 1)

Glickstein, M., Yeo, C., & Stein, J. (Eds.), *Cerebellum and Neuronal Plasticity*. New York: Plenum. (Ch. 7)

Gluck, M. A., & Thompson, R. F. (1987). Modeling the neural substrates of associative learning and memory, a computational approach. *Psychological Review* **94**, 176-191. (Ch. 5)

Golden, R. (1986). The brain-state-in-a-box model is a gradient descent algorithm. *Journal of Mathematical Psychology* **30**, 73-80. (Ch. 4)

Goldman-Rakic, P. S. (1984). Modular organization of prefrontal cortex. *Trends in NeuroSciences* **7**, 419-429. (Ch. 4)

Goldman-Rakic, P. S. (1989). Circuitry of primate prefrontal cortex and regulation of behavior by representational memory. *Handbook of Physiology — The Nervous System, V* (pp. 373-417). (Ch. 8)

Gong, W., & Manry, M. T. (1989). Analysis of non-Gaussian data using a neural network. *International Joint Conference on Neural Networks, Washington, DC, June, 18-22, 1989*. Piscataway, NJ: IEEE (Vol. II, p. 576). (Ch. 6, 8)

Graham, N. (1980). The visual system does a crude Fourier analysis of patterns. In S. Grossberg (Ed.), *Mathematical Psychology and Psychophysiology* (pp. 1-16). Providence, RI: American Mathematical Society. (Ch. 4)

Gray, J. A. & Smith, P. T. (1969). An arousal-decision model for partial reinforcement and discrimination learning. In R. M. Gilbert, & N. S. Sutherland (Eds.), *Animal Discrimination Learning* (pp. 243-272). New York: Academic, 1969. (Ch. 3, 7)

Greenspan, D., & Casulli, V. (1988). *Numerical Analysis for Applied Mathematics, Science, and Engineering*. Reading, MA: Addison-Wesley. (Appendix 2)

Griffith, J. S. (1963a). On the stability of brain-like structures. *Biophysical Journal* **3**, 299-308. (Ch. 2)

Griffith, J. S. (1963b). A field theory for neural nets, I: derivation of the field equations. *Bulletin of Mathematical Biophysics* **25**, 111-120. (Ch. 2)

Griffith, J. S. (1965). A field theory for neural nets, II: properties of the field equations. *Bulletin of Mathematical Biophysics* **27**, 187-195. (Ch. 2)

Grimson, W. E. L. (1983). To have your edge and fill-in too. *The Behavioral and Brain Sciences* **4**, 666-667. (Ch. 4)

Gross, G. W., Wen, W. Y., & Lin, J. W. (1985). Transparent indium tin oxide electrode patterns for extracellular, multisite recording in neuronal cultures. *Journal of Neuroscience Methods* **15**, 243-252. (Ch. 8)

Grossberg, S. (1968a). A prediction theory for some non-linear functional-differential equations, I. Learning of lists. *Journal of Mathematical Analysis and Applications* **21**, 643-694. (Ch. 3, 5)

Grossberg, S. (1968b). A prediction theory for some non-linear functional-differential equations, II. Learning of patterns. *Journal of Mathematical Analysis and Applications* **22**, 490-522. (Ch. 3, 5)

Grossberg, S. (1969a). Embedding fields: a theory of learning with physiological implications. *Journal of Mathematical Psychology* **6**, 209-239. (Ch. 3, 5)

Grossberg, S. (1969b). On the production and release of chemical transmitters and related topics in cellular control. *Journal of Theoretical Biology* **22**, 325-364. (Ch. 3)

Grossberg, S. (1969c). On learning and energy-entropy dependence in recurrent and nonrecurrent signed networks. *Journal of Statistical Physics* **1**, 319-350. (Ch. 3, 5)

Grossberg, S. (1969d). Some networks that can learn, remember, and reproduce any number of complicated space-time patterns, I. *Journal of Mathematics and Mechanics* **19**, 53-91. (Ch. 3)

Grossberg, S. (1970a). Neural pattern discrimination. *Journal of Theoretical Biology* **27**, 291-337. (Ch. 4)

Grossberg, S. (1970b). Some networks that can learn, remember, and reproduce any number of complicated space-time patterns, II. *Studies in Applied Mathematics* **49**, 135-166. (Ch. 3)

Grossberg, S. (1971). On the dynamics of operant conditioning. *Journal of Theoretical Biology* **33**, 225-255. (Ch. 3, 5, 6)

Grossberg, S. (1972a). Pattern learning by functional-differential neural networks with arbitrary path weights. In K. Schmitt (Ed.), *Delay and Functional Differential Equations and Their Applications* (pp. 121-160). New York: Academic. (Ch. 3)

Grossberg, S. (1972b). A neural theory of punishment and avoidance. I. Qualitative theory. *Mathematical Biosciences* **15**, 39-67. (Ch. 2, 3, 5, 7, Appendix 1)

Grossberg, S. (1972c). A neural theory of punishment and avoidance. II. Quantitative theory. *Mathematical Biosciences* **15**, 253-285. (Ch. 2, 3, 5, 7)

Grossberg, S. (1972d). Cerebellar and retinal analogs of cells fired by learnable or unlearned pattern classes. *Kybernetik* **10**, 49-57. (Ch. 4, 6)

Grossberg, S. (1973). Contour enhancement, short term memory, and constancies in reverberating neural networks. *Studies in Applied Mathematics* **52**, 213-257. (Ch. 4, 6, 7)

Grossberg, S. (1975). A neural model of attention, reinforcement, and discrimination learning. *International Review of Neurobiology* **18**, 263-327. (Ch. 4, 5, 7)

Grossberg, S. (1976a). On the development of feature detectors in the visual cortex with applications to learning and reaction-diffusion systems. *Biological Cybernetics* **21**, 145-149. (Ch. 6)

Grossberg, S. (1976b). Adaptive pattern classification and universal recoding: parallel development and coding of neural feature detectors. *Biological Cybernetics* **23**, 121-134. (Ch. 5, 6)

Grossberg, S. (1976c). Adaptive pattern classification and universal recoding: feedback, expectation, olfaction, and illusions. *Biological Cybernetics* **23**, 187-202. (Ch. 6)

Grossberg, S. (1978a). Competition, decision and consensus. *Journal of Mathematical Analysis and Applications* **66**, 470-493. (Ch. 4)

Grossberg, S. (1978b). A theory of human memory: self-organization and performance of sensory-motor codes, maps, and plans. In R. Rosen & F. Snell (Eds.), *Progress in Theoretical Biology* (Vol. 5, pp. 233-374). (Ch. 4, 7)

Grossberg, S. (1978c). Do all neural models really look alike? A comment on Anderson, Silverstein, Ritz, and Jones. *Psychological Review* **85**, 592-596. (Ch. 6)

Grossberg, S. (1980). How does a brain build a cognitive code? *Psychological Review* **87**, 1-51. (Ch. 6, 7)

Grossberg, S. (1982a). Processing of expected and unexpected events during conditioning and attention: a psychophysiological theory. *Psychological Review* **89**, 529-572. (Ch. 5)

Grossberg, S. (1982b). A psychophysiological theory of reinforcement, drive, motivation, and attention. *Journal of Theoretical Neurobiology* **1**, 286-369. (Ch. 5)

Grossberg, S., Ed. (1982c). *Studies in Mind and Brain*. Boston: Reidel. (Ch. 6)

Grossberg, S. (1983). The quantized geometry of visual space: the coherent computation of depth, form, and lightness. *The Behavioral and Brain Sciences* **4**, 625-692. (Ch. 4)

Grossberg, S. (1984). Some normal and abnormal behavioral syndromes due to transmitter gating of opponent processes. *Biological Psychiatry* **19**, 1075-1117. (Ch. 7)

Grossberg, S. (1987a). Cortical dynamics of three-dimensional form, color, and brightness perception, II. Binocular theory. *Perception and Psychophysics* **41**, 117-158. (Ch. 4)

Grossberg, S. (1987b). Competitive learning: from interactive activation to adaptive resonance. *Cognitive Science* **11**, 23-63. (Ch. 4, 6, 7)

Grossberg, S., Ed. (1987c). *The Adaptive Brain, Vols. I and II.* New York: Elsevier. (Ch. 6)

Grossberg, S. (1988). *Neural Networks and Natural Intelligence.* Cambridge, MA: MIT Press. (Ch. 6)

Grossberg, S., & Gutowski, W. (1987). Neural dynamics of decision making under risk: Affective balance and cognitive-emotional interactions. *Psychological Review* **94**, 300-318. (Ch. 7)

Grossberg, S., & Kuperstein, M. (1986). *Neural Dynamics of Adaptive Sensory-motor Control: Ballistic Eye Movements.* Amsterdam: Elsevier/North-Holland. Expanded edition by Pergamon, 1989. (Ch. 3, 7)

Grossberg, S., & Levine, D. S. (1975). Some developmental and attentional biases in the contrast enhancement and short-term memory of recurrent neural networks. *Journal of Theoretical Biology* **53**, 341-380. (Ch. 4, 6)

Grossberg, S., & Levine, D. S. (1987). Neural dynamics of attentionally modulated Pavlovian conditioning: blocking, interstimulus interval, and secondary reinforcement. *Applied Optics* **26**, 5015-5030. (Ch. 3, 5, 7).

Grossberg, S., Levine, D. S., & Schmajuk, N. A. (1991). Associative learning and selective forgetting in a neural network regulated by reinforcement and attentive feedback. In D. S. Levine & S. J. Leven (Eds.), *Motivation, Emotion, and Goal Direction in Neural Networks.* Hillsdale, NJ: Lawrence Erlbaum Associates, in press. (Ch. 5)

Grossberg, S., & Mingolla, E. (1985a). Neural dynamics of form perception: boundary completion, illusory figures, and neon color spreading. *Psychological Review* **92**, 173-211. (Ch. 4, 7, 8)

Grossberg, S., & Mingolla, E. (1985b). Neural dynamics of perceptual grouping: textures, boundaries, and emergent segmentations. *Perception and Psychophysics* **38**, 141-171. (Ch. 4, 8)

Grossberg, S., & Mingolla, E. (1987). Neural dynamics of surface perception: boundary webs, illuminants, and shape-from-shading. *Computer Vision, Graphics, and Image Processing* **37**, 116-165. (Ch. 8)

Grossberg, S., & Pepe, J. (1971). Spiking threshold and overarousal effects in serial learning. *Journal of Statistical Physics* **3**, 95-125. (Ch. 7)

Grossberg, S., & Schmajuk, N. A. (1987). Neural dynamics of attentionally-modulated Pavlovian conditioning: conditioned reinforcement, inhibition, and opponent processing. *Psychobiology* **15**, 95-240. (Ch. 3, 5, 7)

Grossberg, S., & Schmajuk, N. A. (1989). Neural dynamics of adaptive timing and temporal discrimination during associative learning. *Neural Networks* **2**, 79-102. (Ch. 5)

Grossberg, S., & Stone, G. O. (1986). Neural dynamics of word recognition and recall: attentional priming, learning, and resonance. *Psychological Review* **93**, 46-74. (Ch. 7)

Harmon, L. D., & Lewis, E. R. (1968). Neural modeling. *Physiological Reviews* **46**, 513-591. (Ch. 2)

Harth, E. M., Csermely, T. J., Beek, B., & Lindsay, R. D. (1970). Brain functions and neural dynamics. *Journal of Theoretical Biology* **26**, 93-120. (Ch. 2)

Hartline, H. K., & Ratliff, F. (1957). Inhibitory interactions of receptor units in the eye of *Limulus*. *Journal of General Physiology* **40**, 351-376. (Ch. 2, 4)

Hawkins, R. D., Abrams, T. W., Carew, T. J., & Kandel, E. R. (1983). A cellular mechanism of classical conditioning in *Aplysia*: activity-dependent amplification of presynaptic facilitation. *Science* **219**, 400-405. (Ch. 5).

Hawkins, R. D. & Kandel, E. R. (1984). Is there a cell-biological alphabet for simple forms of learning? *Psychological Review* **91**, 375-391. (Ch. 3, 5)

Hebb, D. O. (1949). *The Organization of Behavior.* New York: John Wiley and Sons, 1949. (Ch. 1, 2, 3, 5, 6).

Hebb, D. O. (1955). Drives and the CNS (conceptual nervous system). *Psychological Review* **62**, 243-254. (Ch. 5).

Hecht-Nielsen, R. (1986). Performance limits of optical, electro-optical, and electronic neurocomputers. In H. Szu (Ed.), *Hybrid and Optical Computing* (pp. 277-306). Bellingham, WA: SPIE, Vol. **634**. (Preface, Ch. 1, 8)

Hecht-Nielsen, R. (1987). Counterpropagation networks. *IEEE First International Conference on Neural Networks* (Vol. II, pp. 19-32). San Diego: IEEE/ICNN. (Ch. 1)

Hecht-Nielsen, R. (1988). *Neurocomputer Applications.* In R. Eckmiller, & C. von der Malsburg (Eds.), *Neural Computers* (pp. 445-453). Berlin: Springer-Verlag. (Ch. 1)

Hecht-Nielsen, R. (1990). *Neurocomputing.* Reading, MA: Addison-Wesley. (Ch. 1)

Heiner, R. A. (1983). The origin of predictable behavior. *American Economic Review* **73**, 560-585. (Ch. 7)

Heiner, R. A. (1985). The origin of predictable behavior: Further modeling and applications. *American Economic Review* **75**, 391-396. (Ch. 7)

Hestenes, D. (1991). A neural network theory of manic-depressive illness. In D. S. Levine & S. J. Leven (Eds.), *Motivation, Emotion, and Goal Direction in Neural Networks (pp. 209-257).* Hillsdale, NJ: Lawrence Erlbaum Associates, in press. (Ch. 5, 7, 8)

Hewitt, C. (1986). Concurrency in intelligent systems. *AI Expert* Premier, 1986, 44-50. (Ch. 1, 7)

Hinton, G. E. (1987). Connectionist learning procedures. Technical Report No. CMU-CS-87-115. Pittsburgh: Carnegie-Mellon University, Computer Science Department. (Ch. 5, 7)

Hinton, G. E. & Anderson, J. A. (Eds.) (1981). *Parallel Models of Associative Memory*. Hillsdale, NJ: Lawrence Erlbaum Associates. (Ch. 5)

Hinton, G. E., & Becker, S. (1990). An unsupervised learning procedure that discovers surfaces in random-dot stereograms. *Proceedings of the International Joint Conference on Neural Networks, January, 1990* (Vol. 1, pp. 218-222). Hillsdale, NJ: Lawrence Erlbaum Associates. (Ch. 4)

Hinton, G. E., Sejnowski, T. J., & Ackley, D. H. (1984). Boltzmann machines: constraint satisfaction networks that learn. Technical Report No. CMU-CS-84-119. Pittsburgh: Carnegie-Mellon University, Computer Science Department. (Ch. 7)

Hinton, G. E., & Sejnowski, T. J. (1986). Learning and relearning in Boltzmann machines. In Rumelhart, D. E. & McClelland, J. L. (Eds.), *Parallel Distributed Processing* (Vol. I, pp. 282-317). Cambridge, MA: MIT Press, 1986. (Ch. 4, 7)

Hirsch, H. V. B., & Spinelli, D. N. (1970). Visual experience modifies distribution of horizontally and vertically oriented receptive fields in cats. *Science* **168**, 869-871. (Ch. 4, 6)

Hirsch, M. W. (1982). Systems of differential equations which are competitive or cooperative. I: Limit sets. *SIAM Journal of Mathematical Analysis* **13**, 167-179. (Ch. 4)

Hirsch, M. W. (1984). The dynamical systems approach to differential equations. *Bulletin of the American Mathematical Society* **11**, 1-64. (Ch. 4)

Hirsch, M. W. (1990). On the Amari-Takeuchi theory of category formation. *Proceedings of the International Joint Conference on Neural Networks*, January, 1990 (Vol. I, pp. 297-300). Hillsdale, NJ: Lawrence Erlbaum Associates. (Ch. 4)

Hirsch, M. W. & Smale, S. (1974). *Differential Equations, Dynamical Systems, and Linear Algebra*. New York: Academic Press. (Appendix 2)

Hodgkin, A. L. (1964). *The Conduction of the Nervous Impulse*. Springfield, IL: C. C. Thomas. (Ch. 4)

Hodgkin, A. L., & Huxley, A. F. (1939). Action potentials recorded from inside nerve fiber. *Nature* **144**, 710-711. (Appendix 1)

Hodgkin, A. L., & Huxley, A. F. (1952). A quantitative description of membrane current and its application to conduction and excitation in nerve. *Journal of Physiology* **117**, 500-544. (Appendix 1)

Hogan, N. (1984). An organizing principle for a class of voluntary movements. *Journal of Neuroscience* **4**, 2745-2754. (Ch. 7)

Homa, D. (1984). On the nature of categories. In G. H. Bower (Ed.), *The Psychology of Learning and Motivation* (Vol. 18, pp. 49-94). Orlando: Academic Press. (Ch. 6)

Hopfield, J. J. (1982). Neural networks and physical systems with emergent collective computational abilities. *Proceedings of the National Academy of Sciences* **79**, 2554-2558. (Ch. 1, 4, 7, Appendix 2)

Hopfield, J. J. (1984). Neurons with graded response have collective computational properties like those of two-state neurons. *Proceedings of the National Academy of Sciences* **81**, 3088-3092. (Ch. 4, 7)

Hopfield, J. J., and Tank, D. W. (1985). "Neural" computation of decisions in optimization problems. *Biological Cybernetics* **52**, 141-152. (Ch. 4, 7)

Hopfield, J. J., and Tank, D. W. (1986). Computing with neural circuits: a model. *Science* **233**, 625-633. (Ch. 3, 4, 7)

Houk, J. C. (1987). Model of the cerebellum as an array of adjustable pattern generators. In M. Glickstein, C. Yeo, & J. Stein (Eds.), *Cerebellum and Neuronal Plasticity*. New York: Plenum. (Ch. 7)

Hubel, D. H., & Wiesel, T. N. (1962). Receptive fields, binocular interaction, and functional architecture in the cat's visual cortex. *Journal of Physiology* **160**, 106-154. (Ch. 2, 4, 6, 8)

Hubel, D. H., & Wiesel, T. N. (1963). Receptive fields of cells in striate cortex of very young, visually inexperienced kittens. *Journal of Neurophysiology* **26**, 994-1002. (Ch. 6, 8)

Hubel, D. H., & Wiesel, T. N. (1965). Receptive fields and functional architecture in two non-striate visual areas (18 and 19) of the cat. *Journal of Neurophysiology* **28**, 229-298. (Ch. 2, 4, 8)

Hubel, D. H., & Wiesel, T. N. (1968). Receptive fields and functional architecture of monkey striate cortex. *Journal of Physiology (London)* **195**, 215-243. (Ch. 6, 8)

Hull, C. L. (1943). *Principles of Behavior*. New York: Appleton. (Ch. 2, 3, 5)

Huyser, K. A., & Horowitz, M. A., Generalization in digital functions. *Neural Networks* **1**, Supplement 1, 101. (Ch. 6)

Ingvar, D. (1985). Memory of the future: an essay on the temporal organization of conscious awareness. *Human Neurobiology*, **4**, 124-136. (Ch. 7)

Ito, M. (1982). Cerebellar control of the vestibulo-ocular reflex -- around the flocculus hypothesis. *Annual Review of Neuroscience* **5**, 275-296. (Ch. 7)

Ito, M. (1984). *The Cerebellum and Neural Control*. New York: Raven. (Ch. 7)

Johnson-Laird, P. (1988). *The Computer and the Mind.* Cambridge, MA: Harvard University Press. (Ch. 7)

Jordan, M. I. (1986a). An introduction to linear algebra in parallel distributed processing. In D. E. Rumelhart & J. L. McClelland (Eds.), *Parallel Distributed Processing* (Vol. 1, pp. 365-422). Cambridge, MA: MIT Press. (Ch. 6, Appendix 2)

Jordan, M. I. (1986b). Attractor dynamics and parallelism in a connectionist sequential machine. In *Proceedings of the Eighth Annual Conference of the Cognitive Science Society* (pp. 531-546). Hillsdale, NJ: Lawrence Erlbaum Associates, 1986. (Ch. 7)

Julesz, B. (1960). Binocular depth perception of computer-generated patterns. *Bell System Technical Journal* 37, 1125-1162 (Ch. 4)

Julesz, B. (1971). *Foundations of Cyclopean Perception.* Chicago: University of Chicago Press. (Ch. 4)

Kahneman, D., & Treisman, A. (1984). Changing views of attention and automaticity. In R. Parasuraman, & D. R. Davies, (Eds.), *Varieties of Attention (pp. 29-61).* Orlando, FL: Academic Press. (Ch. 5).

Kahneman, D., & Tversky, A. (1982). Subjective probability. In D. Kahneman, P. Slovic, & A. Tversky (Eds.), *Judgment Under Uncertainty: Heuristics and Biases.* New York Cambridge University Press. (Ch. 7)

Kamangar, F. A., & Cykana, M. J. (1988). Recognition of handwritten numerals using a multilayer neural network. Proceedings of Third Annual Technology Conference (Vol. II, pp. 768-780). Washington, DC: United States Postal Service. (Ch. 6, 8)

Kamin, L. J. (1969). Predictability, surprise, attention, and conditioning. In B. A. Campbell and R. M. Church (Eds.), *Punishment and Aversive Behavior* (pp. 279-296). New York: Appleton-Century-Crofts. (Ch. 5)

Kandel, E. R., & Schwartz, J. H. (Eds.) (1985). *Principles of Neural Science.* New York: Elsevier (Appendix 1)

Kandel, E. R., & Tauc, L. (1965). Heterosynaptic facilitation in neurones of the abdominal ganglion of *Aplysia depilans. Journal of Physiology (London)* 181, 1-27. (Ch. 1, 3, 5)

Kanizsa, G. (1976). Subjective contours. *Scientific American* 234, 48-64. (Ch. 4)

Katchalsky, A., Rowland, V., & Blumenthal, R. (Eds.) (1974). Dynamic patterns of brain cell assemblies. *Neurosciences Research Program Bulletin* 12, 3-187. (Ch. 2)

Katz, B. (1966). *Nerve, Muscle, and Synapse.* New York: McGraw-Hill. (Ch. 2, 4, Appendix 1)

Kawato, M., Furukawa, K., & Suzuki, R. (1987). A hierarchical neural-network model for control and learning of voluntary movement. *Biological Cybernetics* 57, 169-185. (Ch. 7)

Kawato, M., Isobe, M., Maeda, Y., & Suzuki, R. (1988). Coordinates transformation and learning control for visually-guided voluntary movement with iteration: a Newton-like method in function space. *Biological Cybernetics* **59**, 161-177. (Ch. 7)

Kelly, G. (1955). *The Psychology of Personal Constructs.* New York: Norton. (Ch. 7)

Kelso, J. A. S., & Holt, K. G. (1980). Exploring a vibratory systems analysis of human movement production. *Journal of Neurophysiology* **28**, 45-52. (Ch. 7)

Kelso, S. R., & Brown, T. H. (1986). Differential conditioning of associative synaptic enhancement in hippocampal brain slices. *Science* **232**, 85-87. (Ch. 3)

Kernell, D. (1965). The adaptation and the relation between discharge frequency and current strength of cat lumbosacral motoneurones stimulated by long-lasting injected currents. *Acta Physiologica Scandinavica* **65**, 65-73. (Ch. 2)

Khotanzad, A., & Lu, J. H. (1989). Object recognition using a neural network and invariant Zernike features. *Proceedings of the IEEE Computer Society on Computer Vision and Pattern Recognition*, San Diego. (Ch. 6)

Killeen, P. R., & Fetterman, J. G. (1988). A behavioral theory of timing. *Psychological Review* **95**, 274-295. (Ch. 5)

Kilmer, W., McCulloch, W. S., & Blum, J. (1969). A model of the vertebrate central command system. *International Journal of Man-Machine Studies* **1**, 279-309. (Ch. 2, 4, 5, 7)

Kilmer, W., & Olinski, M. (1974). Model of a plausible learning scheme for CA3 hippocampus. *Kybernetik* **16**, 133-143. (Ch. 2, 7)

King, D. L. (1979). *Conditioning, an Image Approach.* New York: Gardner Press. (Ch. 5)

Kirkpatrick, S., Gelatt, C. D., Jr., & Vecchi, M. P. (1983). Optimization by simulated annealing. *Science* **220**, 671-680. (Ch. 4, 7)

Klopf, A. H. (1972). Brain function and adaptive systems: a heterostatic theory. Air Force Cambridge Research Laboratories Research Report AFCRL-72-0164, Bedford, MA. (Ch. 3)

Klopf, A. H. (1979). Goal-seeking systems from goal-seeking components. *Cognition and Brain Theory Newsletter* **3**, 2. (Ch. 7)

Klopf, A. H. (1982). *The Hedonistic Neuron.* Washington, DC: Hemisphere. (Ch. 3, 5, 7)

Klopf, A. H. (1986). A drive-reinforcement model of single neuron function: an alternative to the Hebbian neuronal model. In J. S. Denker (Ed.), *Neural Networks for Computing* (pp. 265-270). AIP Conference Proceedings. New York: American Institute of Physics, Vol. **151**. (Ch. 3)

Klopf, A. H. (1988). A neuronal model of classical conditioning. *Psychobiology* **16**, 85-125. (Ch. 3, 5, 7)

Klopf, A. H., & Morgan, J. S. (1991). The role of time in natural intelligence: implications of classical and instrumental conditioning for neuronal and neural modeling. In J. W. Moore & M. Gabriel (Eds.), *Learning and Computational Neuroscience*. Cambridge, MA: MIT Press, in press. (Ch. 5)

Knapp, A. G., & Anderson, J. A. (1984). Theory of categorization based on distributed memory storage. *Journal of Experimental Psychology: Learning, Memory, and Cognition* **10**, 616-637. (Ch. 6)

Kohonen, T. (1977). *Associative Memory — a System-theoretical Approach*. New York: Springer. (Ch. 3, 6)

Kohonen, T. (1984). *Self-organization and Associative Memory*. Berlin: Springer-Verlag. Reprinted in 1988. (Ch. 3, 5, 6, 7)

Kohonen, T. (1986). Representation of sensory information in self-organizing feature maps, and relation of these maps to distributed memory networks. In H. Szu (Ed.), *Hybrid and Optical Computing* (pp. 248-259). Bellingham, WA: SPIE, Vol. **634**. (Ch. 7)

Kohonen, T. (1987). State of the art in neural computing. *IEEE First International Conference on Neural Networks* (Vol. I, pp. 77-90). San Diego: IEEE/ICNN. (Ch. 7)

Kohonen, T. (1988). The "neural" phonetic typewriter. *Computer* **21**, No. 3, 11-22. (Ch. 8)

Kohonen, T. (1989). A self-learning musical grammar, or "associative memory of the second kind". *International Joint Conference on Neural Networks, Washington, DC, June, 18-22, 1989* (Vol. I, pp. 1-6). Piscataway, NJ: IEEE. (Ch. 8)

Kohonen, T., Lehtio, P., Rovamo, J., Hyvarinen, J., Bry, K., & Vainio, L. (1977). A principle of neural associative memory. *Neuroscience* **2**, 1065-1076. (Ch. 3)

Kohonen, T., & Oja, E. (1976). Fast adaptive formation of orthogonalizing filters and associative memory in recurrent networks of neuron-like elements. *Biological Cybernetics* **21**, 85-95. (Ch. 3, 7)

Kohonen, T., Reuhkala, E., Makisara, K., & Vainio, L. (1976). Associative recall of images. *Biological Cybernetics* **22**, 159-168. (Ch. 3)

Koob, G. F., & Bloom, F. E. (1988). Cellular and molecular mechanisms of drug dependence. *Science* **24**, 715-723. (Ch. 8)

Kosko, B. (1986a). Fuzzy cognitive maps. *International Journal of Man-Machine Studies* **24**, 65-75. (Ch. 7)

Kosko, B. (1986b). Differential Hebbian learning. In J. S. Denker (Ed.), *Neural Networks for Computing* (pp. 265-270). AIP Conference Proceedings. New York: American Institute of Physics, Vol. **151**. (Ch. 3, 5)

Kosko, B. (1987a). Adaptive bidirectional associative memories. *Applied Optics* **26**, 4947-4960. (Ch. 3, 5)

Kosko, B. (1987b). Adaptive inference in fuzzy knowledge networks. *IEEE First International Conference on Neural Networks (Vol. II, pp. 261-268).* San Diego: IEEE/ICNN. (Ch. 7)

Kosko, B. (1987c). Competitive adaptive bidirectional associative memories. *IEEE First International Conference on Neural Networks* (Vol. II, pp. 759-766). San Diego: IEEE/ICNN. (Ch. 3, 5)

Kosko, B. (1987d). Constructing associative memory. *Byte*, September, 1987, 137-144. (Ch. 3)

Kosko, B. (1988). Bidirectional associative memories. *IEEE Transactions on Systems, Man, and Cybernetics* **18**, 49-60. (Ch. 3, 5)

Kuffler, S. (1953). Discharge patterns and functional organization of mammalian retina. *Journal of Neurophysiology* **16**, 37-68. (Ch. 4)

Kuperstein, M. (1988a). An adaptive neural model for mapping invariant target position. *Neuroscience* **102**, 148-162. (Ch. 7)

Kuperstein, M. (1988b). Neural model of adaptive hand-eye coordination for single postures. *Science* **239**, 1308-1311. (Ch. 7, 8)

Kuperstein, M., & Wang, J. (1990). Neural controller for adaptive movements with unforeseen payloads. *IEEE Transactions on Neural Networks*, **1**, 137-142. (Ch. 7)

Lashley, K. (1929). *Brain Mechanisms and Intelligence.* Chicago: University of Chicago Press. (Ch. 2)

LeCun, Y. (1985). Une procédure d'apprentissage pour reseau a seuil assymetrique. *Proceedings of Cognitiva 85* Paris (pp. 599-604). (Ch. 2, 3, 6)

Leven, S. J. (1987). Choice and neural process. Unpublished Ph.D. dissertation, University of Texas at Arlington. (Preface)

Leven, S. J. (1991). Learned helplessness, memory, and the dynamics of hope. In D. S. Levine & S. J. Leven (Eds.), *Motivation, Emotion, and Goal Direction in Neural Networks* (pp. 259-299). Hillsdale, NJ: Lawrence Erlbaum Associates, in press. (Ch. 7)

Leven, S. J. & Levine, D. S. (1987). Effects of reinforcement on knowledge retrieval and evaluation. *IEEE First International Conference on Neural Networks* (Vol. II, pp. 269-279). San Diego: IEEE/ICNN. (Ch. 7)

Levine, D. S. (1975). Studies in the transformation and storage of patterns in reverberating neural networks. Unpublished Ph. D. dissertation, Massachusetts Institute of Technology. (Ch. 4)

Levine, D. S. (1983a). Book review: A. Harry Klopf, *The Hedonistic Neuron.* *Mathematical Biosciences* **64**, 295-297. (Ch. 7)

Levine, D. S. (1983b). Neural population modeling and psychology: a review. *Mathematical Biosciences* **66**, 1-86. (Ch. 1, 2, 3, 6, 7)

Levine, D. S. (1986). A neural network theory of frontal lobe function. In *Proceedings of the Eighth Annual Conference of the Cognitive Science Society (pp. 716-727).* Hillsdale, NJ: Lawrence Erlbaum Associates, 1986. (Ch. 4, 7)

Levine, D. S. (1988). Survival of the synapses. *The Sciences* November-December, 1988, 46-52. (See also the ensuing letters in the July-August, 1989 issue). (Ch. 6)

Levine, D. S. (1989a). Selective vigilance and ambiguity detection in adaptive resonance networks. In W. Webster (Ed.), *Simulation and AI 1989* (pp. 1-7). San Diego: Society for Computer Simulation. (Ch. 6)

Levine, D. S. (1989b). Neural network principles for theoretical psychology. *Behavior Research Methods, Instruments, and Computers* **21**, 213-224 (Ch. 7)

Levine, D. S. (1989c). The third wave in neural networks. *AI Expert* December, 1989, 26-33. (Ch. 1)

Levine, D. S., & Grossberg, S. (1976). Visual illusions in neural networks: line neutralization, tilt after-effect, and angle expansion. *Journal of Theoretical Biology* **61**, 477-504. (Ch. 4)

Levine, D. S., Leven, S. J., & Prueitt, P. S. (1991). Integration, disintegration, and the frontal lobes. In Levine, D. S. & Leven, S. J. (Eds.), *Motivation, Emotion, and Goal Direction in Neural Networks (pp. 301-335).* Hillsdale, NJ: Lawrence Erlbaum Associates, in press. (Ch. 6, 7)

Levine, D. S., & Penz, P. A. (1990). ART 1.5 — a simplified adaptive resonance network for classifying low-dimensional analog data. *Proceedings of the International Joint Conference on Neural Networks, January, 1990* (Vol. II, pp. 639-642). Hillsdale, NJ: Lawrence Erlbaum Associates. (Ch. 6)

Levine, D. S., & Prueitt, P. S. (1989). Modeling some effects of frontal lobe damage: novelty and perseveration. *Neural Networks* **2**, 103-116. (Ch. 3, 5, 6, 7)

Levine, D. S., & Woody, C. D. (1978). Effects of active versus passive dendritic membrane on the transfer properties of a simulated neuron. *Biological Cybernetics* **31**, 63-70. (Ch. 3).

Levy, B. C., & Adams, M. B. (1987). Global optimization with stochastic neural networks. *IEEE First International Conference on Neural Networks* (Vol. III, pp. 681-689). San Diego: IEEE/ICNN. (Ch. 7)

Levy, W. B. (1985). Associative changes in the synapse: LTP in the hippocampus. In W. B. Levy, J. A. Anderson, & S. Lehmkuhle (Eds.), *Synaptic Modification, Neuron Selectivity, and Nervous System Organization* (pp. 5-33). Hillsdale, NJ: Lawrence Erlbaum Associates. (Ch. 3)

Levy, W. B., Brassel, S. E., & Moore, S. D. (1983). Partial quantification of the associative synaptic learning rule of the dentate gyrus. *Neuroscience* **8**, 799-808. (Ch. 3)

Li, D., & Wee, W. G. (1990). A new neocognitron structure modified by ART and back-propagation. *Proceedings of the International Joint Conference on Neural Networks, January, 1990* (Vol. I, pp. 420-423). Hillsdale, NJ: Lawrence Erlbaum Associates. (Ch. 1)

Linsker, R. (1986a). From basic network principles to neural architecture: emergence of spatial-opponent cells. *Proceedings of the National Academy of Sciences* **83**, 7508-7512. (Ch. 6)

Linsker, R. (1986b). From basic network principles to neural architecture: emergence of orientation-sensitive cells. *Proceedings of the National Academy of Sciences* **83**, 8779-8783. (Ch. 6)

Linsker, R. (1986c). From basic network principles to neural architecture: emergence of orientation columns. *Proceedings of the National Academy of Sciences* **83**, 8390-8394. (Ch. 6)

Lippmann, R. P. (1987). An introduction to computing with neural nets. *IEEE ASSP Magazine*, April, 1987, 4-22. (Preface, Ch. 8)

Lorenz, E. (1963). Deterministic nonperiodic flow. *Journal of Atmospheric Science* **20**, 130-141. (Appendix 2)

Lynch, G., & Baudry, M. (1983). The biochemistry of memory: a new and specific hypothesis. *Science* **224**, 1057-1063. (Ch. 3)

Mackintosh, N. J. (1974). *The Psychology of Animal Learning*. London: Academic. (Ch. 5)

Mackintosh, N. J. (1975). A theory of attention: variations in the associability of stimuli with reinforcement. *Psychological Review* **82**, 276-298. (Ch. 5)

Mackintosh, N. J. (1983). *Conditioning and Associative Learning*. New York: Oxford University Press. (Ch. 5)

MacLean, P. D. (1970). The triune brain, emotion, and scientific bias. In F. Schmitt (Ed.), *The Neurosciences Second Study Program*. New York: Rockefeller University Press. (Ch. 7)

Malsburg, C. von der (1973). Self-organization of orientation sensitive cells in the striate cortex. *Kybernetik* **14**, 85-100. (Ch. 2, 5, 6)

Malsburg, C. von der, & Cowan, J. D. (1982). Outline of a theory for the ontogenesis of iso-orientation domains in visual cortex. *Biological Cybernetics* **45**, 49-56. (Ch. 6)

Marko, K. A., James, J., Dosdall, J., & Murphy, J. (1989). Automotive control system diagnostics using neural nets for rapid pattern classification of large data sets. *International Joint Conference on Neural Networks, Washington, DC, June, 18-22, 1989* (Vol. II, pp. 13-16). Piscataway, NJ: IEEE. (Ch. 8)

Marr, D. (1969). A theory of cerebellar cortex. *Journal of Physiology*, **202**, 437-470. (Ch. 7)

Marr, D. (1970). A theory for cerebral neocortex. *Proceedings of the Royal Society of London, Series B*, **176**, 161-234. (Ch. 7)

Marr, D. (1971). Simple memory: a theory for archicortex. *Philosophical Transactions of the Royal Society of London*, **262**, 23-81. (Ch. 7)

Marr, D. (1982). *Vision: a Computational Investigation Into the Human Representation and Processing of Visual Information*. San Francisco: W. H. Freeman. (Ch. 4)

Marr, D., & Hildreth, E. (1980). Theory of edge detection. *Proceedings of the Royal Society of London, Series B* **207**, 187-217. (Ch. 4)

Marr, D., & Poggio, T. (1977a). From understanding computation to understanding serial circuitry. *Neurosciences Research Program Bulletin.* **15**, 470-488. (Ch. 4)

Marr, D., & Poggio, T. (1977b). A theory of human stereo vision. Massachusetts Institute of Technology, A. I. Memo #451. (Ch. 4)

Marr, D., & Poggio, T. (1979). A computational theory of human stereo vision. *Proceedings of the Royal Society of London, Series B* **204**, 301-328. (Ch. 4)

Martin, G. L., & Pittman, J. A. (1989). Recognizing hand-drawn and handwritten symbols with neural nets. In D. Touretzky (Ed.), *Advances in Neural Information Processing Systems 2* (pp. 405-414). San Mateo, CA: Morgan Kaufman. (Ch. 8)

Maslow, A. H. (1968). *Toward a Psychology of Being*. New York: Van Nostrand. (Preface)

Maslow, A. H. (1972). *The Farther Reaches of Human Nature*. New York: Viking. (Preface)

Massone, L., & Bizzi, E. (1989). Generation of limb trajectories with a sequential network. *International Joint Conference on Neural Networks, Washington, DC, June, 18-22, 1989* (Vol. II, pp. 345-349). Piscataway, NJ: IEEE. (Ch. 7)

McClelland, J. L. (1985). Putting knowledge in its place: A scheme for programming parallel structures on the fly. *Cognitive Science* **9**, 113-146. (Ch. 7)

McClelland, J. L., & Rumelhart, D. E. (1981). An interactive activation model of context effects in letter perception: Part 1. An account of basic findings. *Psychological Review* **88**, 375-407. (Ch. 7)

McClelland, J. L., & Rumelhart, D. E. (1985). Distributed memory and the representation of general and specific memory. *Journal of Experimental Psychology: General* **114**, 159-188. (Ch. 6)

McClelland, J. L., & Rumelhart, D. E. (1988). *Explorations in Parallel Distributed Processing*. Cambridge, MA: MIT Press. (Ch. 6)

McCulloch, W. S. (1965). *Embodiments of Mind*. Cambridge, MA: MIT Press, p. xx of Introduction by S. Papert. (Ch. 1)

McCulloch, W. S., & Pitts, W. (1943). A logical calculus of the ideas immanent in nervous activity. *Bulletin of Mathematical Biophysics* **5**, 115-133. (Ch. 1, 2, 4, 5)

Mervis, C. B., & Rosch, E. (1981). Categorization of natural objects. *Annual Review of Psychology* **32**, 89-115. (Ch. 6)

Miller, R. K., & Michel, A. N. (1982). *Ordinary Differential Equations*. New York: Academic Press. (Appendix 2)

Miller, R. K., Walker, T. C., & Ryan, A. M. (1989). *Neural Net Applications and Products*. Madison, GA: SEAI Technical Publications. (Ch. 1, 8)

Miller, T., Sutton, R. S., & Werbos, P. J. (Eds.) (1990) *Neural Networks for Control*. Cambridge, MA: MIT Press, in press. (Ch. 7)

Milner, B. (1963). Effects of different brain lesions in card sorting. *Archives of Neurology* **9**, 90-100. (Ch. 7)

Milner, B. (1964). Some effects of frontal lobectomy in man. In J. M. Warren & K. Akert (Eds.), *The Frontal Granular Cortex and Behavior (pp. 313-334)*. New York: McGraw-Hill. (Ch. 7)

Mingolla, E., & Bullock, D. (1989). Book review: J. A. Anderson & E. Rosenfeld (Eds.), *Neurocomputing: Foundations of Research*. *Neural Networks* **2**, 405-409. (Ch. 7)

Minsky, M. (1986). *The Society of Mind*. New York: Simon and Schuster. (Ch. 1)

Minsky, M., & Papert, S. (1969). *Perceptrons: an Introduction to Computational Geometry*. Cambridge, MA: MIT Press. (Ch. 1, 2, 6)

Mishkin, M., & Appenzeller, T. (1987). The anatomy of memory. *Scientific American* June, 1987, 80-89. (Ch. 7)

Mishkin, M., Malamut, B., & Bachevalier, J. (1984). Memories and habits: Two neural systems. In G. Lynch, J. L. McGaugh, & N. M. Weinberger (Eds.), *Neurobiology of Learning and Memory*. New York, London: Guilford. (Ch. 7)

Montalvo, F. S. (1975). Consensus versus competition in neural networks. *International Journal of Man-Machine Studies* **7**, 333-346. (Ch. 4)

Moore, J. W., Desmond, J. E., Berthier, N. E., Blazis, D. E. J., Sutton, R. S., & Barto, A. G. (1986). Simulation of the classically conditioned nictitating membrane response by a neuron-like adaptive element: response topography, neuronal firing, and interstimulus intervals. *Behavioural Brain Research* **21**, 143-154. (Ch. 5)

Morishita, I., & Yajima, A. (1972). Analysis and simulation of networks of mutually inhibiting neurons. *Kybernetik* **11**, 154-165. (Ch. 4)

Mountcastle, V. B. (1957). Modality and topographic properties of single neurons of cat's somatic sensory cortex. *Journal of Neurophysiology* **20**, 408-434. (Ch. 2, 4)

Mpitsos, G. J., Burton, R. M., Creech, H. C., & Soinila, S. O. (1988). Evidence for chaos in spike trains of neurons that generate rhythmic motor patterns. *Brain Research Bulletin* **21**, 529-538. (Appendix 2)

Nass, M. M., & Cooper, L. N. (1975). A theory for the development of feature detecting cells in the visual cortex. *Biological Cybernetics* **19**, 1-18. (Ch. 3, 6)

Nauta, W. J. H. (1971). The problem of the frontal lobe: A reinterpretation. *Journal of Psychiatric Research* **8**, 167-187. (Ch. 7)

Nauta, W. J. H., & Feirtag, M. (1986). *Fundamental Neuroanatomy*. New York: Freeman. (Appendix 1)

Neumann, J. von (1951). Probabilistic logics and the synthesis of reliable organisms from unreliable components. In L. A. Jeffress (Ed.), *Cerebral Mechanisms in Behavior, the Hixon Symposium*. New York: John Wiley and Sons. (Ch. 2, 7)

Newell, A., & Simon, H. A. (1972). *Human Problem Solving*. Englewood Cliffs, NJ: Prentice-Hall. (Ch. 1)

Nguyen, D., & Widrow, B. (1990). The truck backer-upper: an example of self-learning in neural networks. In T. Miller, R. S. Sutton, & P. J. Werbos (Eds.), *Neural Networks for Control*. Cambridge, MA: MIT Press, in press. (Ch. 7)

Nottebohm, F. (1989). From bird song to neurogenesis. *Scientific American* February, 1989, 74-79. (Ch. 7)

Olds, J. (1955). Physiological mechanisms of reward. In Jones, M. (Ed.), *Nebraska Symposium on Motivation* (pp. 73-142). Lincoln, NE: University of Nebraska Press. (Ch. 8, Appendix 1)

Olds, J., & Milner, P. M. (1954). Positive reinforcement produced by electrical stimulation of septal area and other regions of rat brain. *Journal of Comparative and Physiological Psychology* **47**, 419-427. (Ch. 8)

Papez, J. W. (1937). A proposed mechanism for emotion. *Archives of Neurology and Psychiatry* **38**, 725-743. (Appendix 1)

Parasuraman, R., & Davies, D. R., Eds. (1984). *Varieties of Attention*. Orlando, FL: Academic Press. (Ch. 5).

Parker, D. B. (1985). Learning-logic (TR-47). Cambridge, MA: Massachusetts Institute of Technology, Center for Computational Research in Economics and Management Science. (Ch. 2, 3, 6)

Parks, R. W., Crockett, D. J., & McGeer, P. L. (1989). Systems interpretation of cortical organization: Positron emission tomography and neuropsychological test performance. *Archives of Clinical Neuropsychology* **4**, 335-349. (Ch. 8)

Pavlov, I. P. (1927). *Conditioned Reflexes* (V. Anrep, translator). London: Oxford University Press. (Ch. 2, 3, 5)

Pearce, J. M., & Hall, G. (1980). A model for Pavlovian learning: variations in the effectiveness of conditioned but not of unconditioned stimuli. *Psychological Review* **87**, 532-552. (Ch. 5)

Pellionisz, A. (1986). Tensor network theory of the central nervous system. In: G. Palm & A. Aertsen (Eds.), *Brain Theory*. Berlin: Springer-Verlag. (Ch. 7)

Pellionisz, A., & Llinás, R. (1982). Space time representation in the brain: the cerebellum as a predictive space time metric tensor. *Neuroscience* **7**, 2949-2970. (Ch. 7)

Perez, R., Glass, L., & Shlaer, R. (1974). Development of specificity in the cat visual cortex. *Journal of Mathematical Biology* **1**, 275-288. (Ch. 6)

Piaget, J. (1952). *The Origin of Intelligence in Children*, translated by M. Cook. New York: International University Press. (Ch. 7)

Pinto-Hamuy, T., & Linck, P. (1965). Effect of frontal lesions of performance of sequential tasks by monkeys. *Experimental Neurology* **12**, 96-107. (Ch. 7)

Posner, M. I., & Keele, S. W. (1970). Retention of abstract ideas. *Journal of Experimental Psychology* **83**, 304-308. (Ch. 6)

Posner, M. I., & Snyder, C. R. R. (1975a). Attention and cognitive control. In: R. L. Solso (Ed.), *Information Processing and Cognition: the Loyola Symposium* (pp. 55-85). Hillsdale, NJ: Lawrence Erlbaum Associates. (Ch. 7)

Posner, M. I. & Snyder, C. R. R. (1975b). Facilitation and inhibition in the processing of signals. In: P. Rabbitt & S. Dornic (Eds.), *Attention and Performance* (Vol. 5), pp. 669-683. New York: Academic Press. (Ch. 7)

Powers, W. T. (1973). *Behavior: the Control of Perception*. Chicago: Aldine. (Preface)

Press, W. H., Flannery, B. P., Teukolsky, S. A., & Vetterling, W. T. (1988). *Numerical Recipes: the Art of Scientific Computing*. New York: Cambridge University Press. (Appendix 2).

Pribram, K. H. (1961). A further experimental analysis of the behavioral deficit that follows injury to the primate frontal cortex. *Journal of Experimental Neurology* **3**, 432-466. (Ch. 7)

Pribram, K. H. (1991). *Brain and Perception: Holonomy and Structure in Figural Processing.* Hillsdale, NJ: Lawrence Erlbaum Associates, in press. (Ch. 4)

Prigogine, I. (1969). Structure, dissipation, and life. In *First International Conference, Theoretical Physics and Biology, Versailles* (pp. 23-52). Amsterdam: North-Holland. (Ch. 2)

Pylyshyn, Z. (1984). *Computation: Toward a Foundation for Cognitive Science and Cognition.* Cambridge, MA: MIT Press. (Ch. 4)

Rabbitt, P., & Dornic, S. (Eds.) (1975). *Attention and Performance* (Vol. 5). New York: Academic Press. (Ch. 5)

Rainville, E. D., & Bedient, P. E. (1981). *Elementary Differential Equations.* New York: Macmillan. (Appendix 2)

Rall, W. (1955). Experimental monosynaptic input-output relations in the mammalian spinal cord. *Journal of Cellular and Comparative Physiology* **46**, 413-437. (Ch. 2)

Rapoport, J. L. (1989). The biology of obsessions and compulsions. *Scientific American* March, 1989, 83-89. (Ch. 8)

Rashevsky, N. (1960). *Mathematical Biophysics*, Vol. II. New York: Dover. (Ch. 2)

Ratliff, F. (1965). *Mach Bands: Quantitative Studies of Neural Networks in the Retina.* San Francisco: Holden-Day. (Ch. 4)

Reeke, G. N., & Edelman, G. M. (1984). Selective networks and recognition automata. *Annals of the New York Academy of Sciences* **426**, 181-201. (Ch. 6)

Reeke, G. N., & Edelman, G. M. (1987). Selective neural networks and their implications for recognition automata. *International Journal of Supercomputer Applications* **1**, 44-69. (Ch. 6)

Reilly, D. L., Cooper, L. N., & Elbaum, C. (1982). A neural model for category learning. *Biological Cybernetics* **45**, 35-41. (Ch. 6, 7)

Reilly, D. L., Scofield, C., Elbaum, C., & Cooper, L. N. (1987). Learning system architectures composed of multiple learning modules. *IEEE First International Conference on Neural Networks* (Vol. II, pp. 495-503). San Diego: IEEE/ICNN. (Ch. 6, 7)

Rescorla, R. A., & Wagner, A. B. (1972). A theory of Pavlovian conditioning: variations in the effectiveness of reinforcement and non-reinforcement. In A. H. Black & W. F. Prokasy (Eds.), *Classical Conditioning II: Current Research and Theory* (pp. 64-99). New York: Appleton-Century-Crofts. (Ch. 3, 5, 6)

Ricart, R. (1991). Neuromodulatory mechanisms in neural networks and their influence on interstimulus interval effects in Pavlovian conditioning. In D. S. Levine and S. J. Leven (Eds.), *Motivation, Emotion, and Goal Direction in Neural Networks* (pp. 117-166). Hillsdale, NJ: Lawrence Erlbaum Associates, in press. (Ch. 5)

Robinson, D. A. (1981). Control of eye movements. In J. M. Brookhart, V. B. Mountcastle, V. B. Brooks, & S. R. Geiger (Eds.), *Handbook of Physiology*, Vol. II. Bethesda, MD: American Physiological Society. (Ch. 7)

Robson, J. G. (1975). Receptive fields: neural representation of the spatial and intensive attributes of the visual image. In E. C. Carterette, & M. P. Friedman (Eds.), *Handbook of Perception* (Vol. V, pp. 82-116). New York: Academic Press. (Ch. 4)

Romer, A. S., & Parsons, T. (1977). *The Vertebrate Body*. Philadelphia: Saunders. (Appendix 1)

Rose, D., & Blakemore, C. (1974). Analysis of orientation selectivity in the cat's visual cortex. *Experimental Brain Research* **20**, 1-17. (Ch. 4)

Rosenberg, C. R., & Sejnowski, T. J. (1986). The spacing effect on NETtalk, a massively-parallel network. *Proceedings of the Eighth Annual Conference of the Cognitive Science Society* (pp. 72-89). Hillsdale, NJ: Lawrence Erlbaum Associates. (Ch. 6, 7, 8)

Rosenblatt, F. (1962). *Principles of Neurodynamics*. Washington, DC: Spartan Books. (Ch. 1, 2, 3, 6)

Rosenfield, I. (1988). *The Invention of Memory*. New York: Basic Books. (Ch. 6)

Rosenkilde, C. E., Bauer, R. H., & Fuster, J. M. (1981). Single cell activity in ventral prefrontal cortex of behaving monkeys. *Brain Research* **209**, 375-394. (Ch. 4, 8)

Rumelhart, D. E. (1988). Parallel distributed processing. Unpublished plenary lecture, IEEE International Conference on Neural Networks, San Diego, June, 1988. (Ch. 6)

Rumelhart, D. E., Hinton, G. E., & Williams, R. J. (1986). Learning internal representations by error propagation. In D. E. Rumelhart & J. L. McClelland (Eds.), *Parallel Distributed Processing* (Vol. 1, pp. 318-362). Cambridge, MA: MIT Press. (Ch. 2, 3, 6, 7, Appendix 1, Appendix 2)

Rumelhart, D. E., & McClelland, J. L. (1982). An interactive activation model of context effects in letter perception: Part 2. The contextual enhancement effect and some tests and extensions of the model. *Psychological Review* **89**, 60-94. (Ch. 7)

Rumelhart, D. E., & McClelland, J. L., Eds. (1986). *Parallel Distributed Processing*, Vols. 1 and 2. Cambridge, MA: MIT Press. (Ch. 1, 2, 3, 7)

Rumelhart, D. E., & Zipser, D. (1985). Feature discovery by competitive learning. *Cognitive Science* **9**, 75-112. (Ch. 4, 6, 7)

Ryan, T. W. (1988). The resonance correlation network. *IEEE International Conference on Neural Networks* (Vol. I, pp. 673-680). San Diego: IEEE. (Ch. 6)

Ryan, T. W., & Winter, C. L. (1987). Variations on adaptive resonance. *IEEE First International Conference on Neural Networks* (Vol. II, pp. 767-775). San Diego: IEEE/ICNN. (Ch. 6)

Ryan, T. W., Winter, C. L., & Turner, C. J. (1987). Dynamic control of an artificial neural system: the property inheritance network. *Applied Optics* **26**, 4961-4971. (Ch. 7)

Sakai, H., & Woody, C. D. (1980). Identification of auditory responsive cells in coronal-pericruciate cortex of awake cats. *Journal of Neurophysiology* **44**, 223-231. (Ch. 8)

Sasaki, K., Bower, J. M., & Llinás, R. (1989). Multiple Purkinje cell recording in rodent cerebellar cortex. *European Journal of Neuroscience* **1**, 572-586. (Ch. 8)

Scheff, K., & Szu, H. (1987). 1-D optical Cauchy machine infinite film spectrum search. *IEEE First International Conference on Neural Networks* (Vol. III, pp. 673-679). San Diego: IEEE/ICNN. (Ch. 7)

Schmajuk, N. A. & DiCarlo, J. J. (1991). Neural dynamics of hippocampal modulation of classical conditioning. In M. Commons, S. Grossberg & J. Staddon (Eds.), *Neural Network Models of Conditioning and Action* (pp. 149-180). Hillsdale, NJ: Lawrence Erlbaum Associates. (Ch. 5, 7).

Schmajuk, N. A., & Moore, J. W. (1985). Real-time attentional models for classical conditioning and the hippocampus. *Physiological Psychology* **13**, 278-290. (Ch. 5)

Schneiderman, N., & Gormezano, I. (1964). Conditioning of the nictitating membrane response of the rabbit as a function of the CS-US interval. *Journal of Comparative and Physiological Psychology* **57**, 188-195. (Ch. 5)

Sejnowski, T. J., & Rosenberg, C. R. (1986). NETtalk: A parallel network that learns to read aloud. The Johns Hopkins University Electrical Engineering and Computer Science Technical Report JHU/EECS-86/01 (Ch. 6, 7, 8)

Selfridge, O. G. (1959). PANDEMONIUM: A paradigm for learning. *Proceedings of Symposium on Mechanisation of Thought Processes, National Physics Laboratory, Teddington, England* (pp. 511-529). Her Majesty's Stationery Office, London, 2 vols. (Ch. 2)

Selverston, A. (1976). A model system for the study of rhythmic behavior. In J. C. Fentress (Ed.), *Simpler Networks and Behavior (pp. 83-98).* Sunderland, MA: Sinauer. (Appendix 1, Appendix 2)

Sethares, W. A. (1988). A convergence theorem for the modified delta rule. *Neural Networks* 1, Suppl. 1, 130. (Ch. 6)

Shea, P. M., & Lin, V. (1989). Detection of explosives in checked airline baggage using an artificial neural system. *International Joint Conference on Neural Networks, Washington, DC, June, 18-22, 1989* (Vol. II, pp. 31-34). Piscataway, NJ: IEEE. (Ch. 8)

Shepherd, G. M. (1979). *The Synaptic Organization of the Brain*. New York: Oxford University Press. (Appendix 1)

Shepherd, G. M. (1983). *Neurobiology*. New York: Oxford University Press. (Appendix 1)

Sherrington, C. S. (1947). *The Integrative Action of the Nervous System* (2nd ed.) New Haven, CT: Yale University Press. (Appendix 1)

Sholl, D. A. (1956). *The Organization of the Cerebral Cortex*. London: Methuen. (Ch. 4).

Simon, H. (1969). *The Sciences of the Artificial*. Cambridge, MA: MIT Press. (Ch. 1)

Skarda, C., & Freeman, W. J. (1987). How brains make chaos to make sense of the world. *The Behavioral and Brain Sciences* **10**, 161-195. (Ch. 2, Appendix 2)

Skinner, B. F. (1938). *The Behavior of Organisms*. New York: Appleton-Century. (Ch. 5)

Smital, J. (1988). *On Functions and Functional Equations*. Bristol, England: Adam Hilger. (Appendix 2)

Smith, M. C., Coleman, S. R., & Gormezano, I. (1969). Classical conditioning of the rabbit's nictitating membrane response at backward, simultaneous, and forward CS-US intervals. *Journal of Comparative and Physiological Psychology* **69**, 226-231. (Ch. 5)

Smolensky, P. (1986). Harmony theory. In D. E. Rumelhart & J. L. McClelland (Eds.), *Parallel Distributed Processing* (Vol. 1, pp. 194-281). Cambridge, MA: MIT Press. (Ch. 7)

Soechting, J. F., & Flanders, M. (1989). Errors in pointing are due to approximations in sensorimotor transformations. *Journal of Neurophysiology* **62**, 595-608. (Ch. 7)

Solomon, R. L. & Corbit, J. D. (1974). An opponent-process theory of motivation: I. Temporal dynamics of affect. *Psychological Review* **81**, 119-145. (Ch. 3)

Sontag, E., & Sussmann, H. (1989). Backpropagation separates when perceptrons do. *International Joint Conference on Neural Networks, Washington, DC, June 18-22, 1989* (Vol. I, pp. 639-642). Piscataway, NJ: IEEE. (Ch. 6)

Sperling, G. (1970). Binocular vision: a physical and a neural theory. *American Journal of Psychology* **83**, 461-534. (Ch. 4)

Sperling, G., & Sondhi, M. M. (1968). Model for visual luminance detection and flicker detection. *Journal of the Optical Society of America* **58**, 1133-1145. (Ch. 4)

Staddon, J. E. R. (1983). *Adaptive Behavior and Learning*. London: Cambridge University Press. (Ch. 5)

Stefanis, C. (1969). Interneuronal mechanisms in the cortex. In M. Brazier (Ed.), *The Interneuron* (pp. 497-526). Los Angeles: University of California Press. (Ch. 4)

Stent, G. S. (1973). A physiological mechanism for Hebb's postulate of learning. *Proceedings of the National Academy of Sciences of the USA* **70**, 997-1003. (Ch. 3)

Stevens, K. A. (1983). False dilemmas: confusion between mechanism and computation. *The Behavioral and Brain Sciences* **4**, 675. (Ch. 4)

Stone, G. O. (1986). An analysis of the delta rule and the learning of statistical associations. In D. E. Rumelhart & J. L. McClelland (Eds.), *Parallel Distributed Processing* (Vol. 1, pp. 444-459). Cambridge, MA: MIT Press. (Ch. 2)

Stone, G. O. & Van Orden, G. (1989). Are words represented by nodes? *Memory and Cognition* **17**, 511-524. (Ch. 1)

Stork, D. G. (1989a). Self-organization, pattern recognition, and adaptive resonance networks. *Journal of Neural Network Computing* **1**, 26-42. (Ch. 6)

Stork, D. G. (1989b). Is backpropagation biologicallly plausible? *International Joint Conference on Neural Networks, Washington, DC, June 18-22, 1989* (Vol. II, pp. 241-246). Piscataway, NJ: IEEE. (Ch. 6)

Stornetta, W. S., & Huberman, B. A. (1987). An improved three-layer, back propagation algorithm. *IEEE First International Conference on Neural Networks* (Vol. II, pp. 637-643). San Diego: IEEE/ICNN. (Ch. 6)

Sutton, R. S. (1988). Learning to predict by the methods of temporal differences. *Machine Learning* **3**, 9-44. (Ch. 7)

Sutton, R. S., & Barto, A. G. (1981). Toward a modern theory of adaptive networks: expectation and prediction. *Psychological Review* **88**, 135-170. (Ch. 3, 5, 6, 7, Appendix 2)

Sutton, R. S., & Barto, A. G. (1991). Time-derivative models of Pavlovian reinforcement. In J. W. Moore & M. Gabriel (Eds.), *Learning and Computational Neuroscience*. Cambridge, MA: MIT Press, in press. (Ch. 3, 5, 7)

Swerdlow, N. R., & Koob, G. F. (1987). Dopamine, schizophrenia, mania, and depression. *The Behavioral and Brain Sciences* **10**, 197-245. (Ch. 7, 8)

Swokowski, E. W. (1988). *Calculus With Analytic Geometry* (4th edition). Boston: PWS-Kent. (Appendix 2)

Szentagothai, J. (1967a). *The Anatomy of Complex Integrative Units in the Nervous System.* In K. Lissak (Ed.), *Results in Neuroanatomy, Neurochemistry, Neuropharmacology, and Neurophysiology: Recent Development of Neurobiology in Hungary* (Vol. I, pp. 9-45). Budapest: Akademiai Kiado. (Ch. 4)

Szentagothai, J. (1967b). The "module-concept" in cerebral cortex architecture. *Brain Research* **95**, 475-496. (Ch. 4)

Szu, H. (1986). Fast simulated annealing. In J. S. Denker (Ed.), *Neural Networks for Computing* (pp. 420-425). AIP Conference Proceedings. New York: American Institute of Physics, Vol. **151**. (Ch. 7)

Thagard, P. (1989). Explanatory coherence. *The Behavioral and Brain Sciences* **12**, 435-502. (Ch. 7)

Thomas, G. B., Jr., & Finney, R. L. (1988). *Calculus and Analytic Geometry.* Reading, MA: Addison-Wesley. (Appendix 2)

Thompson, R. F. (1967). *Foundations of Physiological Psychology.* New York: Harper and Row. (Ch. 1, 2, 3, Appendix 1)

Tikhomirov, O. (1983). Informal heuristic principles of motivation and emotion in human problem solving. In R. Groner, M. Groner & W. Bischof (Eds.), Methods of Heuristics (pp. 153-170). Hillsdale, NJ: Lawrence Erlbaum Associates. (Ch. 7)

Truex, R. C., & Carpenter, M. B. (1969). *Human Neuroanatomy.* Baltimore: Williams and Wilkins. (Appendix 1).

Tsukahara, N., & Oda, Y. (1981). Appearance of new synaptic potentials at corticorubral synapses after the establishment of classical conditioning. *Proceedings of the Japanese Academy, Series B (Physical and Biological Sciences)* **57**, 398-401. (Ch. 2, 3)

Turner, C. H. (1981). *Maps of the Mind.* New York: Macmillan. (Preface)

Tversky, A., & Kahneman, D. (1974). Judgment under uncertainty: Heuristics and biases. *Science* **185**, 1124-1131. (Ch. 7)

Tversky, A., & Kahneman, D. (1981). The framing of decisions and the rationality of choice. *Science* **211**, 453-458. (Ch. 7)

Tversky, A., & Kahneman, D. (1982). Belief in the law of small numbers. In D. Kahneman, P. Slovic, & A. Tversky, (Eds.), *Judgment Under Uncertainty: Heuristics and Bias.* New York: Cambridge University Press. (Ch. 7)

Uttley, A. M. (1966). The transmission of information and the effect of local feedback in theoretical and neural networks. *Brain Research* **2**, 21-50. (Ch. 5)

Uttley, A. M. (1970). The informon: a network for adaptive pattern recognition. *Journal of Theoretical Biology* **27**, 31-67. (Ch. 5)

Uttley, A. M. (1975). The informon in classical conditioning. *Journal of Theoretical Biology* **49**, 355-376. (Ch. 5)

Uttley, A. M. (1976a). A two-pathway informon theory of conditioning and adaptive pattern recognition. *Brain Research* **102**, 23-35. (Ch. 5)

Uttley, A. M. (1976b). Simulation studies of learning in an informon network. *Brain Research* **102**, 37-53. (Ch. 5)

Uttley, A. M. (1976c). Neurophysiological predictions of a two-pathway informon theory of neural conditioning. *Brain Research* **102**, 55-70. (Ch. 5)

Walters, E. T., & Byrne, J. H. (1983). Associative conditioning of single sensory neurons suggests a cellular mechanism for learning. *Science* **219**, 405-408. (Ch. 5)

Weathers, J. C. (1988). An adaptive neural network traffic controller. Final project report for graduate class, University of Texas at Arlington. (Ch. 8)

Weidemann, W. E., Manry, M. T., & Yau, H. C. (1989). A comparison of a nearest neighbor classifier and a neural network for numeric handprint character recognition. *International Joint Conference on Neural Networks, Washington, DC, June, 18-22, 1989* (Vol. I, pp. 117-120). Piscataway, NJ: IEEE. (Ch. 6, 8)

Weingard, F. S. (1990). Self-organizing analog fields (SOAF). *Proceedings of the International Joint Conference on Neural Networks, January, 1990* (Vol. II, p. 34). Hillsdale, NJ: Lawrence Erlbaum Associates. (Ch. 6)

Weintraub, E. R. (1979). *Microfoundations.* New York: Cambridge University Press. (Ch. 7)

Wenksay, D. L. (1990). Intellectual property protection for neural networks. *Neural Networks* **3**, 229-236. (Ch. 8)

Werbos, P. J. (1974). Beyond regression: new tools for prediction and analysis in the behavioral sciences. Unpublished Ph. D. dissertation, Harvard University. (Ch. 2, 3, 6).

Werbos, P. J. (1988a). Backpropagation: past, present, and future. *IEEE International Conference on Neural Networks* (Vol. I, pp. 343-353). San Diego: IEEE. (Ch. 5, 6, 7)

Werbos, P. J. (1988b). Generalization of backpropagation with application to a recurrent gas model. *Neural Networks* **1**, 339-356. (Ch. 7)

Wheeler, D. D. (1970). Processes in word recognition. *Cognitive Psychology* **1**, 59-85. (Ch. 7)

White, H. (1987). Some asymptotic results for back-propagation. *IEEE First International Conference on Neural Networks* (Vol. III, pp. 261-266). San Diego: IEEE/ICNN. (Ch. 4, 6)

Widrow, B. (1962). Generalization and information storage in networks of adaline neurons. In: Yovits, M. C., Jacobi, G. T. & Goldstein, G. D. (Eds.), *Self-organizing Systems -- 1962* (pp. 437-461). Washington, DC: Spartan Books. (Ch. 2, 6)

Widrow, B. (1987). ADALINE and MADALINE — 1963. *IEEE First International Conference on Neural Networks (Vol. I, pp. 145-157).* San Diego: IEEE/ICNN. (Ch. 2)

Widrow, B., & Hoff, M. E. (1960). Adaptive switching circuits. Stanford Electronics Laboratories Technical Report 1553-1, Stanford University, Stanford, CA, 1960. (Ch. 2, 3, 6)

Widrow, B., Pierce, W. H., & Angell, J. B. (1961). Birth, life, and death in microelectronic systems. *IRE Transactions on Military Electronics* **4**, 191-201.

Wiener, N. (1948). *Cybernetics.* New York: John Wiley and Sons. (Ch. 2)

Wiener, N. (1954). *The Human Use of Human Beings.* New York: Avon Books. (Ch. 1)

Wilson, H. R. (1975). A synaptic model for spatial frequency adaptation. *Journal of Theoretical Biology* **50**, 327-352. (Ch. 2, 5, 6)

Wilson, H. R., & Bergen, J. R. (1979). A four-mechanism model for spatial vision. *Vision Research* **19**, 19-32. (Ch. 4)

Wilson, H. R., & Cowan, J. D. (1972). Excitatory and inhibitory interactions in localized populations of model neurons. *Biophysical Journal* **12**, 1-24. (Ch. 4)

Wilson, H. R., & Cowan, J. D. (1973). A mathematical theory of the functional dynamics of cortical and thalamic nervous tissue. *Kybernetik* **13**, 55-80. (Ch. 4)

Winograd, T., & Flores, F. (1987). *Understanding Computers and Cognition: a New Foundation for Design.* Norwood, NJ: Ablex. (Ch. 1)

Winston, P. H. (1977). *Artificial Intelligence.* Reading, MA: Addison-Wesley. (Ch. 1)

Winter, C. L., Ryan, T. W., & Turner, C. J. (1987). TIN: a trainable inference network. *IEEE First International Conference on Neural Networks* (Vol. II, pp. 777-785). San Diego: IEEE/ICNN. (Ch. 7)

Wise, R. A. (1988). The neurobiology of craving: implications for the understanding and treatment of addiction. *Journal of Abnormal Psychology* **97**, 118-132. (Ch. 8)

Wong, R., & Harth, E. (1973). Stationary states and transients in neural populations. *Journal of Theoretical Biology* **40**, 77-106. (Ch. 2)

Wood, M. R., Pfenninger, K. H., & Cohen, M. J. (1977). Two types of presynaptic configuration in insect central synapses: an ultrasound analysis. *Brain Research* 130: 25-45. (Appendix 1)

Woody, C. D., Buerger, A. A., Ungar, R. A., & Levine, D. S. (1976). Modeling aspects of learning by altering biophysical properties of a simulated neuron. *Biological Cybernetics* **23**, 73-82. (Ch. 3)

Woody, C. D., Knispel, J. D., Crow, T. J., & Black-Cleworth, P. A. (1976). Activity and excitability to electrical current of cortical auditory receptive neurons of awake cat as affected by stimulus association. *Journal of Neurophysiology* **39**, 1045-1061. (Ch. 8)

Woody, C. D., Vassilevsky, N. N., & Engel, J., Jr. (1970). Conditioned eye blink: unit activity at coronal-pericruciate cortex of the cat. *Journal of Neurophysiology* **33**, 851-864. (Ch. 8)

Yoon, Y. O., Brobst, R. W., Bergstresser, P. R., & Peterson, L. L. (1989). A desktop neural network for dermatology diagnosis. *Journal of Neural Network Computing* **1**, 43-52. (Ch. 8)

Young, J. Z. (1936). Structures of nerve fibers and synapses in some invertebrates. *Cold Spring Harbor Symposia in Quantitative Biology* **4**, 1-6. (Appendix 1)

Zadeh, L. (1965). Fuzzy sets. *Information and Control* **8**, 338-353. (Ch. 7)

Zipser, D. (1986). A model of hippocampal learning during classical conditioning. *Neuroscience* **100**, 764-776. (Ch. 5, 7)

Subject Index

(Pages listed in boldface are those where the concept is either defined or given an extensive treatment.)

Author Index